D1481956

HANDBOOK OF PHYCOLOGICAL METHODS

CULTURE METHODS AND GROWTH MEASUREMENTS

$19.50 in USA

HANDBOOK OF PHYCOLOGICAL METHODS:
CULTURE METHODS
AND GROWTH MEASUREMENTS

Sponsored by the Phycological Society of America Inc.

JANET R. STEIN
Professor of Botany
University of British Columbia

This volume, sponsored by the Phycological Society of America, is the first comprehensive collection of techniques for culturing and measuring the growth of algae. The scope of the work is extensive, and includes special culture techniques, growth measurements, isolation and purification, equipment and methods. The contributors deal with algae from fresh-water and marine habitats as well as with phytoplankton. Emphasis is on methodology rather than theory and, wherever possible, methods based on original research are given.

In many chapters, specific test organisms have been designated so that a 'known' taxon is available upon which to try the method. Sources for these organisms are given in the individual chapters and in the general listing on Culture Collections in the Introduction. Other methods and technique books pertinent to the algae are also listed. The volume also includes author, subject and taxonomic indexes and lists of materials and equipment required for specific methods.

This important work will be of interest to professional workers, researchers, teachers, and students in a wide range of disciplines including phycology, botany, microbiology, plant physiology, biochemistry, genetics and marine biology.

HANDBOOK OF PHYCOLOGICAL METHODS
CULTURE METHODS AND GROWTH MEASUREMENTS

EDITED BY

JANET R. STEIN

PROFESSOR OF BOTANY
UNIVERSITY OF BRITISH COLUMBIA, VANCOUVER, CANADA

SPONSORED BY THE
PHYCOLOGICAL SOCIETY OF AMERICA INC.

CAMBRIDGE
AT THE UNIVERSITY PRESS
1973

Published by the Syndics of the Cambridge University Press
Bentley House, 200 Euston Road, London NW1 2DB
American Branch: 32 East 57th Street, New York, N.Y.10022

Library of Congress Catalogue Card Number: 73-79496

ISBN: 0 521 20049 0

Composed in Great Britain
at the University Printing House, Cambridge
(Brooke Crutchley, University Printer)

Printed in the United States of America

Contents

Contributors

Allen, Mary Mennes, Department of Biological Sciences, Wellesley College, Wellesley, Massachusetts 02181 (Chapter 7)

Blankley, William F., Duke University Marine Laboratory, Beaufort, North Carolina 28516 (Chapters 14, 18)

Carlucci, A. F., Institute of Marine Resources, University of California, San Diego, La Jolla, California 92037 (Chapters 25, 26)

Chapman, A. R. O., Department of Biology, Dalhousie University, Halifax, Nova Scotia, Canada (Chapter 5)

Dunstan, William M., Skidway Institute of Oceanography, University System of Georgia, 55 West Bluff Drive, Savannah, Georgia 31406 (Chapter 20)

Edwards, Peter, Nelson Biological Laboratories, Rutgers University, New Brunswick, New Jersey 08903 (Chapter 17)

Gold, Kenneth, Osborn Laboratories of Marine Sciences, Seaside Park, Coney Island, Brooklyn, New York 11224 (Chapter 27)

Green, Paul B., Department of Biological Sciences, Stanford University, Stanford, California 94305 (Chapter 24)

Guillard, Robert R. L., Woods Hole Oceanographic Institution, Woods Hole, Massachusetts 02543 (Chapters 4, 19)

Hamilton, R. D., Freshwater Institute, Fisheries Research Board of Canada, 501 University Crescent, Winnipeg, Manitoba R3T 2N6, Canada (Chapter 12)

Hansmann, Eugene, Department of Biology, University of New Mexico, Albuquerque, New Mexico 87106 (Chapter 23)

Hemerick, Glen, Route 5, Box 5606, Gig Harbor, Washington 98335 (Chapter 16)

Holm-Hansen, Osmund, Institute of Marine Resources, University of California, San Diego, La Jolla, California 92037 (Chapter 13)

Hoshaw, Robert W., Department of Biological Sciences, University of Arizona, Tucson, Arizona 85721 (Chapter 3)

Lewin, Ralph A., Scripps Institution of Oceanography, University of California, San Diego, La Jolla, California 92037 (Chapter 18)

McLachlan, J., Atlantic Regional Laboratory, National Research Council of Canada, Halifax, Nova Scotia, Canada (Chapter 2)

Menzel, David W., Skidway Institute of Oceanography, University System of Georgia, 55 West Bluff Drive, Savannah, Georgia 31406 (Chapter 20)

Natarajan, K. V., Greater Hartford Community College, 34 Sequassen Street, Hartford, Connecticut 06106 (Chapter 28)

Neilson, Alasdair H., Botaniska Institutionen, Stockholms Universitet, Lilla Fescati, Stockholm 50, Sweden (Chapter 18)

Nichols, H. Wayne, Department of Biology, Washington University, St Louis, Missouri 63130 (Chapter 1)

Page, Joanna Ziegler, Blue Meadow Road, Belchertown, Massachusetts 01007 (Chapter 6)

Parsons, T. R., Institute of Oceanography, University of British Columbia, Vancouver 8, British Columbia, Canada (Chapter 22)

Rosowski, James R., Department of Botany, University of Nebraska, Lincoln, Nebraska 68508 (Chapter 3)

Safferman, Robert S., Environmental Protection Agency, National Environmental Research Center, Cincinnati, Ohio 45268 (Chapter 9)

Sorokin, Constantine, Department of Botany, University of Maryland, College Park, Maryland 20742 (Chapter 21)

Starr, Richard C., Department of Botany, Indiana University, Bloomington, Indiana 47401 (Chapters 10, 11)

Stein, Janet R., Department of Botany, University of British Columbia, Vancouver 8, British Columbia, Canada (Introduction)

Throndsen, Jahn, Institute of Marine Biology, Sect. B, University of Oslo, Blindern, Oslo 3, Norway (Chapter 8)

Ukeles, Ravenna, National Marine Fisheries Service, Experimental Biology Investigations, Milford Laboratory, Milford, Connecticut 06460 (Chapter 15)

Van Baalen, Chase, University of Texas Marine Science Institute, Port Aransas, Texas 78373 (Chapter 17)

Editor's Preface

This *Handbook* originated from the suggestion of a member of the Phycological Society of America concerning a need to present a synthesis of experimental aspects of phycology in a single source book. A committee struck to study the suggestion agreed to its implementation and established guidelines, leading to the appointment by spring 1967 of an Editor and Editorial Committee. Later that year the Editorial Committee considered the topics to be included and decided on potential contributors. The result of this 'meeting of minds' was a projected four volume tome encompassing 'experimental phycology'! The subjects proposed included culture techniques, growth measurements, cytological methods, biochemical techniques, physiological processes, and field-oriented methods. After discussion with the membership of the Society, it was decided to start first with the present volume on culture methods and growth measurements. The rationale being that these topics have the widest appeal and market for those studying or contemplating use of algae. The possibility of covering the remaining subjects in future volumes awaits further consideration depending upon the reception this volume achieves.

Manuscripts have been solicited from active researchers in the fields included. Some usual problems resulting from working with 30 ± contributors (and some unusual, but minor catastrophes – at least to the Editor) have delayed the *Handbook* beyond the original proposed publication date. Thus, some of the chapters have been up-dated by notes added in proof.

As editor, this has been a most enlightening experience. My own laboratory has benefited greatly as many of our standard practices have undergone close scrutiny and revision in light of the methods presented in the *Handbook*. Indeed, this is a most useful tome!

Grateful acknowledgement is made for: the patient support of the Membership of the Phycological Society of America, especially the Executive Committee(s), who have effectively (and financially) kept the concept of the *Handbook* alive; the deliberations and support of the Editorial Consultants (including the original Editorial Committee) who were so very helpful and essential; the efforts of the Contributors without whom this *Handbook* could not have been produced; and, the willing cooperation from the members of the Department of Botany, the University of British Columbia. And finally, thank you to Cambridge University Press for their pleasant cooperation and help.

University of British Columbia JANET R. STEIN
Vancouver
March 1973

Editorial Consultants: (*Editorial Committee)

* Philip W. Cook, University of Vermont, Burlington

Paul R. Gorham, University of Alberta, Edmonton (previously at National Research Council of Canada, Ottawa)

* Paul B. Green, Stanford University, Stanford, California (previously at University of Pennsylvania, Philadelphia)

* Robert R. L. Guillard, Woods Hole Oceanographic Institution, Woods Hole, Massachusetts

O. Holm-Hansen, University of California, San Diego (Scripps Institution of Oceanography)

Robert W. Hoshaw, University of Arizona, Tucson

* Joyce Lewin, University of Washington, Seattle

Ralph A. Lewin, University of California, San Diego (Scripps Institution of Oceanography)

* H. Wayne Nichols, Washington University, St Louis, Missouri

Bruce C. Parker, Virginia Polytechnic Institute, Blacksburg (previously at Washington University, St Louis, Missouri)

Luigi Provasoli, Haskins Laboratory, New Haven, Connecticut

Introduction

JANET R. STEIN
Department of Botany, University of British Columbia, Vancouver 8, British Columbia, Canada

The purpose of the *Handbook of Phycological Methods* is to present a compendium of techniques for culturing and measuring growth of those organisms collectively termed 'algae'. The audience for whom it is intended is diverse and includes experienced researchers, novices, phycologists, and/or non-phycologists. In general, each chapter is concerned primarily with the method rather than theories behind the method. Whenever possible methods based on original research are given. The methods are presented so they can be followed easily, although not necessarily in step-wise fashion. Minor variations are possible depending upon a variety of factors, local circumstances and training of those using the techniques.

Editorial revisions have attempted to present the methods in a consistent fashion. However, as the subject matter is not uniform, this has not always been possible.

The materials and equipment listed serve only for reference, and are not necessarily endorsements of given products or manufacturers. United States and Canadian suppliers are mentioned only as guidelines, but unfortunately these change. Sources of equipment are published annually in the United States in *Science* (American Association for the Advancement of Science, 1515 Massachusetts Ave., N.W., Washington, D.C. 20005) and in Canada in *Research and Development* (Maclean Hunter, 418 University Ave., Toronto 101, Ontario) and *Laboratory Product News* (Southam Business Publications Ltd, 1450 Don Mills Rd, Don Mills, Ontario). These publications should be consulted for up to date addresses of suppliers.

In many chapters, specific test organisms have been designated so that a 'known' taxon is available upon which to try the method. Sources for these organisms are given in the individual chapters and in the general listing on Culture Collections at the end of this Introduction. Other methods and technique books pertinent to the algae are also listed and should be consulted for additional information.

CULTURE COLLECTIONS

(Abbreviations used in text in parentheses)

United States

1. Indiana University Culture Collection (IUCC)
   Department of Botany
   Indiana University
   Bloomington, Indiana 47401
   Starr, R. C. 1964. The culture collection of algae at Indiana University. *Amer. J. Bot.* **51**, 1013–44.
   Starr, R. C. 1966. *Culture collection of algae. Additions to the collection July 1, 1963 – July 1, 1966*, pp. 1–8, Department of Botany, Indiana University, Bloomington, Indiana 47401.
   Starr, R. C. 1971. The culture collection of algae at Indiana University – additions to the collection July 1966 – July 1971. *J. Phycol.* **7**, 350–62.
   Rosowski, J. R. and Parker, B. C. 1971. Sources of algal cultures. *Selected Papers in Phycology*, pp. 816–51. Univ. Nebraska Press, Lincoln.
2. American Type Culture Collection (ATCC)
   12301 Parklawn Drive
   Rockville, Maryland 20852
   American Type Culture Collection 1972. *Catalogue of Strains*, 10th ed. American Type Culture Collection, Washington, D.C.
3. Culture Collection of Algae (WHOI)
   Woods Hole Oceanographic Institution
   Woods Hole, Massachusetts 02543
4. Culture Collection of Thermophilic Bluegreen Algae
   c/o Dr R. W. Castenholz
   Department of Biology
   University of Oregon
   Eugene, Oregon 97403
   Castenholz, R. W. 1970. Laboratory culture of thermophilic cyanophytes. *Schweiz. Z. Hydrol.* **32**, 538–51.
5. Carolina Biological Supply Company
   Burlington, North Carolina 27215
   James, D. E. 1969*a*. Unialgal cultures. *Carolina Tips*, **32**, 33–6.
   James, D. E. 1969*b*. Maintenance and media for marine algae. *Carolina Tips*, **32**, 45–6.

6. Ward's Natural Science Establishment, Inc.
   P.O. Box 1712
   Rochester, New York 14603

United Kingdom

1. Culture Centre of Algae and Protozoa (Cambridge)
   The Director
   36 Storeys Way
   Cambridge, CB3 0DT
   Natural Environment Research Council. 1971. *Culture Collection of Algae and Protozoa – List of Strains*, pp. 1–73. Culture Centre of Algae and Protozoa, Cambridge.
2. Marine Biological Association of the UK (MBA)
   The Laboratory
   Citadel Hill
   Plymouth, PL1 2PB
3. Freshwater Biological Association
   Supply Department
   Windermere Laboratory
   The Ferry House
   Ambleside, Westmorland

Germany

1. Sammlung von Algenkulturen
   Universität Göttingen
   18 Nikolausbergerweg
   Göttingen
   Koch, W. 1964. Verzeichnis der Sammlung von Algenkulturen am Pflanzenphysiologisches Institut der Universität Göttingen. *Arch. Mikrobiol.* **47**, 402–32.

Japan

1. Algal Culture Collection
   Institute of Applied Microbiology
   University of Tokyo
   Tokyo
   Watanabe, A. and Hattori, A. 1966. Cultures and collections of algae. *Proc. U.S.–Japan Conf., Hakone, Japan. Japan Soc. Plant Physiol.* 100 pp.

Czechoslovakia

1. Culture Collection of Algae
   Department of Botany
   Charles University of Prague
   Prague
   Fott, B. and Trucova, E. 1968. List of species in the culture collection of algae at the Department of Botany, Charles University of Prague. *Acta Univ. Carol. Biol.* **1967**, 197–221.

GENERAL REFERENCES

Brunel, L., Prescott, G. W. and Tiffany, L. H., eds. 1950. *The Culturing of Algae*. Charles F. Kettering Foundation, Antioch Press, Yellow Springs, Ohio. 114 pp.

Burlew, J. S., ed. 1953. *Algae Culture, From Laboratory to Pilot Plant*. Carnegie Institution of Washington, Washington, D.C. Publ. 600. 357 pp.

Carr, N. G. 1969. Growth of phototrophic bacteria and blue-green algae. In Norris, J. R. and Ribbons, D. W., eds., *Methods in Microbiology*, **3B**, 53–76. Academic Press, New York.

Droop, M. R. 1969. Algae. In Norris, J. R. and Ribbons, D. W., eds. *Methods in Microbiology*, **3B**, 269–313. Academic Press, New York.

Klein, R. M. and Klein, D. T. 1970. *Research Methods in Plant Science*. Natural History Press, Garden City, New York. 756 pp.

Lapage, S. P., Shelton, J. E. and Mitchell, T. G. 1970. Media for the maintenance and preservation of bacteria. In Norris, J. R. and Ribbons, D. W., eds., *Methods in Microbiology*, **3A**, 1–333. Academic Press, New York.

Lapage, S. P., Shelton, J. E., Mitchell, T. G. and MacKenzie, A. R. 1970. Culture collections and the preservation of bacteria. In Norris, J. R. and Ribbons, D. W., eds., *Methods in Microbiology*, **3A**, 135–228. Academic Press, New York.

Pringsheim, E. G. 1946. *Pure Cultures of Algae*. Cambridge Univ. Press, London. 119 pp.

Venkataraman, G. S. 1969. *The Cultivation of Algae*. Indian Council of Agricultural Research, New Delhi. 319 pp.

Section 1

Isolation and purification

1: Growth media – freshwater

H. WAYNE NICHOLS

Department of Biology, Washington University, St Louis, Missouri 63130

CONTENTS

I. INTRODUCTION

There are perhaps as many media and modifications as there are active phycologists today. Each investigator generally employs his or her own particular medium for successful cultivation of freshwater algae. I have attempted to include in this chapter those media generally employed by several investigators, as well as having general applicability to culturing of a variety of species or algal groups. Further, the media selected represent a wide diversity of pH and dilution. Among the major considerations are: pH; concentration of major nutrients; nitrogen source; possible organic or growth factors for enrichment; and micronutrient composition.

Freshwater media may be designed from several viewpoints. They are generally selected because they possess characteristics similar to the natural environment or they differentially select for a specific algal component of the habitat. Media of an artificial nature, i.e., known chemical composition, are often employed as additives to natural media with an unknown chemical composition, such as lakewater. These then function as enrichment media and are often used to simulate diverse nutritional or physical requirements of a particular species or groups of species, especially when the exact nutritional requirements are unknown.

II. MATERIALS

A. *Chemicals*

The chemical constituents used for the preparation of media should generally be of the highest quality available to the investigator. Quality is determined by the manufacturer and reflects certain specifications established by the Committee on Biological Chemistry of the National Research Council (National Academy of Sciences 1960). The quality is determined by the purity of the compounds and highest purity is confirmed by chromatographic or electrophoretic analysis. These analyses are generally supplied to the investigator routinely or on request. Each manufacturer uses his own code for designation of grade and catalogs should be consulted for code designation and purity of constituents to be used in growth media.

One of the most important points to consider in the preparation of media is the accuracy in weighing the chemicals to be used. Therefore, one should be particularly careful that accurate analytical procedures are followed when weighing small quantities of reagents.

B. *Agar*

Agar is a neutral polymer of galactose which has: (1) low viscosity in water solution; (2) a sharp, stable gelling temperature; and (3) a strong gel structure. It has a melting point *ca.* 95 °C and solidifies *ca.* 45 °C. It is routinely used to solidify growth media and is added after the medium has been mixed, prior to autoclaving. However, as a general rule it is unsafe to autoclave agar in media with very acid pH. Hutner *et al.* (1966) suggests that a slight hydrolysis of the agar liberates acid which accelerates further hydrolysis and liberation of additional acid. They suggest autoclaving the agar and the liquid separately (both at double strength) and mixing them aseptically while hot.

Agar can generally be obtained from a variety of suppliers and is coded by the degree of purity of the agar. The purity is generally based on washing procedures and it has been my experience that a medium to high grade agar (Difco 0560, or 0142, Difco Laboratories, P.O. Box 1058A, Detroit, Michigan 48232) has been satisfactory. The most highly purified agars generally available also may be used not only in solidifying growth media but also for preparing gels for immunodiffusion and electrophoretic techniques.

Agar is generally used at concentrations of 1–2%. It has been successfully used at lower concentrations. However, lower concentrations were generally used in purification procedures (see Chp. 3, VI.B) or in bacteriological work for cultivation of anaerobes.

1. Heat to *ca.* 95 °C 1 l of medium in a 2 l flask in a water bath. (The medium may be heated directly, then placed in the water bath.)
2. Slowly, add the desired quantity of agar while stirring continuously.
3. When all the agar is dispersed and in solution, quickly dispense into containers as desired.
4. Sterilize at 121 °C, 20 lb/in² for 20–35 min (see Chp. 12, II.C.2).
5. Sterile petri plates should be filled aseptically after sterilization. Test tubes and flasks may be filled before sterilization. Petri plates should be poured when the temperature of the medium is *ca.* 50 °C (warm to the inside wrist), otherwise large amounts of condensation occur (see C following).

C. *Glassware*

A variety of sizes and types of glassware is available, including disposable glassware. Glassware is generally prepared from either alumina silicate glass or borosilicate. Either of these types of glass is satisfactory for initial use in preparation and dispensing of growth media. A variety of special glassware to meet particular investigators' problems can be obtained from Corning Glass Works (Science Products Division, Corning, New York 14830; 135 Vanderhoof Ave., Toronto, Ontario). Media are generally dispensed into test tubes of a variety of sizes, depending upon the specific future uses of the media (i.e., stock cultures, nutritional experiments or the like). The tubes are plugged with cotton, plastic closures, or screw caps and sterilized. Screw cap tubes or flasks are generally used to maintain sterile cultures for a long period of time and appear to retard evaporation, although they are generally more expensive. More rapid growth can often be obtained using the media given here by dispensing the media in petri plates. However, media used in petri plates are often difficult to maintain in sterile condition unless given particular types of treatment.

To prevent chemical contamination of media during sterilization, glassware may be washed or cleaned in particular ways. Glassware is often washed in a concentrated acid solution (see Chp. 2, II.C.2) after first washing with a detergent, such as 'Alconox' (Alconox Inc., 215 Park Ave. South, New York, New York 10003). A typical washing procedure is as follows: detergent washing in tap water; rinsing in tap water; washing with acid solution; rinsing in tap water; three rinses in glass-distilled, deionized water. Potassium dichromate is often difficult to remove from glassware by repeated washings. Thus, many investigators have discontinued the use of this chemical cleaning agent since it is toxic often in small quantities to most algal species (see Chp. 2, II.C.2).

D. *Water*

Water generally employed for freshwater growth media is as follows: (1) copper-distilled water; (2) single glass-distilled water; (3) double glass-distilled water; (4) membrane filtered water; and (5) deionized water.

Copper-distilled water has the disadvantage of residual copper which is toxic to many algal species. Single or double glass-distilled water is generally used in most laboratories and can be deionized by a simple attachment of a prepacked deionizing column. Deionized glass-distilled water is generally used, as glass stills (unless frequently and carefully cleaned) may impart undesirable contaminants to the medium and adversely affect the growth of algae to be cultured.

E. *Soil*

There is often some mystique regarding sources of soil for inclusion into those media which require it. Soils which have been successfully employed for the growth of algae have been obtained from Tennessee, Vermont, California and a variety of other locations. A satisfactory soil is one containing small quantities of clay and one which will settle after it has been added to liquid. Pringsheim (1946; also see IV.B.4, 5, following) suggests that soil should not have too great a humus content and should not have commercial fertilizers added recently. It has been my experience that the selection of a suitable soil within these parameters is generally an experimental procedure which requires trial for successful cultivation of certain species by the investigator.

III. METHODS

Media are generally prepared from premixed stock solutions. Aliquots from these stocks are measured and added to a given volume of water. However, some media must be prepared by weighing or measuring the desired components and adding them directly to a given volume of liquid. Accuracy in measuring liquid aliquots from stock solutions or water, and weighing of chemicals is essential. Improper procedures may result in precipitation of one or more of the components of the medium (especially nitrates and phosphates) or a failure of some of the constituents to go into solution.

Stock solutions can be prepared and stored at low temperatures for lengthy periods of time. However, they should be stored, if possible, in tightly sealed glassware since evaporation alters initial concentrations. If crucial and accurate experiments are to be performed, it is best to mix fresh stock solutions of both macro- and micronutrients. All solutions are made using deionized or distilled water.

IV. MEDIA

Table 1-1 indicates the classes of algae that have been successfully grown in the media included in this chapter.

A. *Defined* (see Tables 1-2, 1-3)

1. Beijerinck pH 6.8 (Stein 1966). This medium is made from three stock solutions and the micronutrient solution.

a. Stock I – use 100 ml/l

NH_4NO_3	1.5 g/l
K_2HPO_4	0.2
$MgSO_4 . 7H_2O$	0.2
$CaCl_2 . 2H_2O$	0.1

TABLE 1-1. *Application of media*

Medium	Classes cultured[a]
Defined	
Beijerinck	C
Bold Basal	C, CR, CY, R
Bozniak Community	C, CR, CY, R (mixed)
Cg 10	CY
Chu no. 10	BA, C, CR, CY
Rodhe VIII	BA, C, CR, CY
Volvox	C
Waris	C
Woods Hole MBL	BA, C, CR, CY
Enrichment	
Modified *Porphyridium*	R
Polytomella	CO, E
Proteose	C, CR, CY, E
Soil extract agar	C, CR, CY, R
Soil water	BA, C, CO, CR, CY, E, R
Trebouxia agar	C

[a]BA Bacillariophyceae
C Chlorophyceae
CO Colorless flagellates
CR Chrysophyceae
CY Cyanophyceae
E Euglenophyceae
R Rhodophyceae

TABLE 1-2. *pH and concentration of defined media*

		Concentration (total mol/l)		
	pH	Macro	Micro	Macro + micro
Beijerinck	6.8	5×10^{-3}	4×10^{-4}	5×10^{-3}
Bold Basal	6.6	4×10^{-2}	5×10^{-5}	4×10^{-2}
Bozniak Community	8.0	7×10^{-6}	1×10^{-6}	8×10^{-6}
Cg 10	8.0	1×10^{-3}	1×10^{-4}	1×10^{-3}
Chu no. 10	6.5–7.0	9×10^{-4}	—	9×10^{-4}
Rodhe VIII	7.0–7.5	2×10^{-3}	—	2×10^{-3}
Volvox	7.0	5×10^{-2}	3×10^{-3}	5×10^{-2}
Waris	6.0	1×10^{-2}	—	1×10^{-2}
Woods Hole MBL	7.1–7.3	2×10^{-2}	—	2×10^{-2}

b. Stock II – use 40 ml/l

KH_2PO_4	9.07 g/l

c. Stock III – use 60 ml/l

K_2HPO_4	11.61 g/l

d. Micronutrients – use 1 ml/l

H_3BO_3	1.0 g/100 ml
$CuSO_4 . 5H_2O$	0.15
EDTA	5.0
$ZnSO_4 . 7H_2O$	2.2
$MnCl_2 . 4H_2O$	0.5
$FeSO_4 . 7H_2O$	0.5
$CoCl_2 . 6H_2O$	0.15
$(NH_4)_6Mo_7O_{24} . 4H_2O$	0.10

The micronutrients are dissolved one at a time in 100 ml warm water. Following the addition of each, the pH is adjusted to 5 with KOH pellets. The final solution should have a pH *ca.* 6.5 (iron precipitates at 7.0).

2. *Bold Basal pH 6.6* (Nichols and Bold 1965; also known as Bristol solution). Six individual stock solutions are made of the macronutrients. The three minor constituents are also separate stock solutions as is the micronutrient solution.

a. Macronutrients – use 10 ml each/940 ml

$NaNO_3$	10 g/400 ml
$CaCl_2 . 2H_2O$	1
$MgSO_4 . 7H_2O$	3
K_2HPO_4	3
KH_2PO_4	7
NaCl	1

b. EDTA – use 1 ml/l

EDTA	50 g/l
KOH	31

c. Iron – use 1 ml/l

$FeSO_4 . 7H_2O$	4.98 g/l
H_2SO_4	1.0 ml/l

d. Boron – use 1 ml/l

H_3BO_3	11.42 g/l

e. Micronutrients – use 1 ml/l

$ZnSO_4 . 7H_2O$	8.82 g/l
$MnCl_2 . 4H_2O$	1.44

MoO_3	0.71
$CuSO_4.5H_2O$	1.57
$Co(NO_3)_2.6H_2O$	0.49

3. Bozniak Community pH 8.0 (Bozniak 1969). Prior to preparation, all glassware is ashed at 600 °C for 30 min. The autoclave is scrubbed inside with 95% ethanol to minimize organic contamination; adjust pH of water to 8.0. Separate stock solutions of each macronutrient, the micronutrient solution and the vitamin solution may be prepared and stored frozen. The $Ca(NO_3)_2$ is autoclaved separately and added aseptically when cool to avoid precipitation of calcium phosphate. The ferric citrate–citric acid and vitamin stocks are membrane filtered (0.22 μm GS Millipore Filter, Millipore Corp., Ashby Rd, Bedford, Massachusetts 01730; 55 Montpellier Blvd, Montreal 379, Quebec). These are added after sterilization of the other constituents to the cooled medium.

a. Macronutrients – use 1 ml each/l

$Ca(NO_3)_2.4H_2O$	24.4 g/l
$MgSO_4.7H_2O$	1.0
$NaHCO_3$	16.5
$K_2HPO_4.3H_2O$	0.8
KH_2PO_4	0.8
Na_2SiO_3	5.8
H_3BO_3	0.5
Na_2EDTA	10.0

b. Citrate–citric acid – use 1 ml each/l

Ferric citrate	1.0 g/l
Citric acid	1.0

c. Vitamins – use 1 ml each/l

Thiamine.HCl	1×10^{-6} g/l
Cyanocobalamin	10×10^{-6}

d. Micronutrients

K_2CrO_4	0.0037 g/l
$CoCl_2.6H_2O$	0.020
$MnCl_2.4H_2O$	0.890
$ZnCl_2$	0.0104
$VOSO_4.2H_2O$	0.0039
MoO_3	0.0075
$CuSO_4$	0.00125

4. *Cg 10 pH 8.0* (Van Baalen 1967). Major nutrients are added as salts, or can be dissolved in stock solutions in appropriate concentrations. Aliquots of these stock solutions are then added to the final medium before bringing the medium up to volume. The micronutrient solution ('A5') is made as a single stock solution, and adjusted to pH 8.5.

a. Macronutrients

$MgSO_4.7H_2O$	0.25 g/l
K_2HPO_4	1.0
$Ca(NO_3)_2.4H_2O$	0.025
KNO_3	1.0
$Na_2.EDTA$	0.010
$Fe_2(SO_4)_3.6H_2O$	0.004
Glycylglycine	1.0

b. Micronutrient 'A5' – use 1 ml/l

H_3BO_3	2.86 g/l
$MnCl_2.4H_2O$	1.81
$ZnSO_4.7H_2O$	0.222
MoO_3 (85%)	0.018
$CuSO_4.5H_2O$	0.079
$CoCl_2.6H_2O$	0.010

5. *Chu no. 10 pH 6.5–7.0* (Chu 1942). Prior mixing of stock solutions is suggested to avoid repetitive weighing.

$Ca(NO_3)_2$	0.04 g/l
K_2HPO_4	0.01
$MgSO_4.7H_2O$	0.025
Na_2CO_3	0.02
Na_2SiO_3	0.025
$FeCl_3$	0.8 mg/l

6. *Rodhe VIII pH 7.0–7.5* (Rodhe 1948). Prior mixing of stock solutions is suggested to avoid repetitive weighing.

$Ca(NO_3)_2$	60 mg/l
$MgSO_4$	5
Na_2SiO_3	20
K_2HPO_4	5
Ferric citrate	1.0
Citric acid	1.0
$MnSO_4$	0.03

7. Volvox pH 7.0 (Darden 1966; Starr 1969). Stock solutions of the macronutrients are prepared separately. The micronutrient solution is prepared separately. Glycylglycine, glycerophosphate and the vitamins are considered as macronutrients. Starr (1969) suggests increasing the vitamins (biotin, 2.5 increase; cyanocobalamin, 0.5 increase).

a. Macronutrients

$Ca(NO_3)_2.4H_2O$	0.118 g/l	use 10 ml/l
$MgSO_4.7H_2O$	0.04	use 10 ml/l
Na_2.glycerophosphate.$5H_2O$	0.05	use 10 ml/l
KCl	0.05	use 10 ml/l
Glycylglycine	0.5	use 10 ml/l
Biotin	1.0 μg	use 1 ml/l
Cyanocobalamin	1.0 μg	use 1 ml/l

b. Micronutrients – use 3 ml/l

$FeCl_3.6H_2O$	0.097 g/l
$McCl_2.4H_2O$	0.041
$ZnCl_2$	0.005
$CoCl_2.6H_2O$	0.002
Na_2MoO_4	0.004
Na_2.EDTA	0.750

Adjust pH to 7.0 with 1N NaOH

8. Waris pH 6.0 (Waris 1953). Separate stocks of each macronutrient and the iron sequestrine solution may be prepared.

a. Macronutrients

KNO_3	0.1 g/l
$MgSO_4.7H_2O$	0.02
$(NH_4)_2HPO_4$	0.02
$CaSO_4$	0.05

b. Iron sequestrine

EDTA	1.30 g/l
$FeSO_4.7H_2O$	1.25
1N KOH	13.5 ml/l

pH is adjusted to 6.0

FeNaEDTA (ferric sequestrine) may be substituted for the solution (use 2.49 g FeNaEDTA/liter)

9. Woods Hole MBL pH 7.2 (R. R. L. Guillard, personal communication). Stock solutions of each macronutrient are prepared at 1000-fold concentrations thus 1 ml of each stock is added to 1 l medium. Stock solutions of each micronutrient may also be prepared at 1000-fold concentration. Vitamin stocks should be membrane filtered (or autoclaved) and stored frozen. The tris solution, which is also prepared as a stock solution, should have pH adjusted to 7.2.

Guillard offers the following comments:

Experience has shown that many freshwater algae will grow in this medium even if no buffer is used. However, most experience in this connection has been had using ferric citrate–citric acid and 3 mg/liter each, as the chelated iron source, which may be a significant factor.

Note that the nitrate and phosphate levels can be reduced at least fivefold without much reducing the crop of algae, though survival in old cultures may be shorter. For sensitive freshwater algae, tenfold reduction of nitrate and phosphate may be essential. For species that cannot use nitrate, add *ca.* 100 μM ammonium chloride.

The bicarbonate content of the medium can be increased if desired. For the most careful work it is probably best to filter sterilize the bicarbonate solution, but a concentrated solution or the dry powder may be autoclaved. The pH should be checked; if it has not gone up very much, it indicates that no significant decomposition to CO_2 and carbonate has occurred.

a. Macronutrients – use 1 ml each/l	
$CaCl_2.2H_2O$	36.76 g/l
$MgSO_4.7H_2O$	36.97
$NaHCO_3$	12.60
K_2HPO_4	8.71
$NaNO_3$	85.01
$Na_2SiO_3.9H_2O$	28.42
b. Micronutrients – use 1 ml each/l	
$Na_2.EDTA$	4.36 g/l
$FeCl_3.6H_2O$	3.15
$CuSO_4.5H_2O$	0.01
$ZnSO_4.7H_2O$	0.022
$CoCl_2.6H_2O$	0.01
$MnCl_2.4H_2O$	0.18
$Na_2MoO_4.2H_2O$	0.006
c. Vitamins	
Thiamine.HCl	0.1 mg/l
Biotin	0.5 μg/l
Cyanocobalamin	0.5 μg/l

d. Tris – use 2 ml/l
Tris(hydroxymethyl)-
aminomethane 50 g/200 ml
Adjust pH to 7.2 with HCl

e. Glycylglycine may be substituted for tris if bacteria are absent

B. *Enrichment* (see Table 1-1)

Most of the media previously described as defined media can be used for additional algal groups by adding a variety of other components or modifying the amounts of certain reagents. Enrichment media often have the advantage of supporting growth of large numbers of algal species. However, if these media are highly organic, they have a distinct disadvantage of encouraging considerably more bacterial growth than strictly inorganic media. Thus, the previously described defined media may become undefined enrichment media when certain modifications are made.

One of the chief elements which has been successfully manipulated in media is the nitrogen source (see A.9, preceding). Nitrogen is generally supplied as nitrate, although ammonia, urea and certain amino acids have been used. Nitrogen levels can often be elevated to three times normal level or reduced by one-half or lower. The use of one-tenth the normal amount of nitrogen has been used for observing rapid color changes in cultures due to nitrogen depletion.

Urea has been employed very successfully as a nitrogen source and is particularly good in a medium such as Bold Basal medium (A.2, preceding). A stock of urea is prepared by dissolving 54 g/l in distilled water. Ten ml of this stock is used as a nitrogen source for the solution to replace nitrate. Often casamino acids (a mixture containing a variety of amino acids) may be added at very low concentrations to a growth medium when amino acid requirements of a specific nature are unknown.

A variety of plant extracts have also been used to enrich certain media. For example, peat moss has been used in conjunction with soil in soil water cultures for successful cultivation of Charophyceae or the desmids (Chlorophyceae).

Additionally, yeast extract may be used as a source of vitamins when a precise requirement is unknown. Yeast extract is particularly useful when attempting to purify cultures as it greatly enhances the growth of bacteria. Vitamins, generally cyanocobalamin, biotin and thiamine, are required by many algae when grown in axenic culture. They are generally used in very low concentrations and stored in a crystalline form or in a stock solution at sub-freezing temperature. They may be autoclaved or membrane filter sterilized and added aseptically to the sterile medium. Vitamins are particularly important to supply when the cultures are free of bacteria since the bacteria which were initially present may have supplied appropriate vitamins to the growing culture. It should be noted certain filters may retain large ions with positive charges, thus retaining a portion of the vitamin within the filter.

Further, enrichments such as the addition of large quantities of calcium carbonate to an inorganic medium as Bold Basal (A.2, preceding) or Beijerinck (A.1, preceding) have been successful for growing certain Rhodophyceae. Two particular enrichment media have been successful in my own laboratory, the first being a combination of deep layer (7 cm) of Bold Basal plus agar in petri plate overlain with 2–3.5 cm liquid combination of six parts Bold Basal and four parts soil extract (4, 5, following). The second medium which has been successfully used is Beijerinck (A.1, preceding) with the addition of excess calcium carbonate, the latter being particularly successful in the cultivation of limestone spring organisms.

1. Modified Porphyridium (Sommerfeld and Nichols 1970). For each 1000 ml of medium required:

Soil extract (see 4 or 5, following)	100.0 ml
Yeast extract	1.0 g
Tryptone	1.0 g

2. Polytomella (Starr 1964). For each 1000 ml of medium required:

Sodium acetate	2.0 g
Yeast extract	1.0
Tryptone	1.0

3. Proteose agar (Starr 1964). For each 1000 ml of medium required:

Bold Basal medium (see A.2, preceding)	1000.0 ml
Proteose peptone	1.0 g
Agar	15.0 g

4. Soil extract agar (Starr 1964). This is prepared by adding 40 ml of soil water supernatant to 960 ml of an inorganic medium (see A, preceding). The medium is solidified with 15 g agar/liter.

Soil water supernatant is prepared by covering the bottom of a large flask with soil, adding distilled water over the soil until the flask is two-thirds full. The flask is then steamed (tyndallization) for 2 h (see Chp. 12,

TABLE 1-3. *Concentrations of additives in defined media*

Additive	Concentration/liter								
	Beijerinck	Bold	Bozniak	Cg 10	Chu no. 10	Rodhe	Volvox	Waris	Woods Hole
Macronutrients (in mM)									
$(NH_4)_2 . HPO_4$	—	—	—	—	—	—	—		0.15 —
NH_4NO_3	1.87	—	—	—	—	—	—	—	—
K_2HPO_4	4.11	0.43	0.03	0.57	0.06	0.029	—	—	0.05
KH_2PO_4	2.67	1.29	0.03	—	—	—	—	—	—
KNO_3	—	—	—	9.89	—	—	—	0.99	—
KCl	—	—	—	—	—	—	0.67	—	—
$NaNO_3$	—	2.94	—	—	—	—	—	—	—
NaCl	—	0.43	—	—	—	—	—	—	—
$NaHCO_3$	—	—	0.72	—	—	—	—	—	0.15
Na_2CO_3	—	—	—	—	0.19	—	—	—	—
Na_2SiO_3	—	—	0.21	—	0.20	0.164	—	—	0.1
$NaNO_3$	—	—	—	—	—	—	—	—	1.0
$MgSO_4$	0.08	0.30	0.04	1.01	0.10	0.042	0.16	0.08	0.15
$MnSO_4$	—	—	—	—	—	0.0002	—	—	—
$CaCl_2$	0.07	0.17	—	—	—	—	—	—	0.25
$Ca(NO_3)_2$	—	—	0.61	0.11	0.24	0.366	0.5	—	—
$CaSO_4$	—	—	—	—	—	—	—	0.37	—
$Fe_2(SO_4)_3$	—	—	—	0.008	—	—	—	—	—
Fe citrate	—	—	0.018	—	—	0.0041	—	—	—
Citric acid	—	—	0.005	—	—	0.0052	—	—	—
Na_2glycerophosphate	—	—	—	—	—	—	0.16	—	—
Glycylglycine	—	—	—	7.57	—	—	3.78	—	3.3[a]
Tris	—	—	—	—	—	—	—	—	4.13[a]

Vitamins (in nM)									
Biotin	—	—	—	—	—	—	0.41[b]	—	2.047
Cyanocobalamin	—	—	7.4	—	—	—	0.74[b]	—	0.368
Thiamine.HCl	—	—	3.0	—	—	—	—	—	297
Micronutrients (in μM)									
H_3BO_3	16.17	184.7	46.0	46.3	—	—	—	—	—
EDTA	7.65	171.09	—	—	—	—	—	17.86	—
Na_2.EDTA	—	—	26.86	26.86	—	—	12.09	—	11.71
$CoCl_2$	0.63	—	0.084	0.042	—	—	0.05	—	0.042
$Co(NO_3)_2$	—	16.8	—	—	—	—	—	—	—
$CuSO_4$	0.601	62.9	0.008	0.316	—	—	—	—	0.040
$FeCl_3$	—	—	—	—	4.9	—	2.15	—	11.65
$FeSO_4$	1.798	17.9	—	—	—	—	—	17.91	—
K_2CrO_4	—	—	0.002	—	—	—	—	—	—
$(NH_4)_6Mo_7O_{24}$	0.081	—	—	—	—	—	—	—	—
MoO_3	—	4.9	0.052	0.147	—	—	—	—	—
Na_2MoO_4	—	—	—	—	—	—	0.12	—	0.025
$MnCl_2$	2.56	7.3	4.498	9.147	—	—	1.24	—	0.910
$VOSO_4$	—	—	0.019	—	—	—	—	—	—
$ZnCl_2$	—	—	0.076	—	—	—	0.22	—	—
$ZnSO_4$	7.65	30.7	—	0.772	—	—	—	—	0.077
KOH	—	553	—	—	—	—	—	54.0	—

[a] Use either glycylglycine or tris for buffer (see IV.A.9).

[b] Starr (1969) suggests: biotin 1.023 nM; cyanocobalamin 0.11 nM.

II.C.2.b). The flask is allowed to stand until the soil has settled out (this may take 1–7 days depending on the soil used). The supernatant is drawn off, taking care not to resuspend the soil; filtered through a no. 1 or no. 2 Whatman filter. If soil water is to be left to stand following filtration, it should be autoclaved as it will readily become contaminated.

Supernatant from soil water (5, following) may be used as soil extract.

5. *Soil water* (Pringsheim 1946; Starr 1964). Variations of this medium are for non-axenic culture, especially for isolation purposes and for growing algae in order to secure 'normal' growth forms. Success with soil water media depends on the selection of a suitable garden soil. This soil should be of medium, but not too great, humus content, and should not have been recently fertilized with commercial fertilizers. Soils with a high clay content are usually not the most suitable for most organisms, although some clay should be present (see II.E, preceding).

a. Place 0.65–1·5 cm garden soil in bottom of container (test tube, 8 oz bottle, etc.).

b. Add distilled water until container is three-fourths full.

c. Cover. Use cotton plugs, metal or plastic caps, beakers, etc.

d. Steam (see Chp. 12, II.C.2.b) for 1–1½ h on three consecutive days. It is important that the temperature in the steamer be at 100 °C for the full time period (see Chp. 12, II.C.3). Also, it is essential that the containers cool to room temperature for at least 18 h between steaming. Thus, it may be necessary to wait two days between the second and third steaming periods.

A few algae such as *Spirogyra* (Chlorophyceae) grow well in the near neutral medium. For most presumptively phototrophic algae which thrive in an alkaline medium, a small pinch of powdered $CaCO_3$ is placed in the bottom of the container before the soil and water are added.

Some algae like the Euglenophyceae, *Astasia* and *Euglena*, and the Chlorophyceae, *Pyrobotrys*, *Polytomella* and *Polytoma*, require additional complex nitrogenous or carbon compounds not present in the basic formula. For *Euglena* and *Pyrobotrys*, the best results have been obtained by adding one-quarter of a garden pea cotyledon (*Pisum sativum*) to the basic medium (including $CaCO_3$) before steaming. For the colorless forms, the addition of a barley grain (fruit of *Hordeum*) before steaming supplies the necessary carbon source. A few species, such as *Botryococcus* (Chlorophyceae), grow best when a pinch of sterile NH_4MgPO_4 is added after the steaming of the basic medium (including $CaCO_3$).

6. *Trebouxia agar* (Starr 1964). For each 1000 ml of medium required:

Bold Basal medium (see A.2, preceding)	850.0 ml
Soil extract (see 4 or 5 preceding)	140.0
Proteose peptone	10.0 g
Glucose	20.0
Agar	15.0

V. REFERENCES

Bozniak, E. 1969. Laboratory and Field Studies of Phytoplankton Communities. *Ph.D. Dissertation.* Washington University, St Louis, Missouri. 106 pp.

Chu, S. P. 1942. The influence of the mineral composition of the medium on the growth of planktonic algae. *J. Ecol.* **30**, 284–325.

Darden, W. H. 1966. Sexual differentiation in *Volvox aureus*. *J. Protozool.* **13**, 239–55.

Gerloff, G. C., Fitzgerald, G. P. and Skoog, F. 1950. The isolation, purification, and nutrient solution requirements of blue-green algae. In Brunel, J., Prescott, G. W. and Tiffany, L. H., eds., *The Culturing of Algae*, pp. 27–44. C. F. Kettering Foundation, Yellow Springs, Ohio.

Hutner, S. H., Zahalsky, A. C., Aaronson, S., Baker, H. and Frank, O. 1966. Culture media for *Euglena gracilis*. In Prescott, D. M., ed., *Methods in Cell Physiology*, **2**, 217–28. Academic Press, New York.

National Academy of Sciences – National Research Council. 1960. *Specifications and Criteria for Biochemical Compounds*. Publ. 719. Washington, D.C. (Suppl. 719-1, 1963).

Nichols, H. W. and Bold, H. C. 1965. *Trichosarcina polymorpha* gen. et sp. nov. *J. Phycol.* **1**, 34–8.

Pringsheim, E. G. 1946. *Pure Cultures of Algae*. Cambridge Univ. Press, London. 119 pp.

Provasoli, L. and Pintner, I. J. 1960. Artificial media for freshwater algae; problems and suggestions. In Tryon, C. A., Jr and Hartman, R. T., eds., *The Ecology of Algae*, pp. 84–96. Special Publ. No. 2, Pymatuning Laboratory of Field Biology, Univ. Pittsburgh.

Rodhe, W. 1948. Environmental requirements of freshwater plankton algae. *Symbol. Bot. Upsaliensis* **10** (1), 5–145.

Sommerfeld, M. R. and Nichols, H. W. 1970. Comparative studies in the genus *Porphyridium* Naeg. *J. Phycol.* **6**, 67–78.

Starr, R. C. 1964. The culture collection of algae at Indiana University. *Amer. J. Bot.* **51**, 1013–44.

Starr, R. C. 1969. Structure, reproduction and differentiation in *Volvox carteri* f. *nagariensis* Iyengar, strains HK9 and 10. *Arch. Protistenk.* **111**, 204–22.

Stein, J. R. 1966. Growth and mating of *Gonium pectorale* (Volvocales) in defined media. *J. Phycol.* **2**, 23–8.

Van Baalen, C. 1967. Further observations on growth of single cells of coccoid blue-green algae. *J. Phycol.* **3**, 154–7.

Waris, H. 1953. The significance for algae of chelating substances in the nutrient solutions. *Physiol. Plant.* **6**, 538–43.

2: Growth media – marine*

J. McLACHLAN

Atlantic Regional Laboratory National Research Council of Canada, Halifax, Nova Scotia, Canada

CONTENTS

* NRCC No. 12916.

I. INTRODUCTION

Seawater is an ideal medium for growth of marine plants, and has not been improved upon. It is, however, an intrinsically complex medium, and contains over fifty known elements in addition to a large but variable number of organic compounds. Unfortified seawater is adequate for some purposes such as spore germination and isolation of various species. It is usually necessary, however, to fortify natural water with several plant nutrients especially nitrogen, phosphorus and perhaps iron. Soil and other extracts are generally beneficial, although it is now possible in many instances to substitute known additives.

Synthetic media have been designed primarily to provide simplified, defined media. Growth and development may be as satisfactory as in enriched seawater, but commonly these media are disappointing or fail completely. A better understanding of micronutrients and growth factors has led to improved media, yet much remains to be accomplished especially with regard to benthic species.

Precipitation resulting from steam sterilization has been a major problem with marine media. Introduction of several compatible, non-toxic pH buffers and use of metal chelators has alleviated some of these difficulties. These additives are included in most recent synthetic media in which there has been a tendency to lower the salinity and especially the calcium concentration. The task has, therefore, been to develop media within a restricted pH range capable of being autoclaved. It is not, however, necessary to heat sterilize media! Filters and filtration apparatus are available commercially, and for most purposes media can be thus filter sterilized easily and practically. More attention should be focused on the organisms (and less on the autoclave) which may permit the culture of more difficult and fastidious species. (See Chp. 12 for a discussion of sterilization methods.)

Numerous enriched and synthetic media have been formulated, which together with generally trivial modifications, almost equal the number of investigators. Most media, and especially the synthetics, were developed for unicellular algae although benthic species have been cultured successfully in some of them. It is commonly stated that a particular medium is best suited for certain species, but this is rarely supported by comprehensive, quantitative comparisons. Marine algae generally have fairly wide tolerances, and difficulties attributed to media can frequently be related to problems of isolation, conditions of manipulation, and incubation and physiological state of the organism. A single medium will generally serve most needs of an investigator. The 'f/2' and ASP-6 media are excellent examples of the potential usefulness of a single medium in which a wide selection of organisms have been grown (see IV, A, B, following). At the Atlantic Regional Laboratory, numerous unicellular and benthic species have been grown in one basal medium. The synthetic medium is employed in the same manner as the enriched seawater medium with a formulation of conservative elements replacing seawater.

The media given in this section have had wide use and applicability. Although only a few media are presented, these are examples of possible *major* variations. Undoubtedly numerous other media are equally effective. The purpose of marine media is, however, to grow marine algae, and it is imperative to emphasize that media are only a means to this end. While fulfilling this purpose, a medium should be as simple as possible in composition and preparation.

Included in this book are chapters concerned with aspects which have a direct bearing on media such as sterilization (Chp. 5, 12), carbon sources (Chp. 18), sterility testing (Chp. 12), and agar (Chp. 1, II.B). Documentation in this chapter is minimal; the references selected may be used as a guide to some of the more recent publications on marine media. Readers are also referred to a comprehensive article on the culture of algae written by Droop (1969) for *Methods in Microbiology*.

II. MATERIALS

A. *Chemicals*

Most chemicals employed in marine media are common and inexpensive. Reagent grade is used whenever possible, although a few compounds are available commercially only in the technical grade. Some of the organic chemicals, such as vitamins or buffers, may not be generally available, but there are a number of reliable vendors (e.g., Calbiochem, P.O. Box 12087, San Diego, California 95112; Schwarz/Mann, Orangeburg, New Jersey 10962; ICN Nutritional Biochemicals Division, International Chemical and Nuclear Corp., 26201 Miles Rd, Cleveland, Ohio 44128; Sigma Chemical Co., P.O. Box 14508, St Louis, Missouri 63178). Digests, extracts, etc., may be purchased from microbiological supply houses (e.g., BBL, BioQuest Division, Becton, Dickinson and Co., P.O. Box 243, Cockeysville, Maryland 21030; Difco Laboratories, P.O. Box

1058A, Detroit, Michigan 48232; Sheffield Chemicals, Division of National Dairy Products, 2400 Morris Ave., Union, New Jersey 07083). It is often less expensive to obtain chemicals in relatively small lots to avoid the common hazard of hydration in humid marine laboratories. Containers must always be kept tightly closed.

B. *Equipment*

Minimum equipment is required for preparation of media. The autoclave is the single most expensive item. Additional requirements include an analytical balance (1 mg sensitivity), line operated pH meter, hot plate-magnetic stirrer, automatic pipette washer (plastic or stainless steel), good quality stainless steel pails of 13–15 qt (US) capacity (milk pails are adequate and cheap) and a refrigerator with a freezer compartment (a domestic type is entirely satisfactory). A top-loading balance (2 kg capacity) is useful but not essential except for large-scale work.

C. *Glassware*

Borosilicate glassware should be used exclusively in sizes and quantities as required. Essential items include: membrane filter apparatus and syringe adapter (Millipore Corp., Ashby Rd, Bedford, Massachusetts 01730; 55 Montpellier Blvd, Montreal 379, Quebec: Gelman Instrument Co., 600 S. Wagner Rd, Ann Arbor, Michigan 48106), syringes, graduated cylinders (some with stoppers), beakers, Erlenmeyer flasks, reagent bottles, plastic bottles with closures, pipettes (either Mohr or serological type), wash bottles, flasks and tubes with teflon lined screw-caps, ampules, stirring rods, buret (25 or 50 ml capacity), petri dishes, thermometers (−20 to 110 °C), spatulas, funnels, Büchner funnel (glazed porcelain), and filter flasks. Many of these items are available in plastics or teflon.

All glassware must be scrupulously clean, and *general purpose glassware should not be used in media preparation.* New glassware is soaked in dilute HCl to remove any free alkali present.

1. Detergents. Domestic detergents leave a residual film on glassware, but there are a number of satisfactory laboratory detergents such as 'Haemo-Sol' (American Hospital Supply, 2020 Ridge Ave., Evanston, Illinois 60201). Other detergents are available through scientific supply companies. The detergent should remove organic matter below 100 °C, and leave the glassware neutral after rinsing.

a. Rinse the glassware in a stream of warm tap water immediately after use.

b. Soak overnight in a detergent solution prepared in distilled water and contained in a stainless steel receptacle.

c. Heat or boil gently for approximately one hour, rinse in tap water followed by several rinses in distilled water.

d. Invert to dry (never use a pegboard) and then cover.

The same procedure is used for washing up pipettes which are rinsed in an automatic washer; the washer should be inverted and dried between uses, and cleaned periodically with hot detergent.

2. Acids. Glassware which fails to respond to detergent cleaning may be soaked in concentrated H_2SO_4 saturated with $NaNO_3$ or in hot HNO_3 (in a fume hood) followed by thorough rinsing with distilled water. Chromate cleaning solutions should never be used as the chromium ion, which is quite toxic, adsorbs onto glass and is difficult to remove.

3. Baking. Minute amounts of organic matter can be destroyed by baking the glassware at 350–400 °C for 3–4 h. After baking exercise care to prevent dust and other airborne contamination; dirty fingers are a common source of contamination.

Distilled water is discussed by Nichols in Chp. I, II.D. For the media and stock solutions included here, distilled water for preparation of stock solutions and media implies glass- or double-distilled water. There is, however, little value in adding quality water to media containing soil extract or unwashed agar.

D. *Seawater*

To obtain seawater free from gross pollution, it may be necessary to collect offshore. Offshore water is, in addition, relatively free of fine sediments and if large volumes are required more easily filtered.

1. Collection. Inshore water may be collected either from the shore or a small boat. Water around piers is usually contaminated. To avoid contamination from a vessel, it is necessary to pump water from below the surface by means of a small, self-priming (stainless steel or plastic) gear pump. Large volumes of seawater can be pumped through a 293 mm diameter membrane filter contained in a plastic or stainless steel holder (Millipore or Gelman). The pump is driven by a small gasoline or electric motor. The entire apparatus is portable, and water may be collected, filtered and bottled in a single operation. Stainless steel comes in a number of varieties, some of which are relatively soluble in seawater. Caution should therefore be exercised in the use of this material, and if problems arise plastic is the obvious alternative.

2. *Filtration.* Seawater always contains some particulate matter. Much is removable by paper filters, e.g., Whatman no. 1, although membrane filters are necessary for complete removal. A 0.45 μm pore-size filter, which has a reasonable rate of flow, is commonly used to eliminate all but the finest particles. Large diameter filters will save considerable time and effort, and it is often practical to coarse-filter the water before membrane filtration.

a. Prepare a slurry of 'Celite' (acid-washed diatomaceous earth; Johns Manville Corp. Celite Division, 22 E. 40 St., New York, New York 10016) in distilled water, and pour onto a filter paper (Whatman no. 3) in a Büchner funnel.

b. Remove most of the water by gentle vacuum, but do not permit the Celite to dry.

c. Transfer the filter to the receiving vessel, lay a second sheet of filter paper (Whatman no. 1) over the Celite bed and gently add seawater to the funnel.

d. If the rate of filtration slows appreciably, replace the top filter disc, and remove a thin layer of Celite from the filter bed with a spatula. This water may be used directly, or it may be membrane filtered.

Small quantities of water can be filtered quickly by preparing a Celite bed directly on the membrane filter. Alternatively, the membrane filter is overlain with a glass-fiber filter (obtainable from the same source as the membrane filter and in the same diameter). For many purposes filtration through the glass-fiber filter only is sufficient.

3. *Removal of dissolved organic matter.* It may be desirable to remove dissolved organic matter because of suspected contamination or for use in bioassays such as vitamins. This material can be removed by adsorption onto activated charcoal. This procedure also tends to standardize the seawater. The charcoal is first washed exhaustively with ethanol, benzene, methanol, 50% ethanol and distilled water (see Craigie and McLachlan 1964). Small quantities of seawater are treated by the addition of 2 g of powdered, washed charcoal per liter of water, and agitated for about one hour, followed by membrane filtration. Large volumes of seawater are passed through charcoal contained in a glass column or fritted filter. The charcoal may be washed in the column or filter, but it must not be allowed to dry before addition of the seawater.

An alternative method has been described by Hamilton and Carlucci (1966) which involves destruction of organic material by means of high-energy ultraviolet radiation. This procedure also sterilizes the water.

4. Containers, storage. Seawater is stored in either glass or plastic (linear polyethylene) containers fitted with grips to facilitate handling when wet. Rectangular containers require minimal storage space and can be stacked. New containers should be leached for several days with dilute HCl and then rinsed thoroughly. Raw seawater must not be stored for prolonged periods. Use of 'aged' seawater is frequently recommended, but media prepared from recently collected and properly filtered seawater are entirely satisfactory. Water should be refrigerated during storage if possible, or kept cool and in the dark.

III. PREPARATION OF MEDIA, STOCK SOLUTIONS

A. *Media*

Reference to specific media is given in IV, following, and Tables 2-2 to 2-5. Procedures for preparation frequently depend upon the method of sterilization used.

1. Enriched seawater. Only 'plant' nutrients are added to seawater. The salinity of the water is determined, and any necessary adjustments are made before addition of nutrients. A salinity of 35‰ is considered normal seawater. The salinity is lowered by dilution with distilled water if required. Increased salinities are achieved by concentrating the seawater in a rotary or flash evaporator. An alternative procedure which involves freezing-out has been described by Shapiro (1961). Rather than concentrating seawater, osmotic agents such as NaCl, sucrose, glycerol, mannitol, etc. may be added. It must be remembered that normal seawater has an ionic strength of *ca.* 1 M which is equivalent to, for example, 0.5 M NaCl or 1 M sucrose. Salinity is not an inherent feature of any medium; most algae will grow well around 30‰, and many unicells can be cultured at much lower salinities (see Chp. 4, IV.A.3).

a. Basic method

i. Add approximately 80–90% of the required volume of seawater to a beaker.

ii. Dispense the appropriate nutrients from previously prepared stock solutions; stir thoroughly after each addition.

iii. Check the pH and make any necessary adjustments with either 1 N HCl or NaOH; ensure that the medium is well mixed. The pH is generally adjusted to around 7.5 whereas the pH of seawater is 8.2–8.4.

iv. Dilute the medium to volume with seawater and dispense into appropriate containers preparatory to sterilization.

b. Alternatives – several alternatives may be employed following sterilization of the seawater, or partially prepared medium, by autoclaving or tyndallization (see Chp. 12, II.C.2).

i. Sterilize the nutrients separately by filtration or autoclaving. Nitrogen and phosphorus compounds may be autoclaved together; autoclave trace metals separately in a slightly acidified (pH 5.5–6.0) solution; dispense the vitamins from a previously sterilized solution (see E.1, following); soil extract should be autoclaved for at least 30 min.

ii. Add the sterile nutrient solutions aseptically to the cooled seawater after it has been shaken or aerated aseptically.

Precipitation in an unbuffered, autoclaved medium can be minimized by addition of phosphate and trace metals after sterilization, and by dilution of the seawater with 5% or more of distilled water. Precipitates are, however, generally not detrimental to growth. Little or no precipitate will form if the medium is buffered with 4–5 mM of either tris or glycylglycine (see B.2.g, following).

2. Synthetic

a. Dissolve appropriate quantities of weighed NaCl, $MgSO_4$, KCl and Na_2SO_4 (if required) in approximately 80% of the final volume of distilled water; these salts are easily soluble with minimal stirring.

b. Dispense $MgCl_2$ and $CaCl_2$ from previously prepared stock solutions (see B.2.e, f, following).

The additions in a and b form the 'basal seawater'.

c. Addition of the other nutrients is as in the preparation of enriched seawater media.

d. Check the pH, which is usually also adjusted to *ca.* 7.5, and dilute to volume with distilled water.

The same procedures used for sterilization of enriched seawater media are employed for the synthetic media.

3. Commercial preparations. These preparations (Table 2-1) are synthetic mixes and not dehydrated seasalts. Most are sold in 100 lb units which require large mixing containers and sufficient storage space. A portion of the salts may be weighed for small-scale preparations, but there is no assurance that the components, especially the minor ones, are evenly distributed in the dry mix. At least one preparation is packaged in small quantities sufficient for 5 gal (US) of solution. Usually some of the trace metals are provided in liquid form contained in small vials. Do not allow the salts to hydrate during storage.

TABLE 2-1. *Commercial preparations, conc.* μM/l (*sp. gr.* 1.025)

Additive	Instant ocean[a] Conc.	Additive	Dayno[b] Conc.
Cl	519×10^3	NaCl	471×10^3
Na	444×10^3	$MgCl_2$	28.5×10^3
SO_4	26×10^3	$MgSO_4$	57.5×10^3
Mg	49.4×10^3	$CaCl_2$	12.4×10^3
K	9.5×10^3	KCl	9.84×10^3
Ca	9.2×10^3	$NaHCO_3$	2.5×10^3
HCO_3	2.3×10^3	$SrCl_2$	74
H_3BO_3	404	$MnSO_4$	23.4
Br	250	Na_2HPO_4	12.3
Sr	91.3	Na_2MoO_4	4.1
PO_4	10.5	LiCl	23.4
Mn	18	$Ca(C_6H_{11}O_7)_2$	1.53
MoO_4	4.4	KI	0.57
S_2O_3	3.6	KBr	240
Li	28.8	Cu (as Cl)	0.028
Rb	1.2	Al (as SO_4)	0.30
I	0.55	Co (as SO_4)	0.34
EDTA	0.13	Rb (as SO_4)	1.31
Al	1.5	Zn (as SO_4)	0.24
Zn	0.31	Fe (as SO_4)	0.42
V	0.39		
Co	0.17		
Fe	0.18		
Cu	0.047		

[a] Aquarium Systems, Inc., 33208 Lakeland Blvd, Eastlake, Ohio 44094.
[b] Dayno Sales Co., 678 Washington St., Lynn, Massachusetts 01901.

a. Add the salts to 80–90% of the final volume of distilled water and stir; a mechanical stirrer is useful. The solution will be cloudy initially, but should clear within several hours. Aeration will frequently facilitate solution.

b. After the salts have dissolved, membrane filter, and dilute to the desired specific gravity (1.025 is equivalent to *ca.* 35‰).

c. Add the liquefied trace metal solution, or alternatively dilute this solution and add during preparation of the medium.

d. Store the solution in the same manner as seawater.

Procedures for preparation of media are as for enriched seawater media (1, preceding). Additional trace metals are not required, but a suitable metal chelator is indicated.

B. *Stock solutions*

1. General. Stock solutions are prepared as a convenience. The concentration largely depends upon the purpose for which they will be used and the volume of media routinely prepared. It is important to ensure that in preparation of enriched seawater media, the addition of nutrients does not alter the salinity significantly. Most salts are sufficiently soluble to permit a 500–1000 dilution of the stock solution. In general cations are supplied as chlorides or sulphates and anions as sodium or potassium salts. Many salts are hygroscopic and some are impossible to weigh. Generally the inaccuracies are minor due to the dilution factor. The concentration of some stock solutions is very low, viz. some trace metals and vitamins, and primary stock solutions are prepared for dilution to obtain the working stock solutions.

a. Dissolve the required amount of salt in 80–90% of the final volume of distilled water. Gentle heating will often expedite solution, and a magnetic stirrer is used for prolonged mixing.

b. After complete dissolution and cooling, make any necessary pH adjustments. Dilute to volume in a graduated cylinder; a volumetric flask is rarely required.

In preparation of a stock solution containing a mixture of compounds, dissolve each individually in a minimal volume of water. Then combine these solutions, and dilute to volume.

c. Store stock solutions in glass reagent bottles or plastic containers and refrigerate. If the media are to be autoclaved, a volatile preservative may be used. Add a few drops of a solution containing 1 part (v/v) chlorobenzene, 1 part 1,2-dichloroethane and 2 parts *n*-butylchloride (1-chlorobutane) (Hutner, Cury and Baker 1958) and shake vigorously.

2. Specific additives. Some additives require either special procedures or precautions not indicated above.

a. A few salts are usually not prepared as stock solutions because of low solubilities or excessive efflorescence. These include NaCl, $MgSO_4$, KCl, and Na_2SO_4.

b. Nitrogen is usually added as nitrate. Ammonia, as NH_4Cl, is toxic to many species in alkaline media, and the maximum concentration tolerated may be as low as 0.1 mM/l. Considerable quantities may be lost from the medium through volatilization during autoclaving, and sterile solutions are added aseptically after the medium has cooled. Urea will decompose when heated, and must be added by sterile filtration. Some amino acids will decompose when autoclaved with the medium.

c. Phosphorus is usually added as orthophosphate. Sodium glycerophosphate may precipitate as the calcium salt at elevated temperatures. If autoclaved with the medium, equimolar concentrations of an acid whose calcium salt is soluble may be added. To stabilize glycerophosphate, nitrilotriacetic acid (see D.3, following) is commonly used and some organic acids, such as citric, are equally effective.

d. Silicon can probably be omitted from media, except for the Chrysophyta (especially the Bacillariophyceae, the diatoms). This element is most often added as sodium metasilicate (Na_2SiO_3) from prepared stock solutions. The salt is easily dissolved in minimum quantities of warm water. These solutions, which are strongly alkaline, precipitate (or polymerize) when added to media or seawater, and furthermore it is difficult to sterile filter media with added silicate. These problems can be overcome if the stock solution is acidified to pH 2 with concentrated HCl; the solution is stable. Diatoms grow equally well as in media prepared from alkaline silicate solutions.

e. Calcium in synthetic media is provided as $CaCl_2$ which is impossible to weigh accurately because of deliquescence.

i. Weigh an approximate amount of $CaCl_2$; dissolve in about 80% of the final volume of distilled water.

ii. Filter through a 0.45 μm membrane filter.

iii. Determine the chloride concentration by titration with $AgNO_3$.

iv. Dilute to the required volume to obtain the desired concentration.

Alternatively, prepare the stock solution from $CaCO_3$.

i. Weigh accurately sufficient $CaCO_3$ to give the desired calcium concentration.

ii. Place in a beaker; slowly add concentrated HCl until the $CaCO_3$ has dissolved (160–170 ml will dissolve 1 mole of $CaCO_3$); there will be considerable effervescence.

iii. Warm the solution; stir for *ca.* 30 min to drive off excess CO_2.

iv. Cool; membrane filter as above and dilute to volume.

f. Some or all of the magnesium in synthetic media may be added as $MgCl_2$. This salt is also extremely deliquescent. Stock solutions are prepared and the concentration determined as for $CaCl_2$. Magnesium carbonate cannot be used since the magnesium content is only approximate.

g. Buffers – only two buffers, tris (hydroxymethyl)aminomethane (known as tris) and glycylglycine, are used extensively in marine media. Both buffer well in the pH range 7.5–8.5.

i. Tris buffer (also referred to as 2-amino-2-(hydroxymethyl)-1,3-propanediol) may be purchased as preset crystals which give the desired pH upon dissolution. Also available are Trisma.HCl and Trisma.Base which, when combined in specified proportions, give the desired pH. There are reports that stock solutions will deteriorate upon standing, and preparation of a fresh solution prior to use is recommended. Tris is strongly alkaline and to avoid precipitation, it is not added directly to the medium. Dissolve the required amount of tris in distilled water; if added to enriched seawater media use minimal volumes. Adjust the pH with concentrated HCl to the desired pH of the medium; usually no further adjustment of the pH will be required.

ii. Glycylglycine is readily soluble in water, and the powder is added directly to the medium. It is slightly acidic, and it may be necessary to make a small adjustment of the pH with several drops of 1N NaOH.

Glycylglycine is rapidly metabolized by microbes and can be used only with axenic cultures, whereas tris may be used with xenic or agnotobiotic cultures. There have been no reports of glycylglycine toxicity, but inhibition of growth of some organisms has been noted with tris. Minimal quantities should therefore be employed, and for most purposes 1–5 mM is sufficient. Apparently neither tris nor glycylglycine can serve as a nitrogen source for algal growth.

C. *Inorganic micronutrients*

Inorganic micronutrients fall into two groups: (1) those added to all media; and (2) those added to synthetic media only. Frequently the latter are omitted, but there is mounting evidence that at least some of these elements have a marked influence on the growth of benthic species. This may explain in part why synthetic media have often been unsatisfactory, although most unicellular species investigated do not appear to require these nutrients.

1. Trace metals – I (Table 2-2). These elements are prone to precipitate in alkaline media and should be added in a chelated form. All salts, with the exception of boron, are sufficiently soluble to permit preparation of a concentrated stock solution for 1000-fold dilution. Most of these elements are obtainable from Geigy Industrial Chemicals Division (Saw Mill River Rd, Ardsley, New York 10502; 630 Evans Ave., Toronto 520, Ontario). The manufacturer's recommended concentrations of chelates may differ slightly from the amounts specified in Table 2-2.

a. It may be desirable to prepare a separate stock solution of iron. Crystals of $FeSO_4.7H_2O$ become hydrated easily and should not be used. Iron chelated with EDTA prepared from either $FeCl_3$ or $Fe_2(SO_4)_3$ be-

TABLE 2-2. *Trace metals – I (see text,* III.C.I)

	Conc. μM/l of medium					
Element	P-II[a]	P-8[a]	TM-2[b]	'f/2'[c]	M-I[d]	TMS-I[e]
Zinc	0.8	7.7	35.2	0.08	14	35
Manganese	7.3	18	11.8	0.9	3.8	10
Molybdenum	—	5.2	2.1	0.03	0.83	5
Cobalt	0.17	0.17	0.11	0.05	0.04	0.3
Copper	—	0.32	0.02	0.04	0.008	0.3
Iron	1.8	35.8	1.8	11.7	1.2	2
EDTA	26.9	—	53.7	11.7	53.7	48
Versonal	—	87.2	—	—	—	—
Chelate:metal	2.7:1	1.3:1	0.8:1	0.9:1	2.7:1	2:1
Boron	185	185	—	—	32	400

[a] Provasoli 1963*a*; Provasoli *et al.* 1957.
[b] Droop, in Provasoli *et al.* 1957.
[c] Guillard, unpublished 1968.
[d] Müller 1962.
[e] McLachlan, unpublished.

comes turbid if titrated above pH 3. Solutions prepared from $FeSO_4$ $(NH_4)_2SO_4$ are more stable, but this compound must be dissolved in 0.05% (w/v) sulphosalicylic acid. Iron-EDTA prepared from ferric citrate is stable even when titrated to pH 8, and at 20 μM/l will not precipitate in buffered seawater autoclaved for 30 min.

i. Dissolve the required amount of Fe.citrate in a solution containing 2–3 molar equivalents of Na_2EDTA.

ii. Boil for 5–10 min to hasten complexing. The solution will be acidic (*ca.* pH 2), and when cooled the pH may be adjusted to neutral with 1N NaOH.

iii. Dilute to volume with distilled water, and store under refrigeration. This solution may be used in cultures with bacteria.

b. Cobalt, copper, manganese, molybdenum, zinc – separate primary stock solutions of these elements are prepared, and diluted to form the working stock solution. Either the chlorides or sulphates of Co, Cu, Mn and Zn may be used; Mo is prepared from Na_2MoO_4.

Dissolve the required amount of salt in a solution of 2 molar equivalents of Na_2EDTA, boil for 5–10 min and dilute to volume with distilled water. Working stock solutions are prepared by making the appropriate dilution. These solutions are stable under alkaline conditions. If combined into a single solution, Fe.EDTA can also be added.

TABLE 2-3. *Trace metals – II (see text,* III.C.2)

	Conc. μM/l of medium			
Element	S-II[a]	TM-2[b]	M-II[c]	TMS-II[d]
Bromine	125	275	185	500
Strontium	23	44	14	100
Aluminium	—	1	0.21	—
Rubidium	2.3	0.7	0.13	2
Lithium	28.8	0.87	0.14	10
Iodine	0.08	0.16	0.12	0.2
Molybdenum[e]	5.2	—	—	—

[a] Provasoli 1963*a*; Iwasaki 1967.
[b] Droop, in Provasoli *et al.* 1957.
[c] Müller 1962.
[d] McLachlan, unpublished.
[e] It is more convenient to supply this element with those in group I (Table 2-2).

c. There is sufficient boron in seawater so that it is unnecessary, and undesirable, to add this element to enriched seawater media. Boron is routinely added to synthetic media.

Dissolve the required amount of H_3BO_3 in distilled water with gentle heating. If synthetic media only are to be prepared, H_3BO_3 may be combined into a single stock solution containing the elements listed earlier. However, H_3BO_3 is relatively insoluble in cold water, and stock solutions for 500–1000-fold dilutions cannot be prepared.

2. *Trace metals – II* (Table 2-3). These metals are added to synthetic media only. They are stable under alkaline conditions, and a single stock solution may be prepared for 1000-fold dilution.

Bromine is added as NaBr; iodine as either NaI or KI; the remaining elements as chlorides. It is necessary to prepare a primary stock solution of iodine; this solution and the working solution should be slightly alkaline to prevent oxidation.

D. *Chelates*

Theoretically a chelate should combine with a metal in a 1:1 molar ratio, but in practice it is necessary to add an excess of chelate, from 10 to 100%, to ensure adequate chelation. Chelate-metal ratios of 1.5–3:1 are commonly used which results in adequate chelation of the added metals in addition to those present as impurities.

1. Ethylenediaminetetraacetic acid (EDTA). This is the most widely used chelate in marine media, and is not readily metabolized by microbes. The free acid is insoluble in water, but freely soluble in a solution containing 2 molar equivalents of NaOH. The disodium salt ($Na_2EDTA.2H_2O$) is readily soluble in water; solutions have a pH of 4.5–5.0. It is marketed as Versene (Fisher Scientific Co., 711 Forbes Ave., Pittsburgh, Pennsylvania 15219; 8555 Devonshire Rd, Montreal 307, Quebec).

2. Na_3hydroxyethylethylenediaminetriacetate. This chelate is obtainable as a liquid (Versonal; Fisher Scientific Co.) and has a pH of 11.4. Under alkaline conditions, iron chelated with this is reported to be more stable than Fe.EDTA, but it forms relatively weak chelates with other heavy metals.

3. Nitrilotriacetic acid (NTA). This chelate is less effective than EDTA for most purposes. Its use in several media (Table 2-6) is recommended in combination with glycerophosphate, and it has been used in place of Versonal in the P-8 metal mix (Fries 1963). The free acid is insoluble in water, and is dissolved by titrating to *ca.* pH 7.5 with NaOH.

4. Citric acid. Citric acid can replace NTA. In alkaline solutions it is probably slowly metabolized, and may be used in non-sterile media with caution. The acid and sodium salt are both soluble in water. If the salinity of the medium is to be varied, Droop (1961) recommends the use of citrate. It has a low affinity for calcium and hence is little influenced by changes in salinity.

E. *Organic micronutrients*

1. Cyanocobalamin, biotin, thiamine. Only three vitamins, cyanocobalamin, biotin and thiamine have been found necessary for growth of marine algae, and are added to most media. Some media, in addition, contain other vitamins. Although a few species do not require an exogenous source of vitamins, the majority remain to be investigated.

a. Cyanocobalamin is stable in powder form, and contains *ca.* 12% water of hydration. It may be purchased in ampules which contain small quantities (e.g., 10 mg) of the crystalline compound.

i. Prepare a primary stock solution by dissolving the contents of an ampule in distilled water.

ii. Adjust the pH to 4.5–5.0 with HCl; dilute to volume.

iii. Either membrane filter or autoclave (the acidified solution is stable); store in a deep-freezer at −20 °C. This solution may be refrozen after each use.

b. Crystalline biotin is stable and contains *ca.* 4% water of hydration. Primary stock solutions are prepared as for cyanocobalamin except that the required amount of vitamin is weighed. The acidified solution is stable to autoclaving.

c. It is unnecessary to prepare a primary stock solution of thiamine.HCl because of the quantities usually added to media. Acidified (*ca.* pH 5.5) solutions may be autoclaved.

d. These three vitamins may be combined into a single working stock solution for 1000-fold dilution.

i. Dispense sufficient volumes for general use into glass ampules or screw-cap tubes (*ca.* 5 ml).

ii. Sterilize; store in a deep-freezer.

These vitamins are frequently autoclaved with the medium. This undoubtedly results in some decomposition, but the moieties in many instances are apparently equally effective. It is, however, recommended that vitamins be added aseptically to the medium from previously sterilized solutions.

2. Other vitamins. Additional vitamins are added to some media in the form of mixes (Table 2-4). Stock solutions, which can be prepared for 1000-fold dilution, are stored frozen in ampules or screw-cap tubes as indicated earlier. Most of these compounds can be autoclaved if the solution is acidic (pH 4.5–5.0).

a. The calcium salt (Ca.leucovorin) of folinic acid is more soluble than the free acid. It is more stable in neutral or alkaline solution.

b. Riboflavin decomposes rapidly in alkaline media with decomposition accelerated by light. It is more soluble in solutions of NaCl.

c. Folic acid is dissolved in a solution of $NaHCO_3$ or in boiling water.

d. Ca.pantothenate is unstable to autoclaving.

3. Growth factors. Several growth factors, not normally added to marine media, have been indicated for normal development of benthic species in axenic culture.

a. Adenine and kinetin are only slightly soluble in water; the latter is freely soluble in dilute solutions of NaOH or HCl.

b. Gibberellic acid is slightly soluble in water, but more soluble in dilute solutions of $NaHCO_3$.

TABLE 2-4. *Vitamin mixes*

Additive	Conc./l of medium S-3[a]	8A[a]
Thiamine.HCl	0.5 mg	2.0 mg
Nicotinic acid	0.1 mg	1.0 mg
Ca.pantothenate	0.1 mg	1.0 mg
p-Aminobenzoic acid	10 μg	0.1 mg
Biotin	1.0 μg	5.0 μg
i-Inositol	5 mg	10 mg
Folic acid	2.0 μg	25 μg
Cyanocobalamin	1.0 μg	0.5 μg
Thymine	3.0 mg	8.0 mg
Pyridoxine.HCl	—	0.4 mg
Pyridoxamine.2HCl	—	0.2 mg
Putrescine.2HCl	—	0.4 mg
Riboflavin	—	50 μg
Choline-H_2-citrate	—	5.0 mg
Orotic acid	—	2.6 mg
Folinic acid	—	2.0 μg

[a] Provasoli 1963*a*; Provasoli *et al.* 1957.

c. Indoleacetic acid (IAA) is only slightly soluble in hot water; it may be added to the medium in minimal volumes of ethanol. This compound is photolabile.

4. Extracts. Many extracts and digests are rich in vitamins and other undetermined factors, and are added to the medium at concentrations from 0.05–0.5% (w/v). These include extracts of algae, liver, protein, yeast, etc., and digests from such materials as milk, protein and blood. Most provide a good substrate for microbes, and minimum quantities should be used in cultures with bacteria.

5. Soil extract. In maritime areas it is frequently difficult to obtain good quality loamy soil although commercial nurseries and greenhouses are a possible source. Precautions should be taken to ensure that fungicides, insecticides, etc. have not been added to the soil (see Chp. 1, II.E).

The most common procedure for preparation of soil extract consists of mixing 1 vol. of soil with 2 vol. distilled water and steaming for one hour or autoclaving for a few minutes; after cooling the liquid is decanted, allowed to stand overnight (or centrifuged) and filtered clear.

Alternatively, an extract may be prepared by alkaline extraction (Droop 1969; Provasoli, McLaughlin and Droop 1957). One vol. of soil is autoclaved for one hour with 1–2 vol. of water and 3 g NaOH. After standing overnight the liquid is decanted and filtered clear. This extract is added to media at 0.1–0.2 the recommended concentration.

Care should be exercised in dealing with soil in the laboratory. It is a potent source of contaminants, and moist soil should never be stored in an area where culture work is being done.

IV. MEDIA

A. *Enriched* (Tables 2-2 to 2-6)

The number of variations in media based on natural seawater is limited. In the simplest media only a source of nitrogen, phosphorus and usually soil extract, are added to seawater; whereas, the more sophisticated media contain in addition mixtures of chelated trace metals and vitamins.

1. Erdschreiber (Schreiber 1927; Føyn 1934; Gross 1937; M. Parke, personal communication). This medium has been used successfully for many years and for a wide variety of organisms including both unicellular and benthic species. Many species have not been grown in other media. It is employed extensively at the Marine Biological Association Laboratory (Plymouth, England) for both isolation and maintenance.

The seawater and additives are sterilized separately and combined aseptically. The concentration of nitrogen and phosphorus is high, especially for an unbuffered medium. In many instances addition of iron and other trace metals and vitamins can replace soil extract.

2. Grund (von Stosch 1963, 1969). Von Stosch has cultured and completed the life histories of a number of Rhodophyceae, including Ceramiales and coralline species (Cryptonemiales), in this medium.

The seawater and additives are sterilized separately and combined aseptically. Soil extract or ashed soil extract is necessary for some species.

Inhibition due to a high phosphorus content has been noted, and reproduction may be induced in some species by lowering the nitrogen concentration 10–20 times and the phosphorus concentration 2.5–5 times. It is assumed that the nitrogen concentration given by von Stosch (1963) as 115 mg/l is in error.

The medium is overchelated which could induce a trace metal deficiency, and possibly this is why ashed soil extract is effective.

TABLE 2-5. *Enriched seawater media*

Additive	Concentration/liter				
	Erdschreiber	Grund	ES	'f/2'	SWM
$NaNO_3$	1.18–2.35 mM	0.5 mM	0.66 mM	0.88 mM	0.5–2.0 mM
Na_2HPO_4	56–140 μM	30 μM	—	—	—
NaH_2PO_4	—	—	—	36.3 μM	50–100 μM
Na_2.glycerophosphate	—	—	25 μM	—	—
Na_2SiO_3	—	—	—	0.054–0.107 mM	0.2 mM
Fe.EDTA	—	—	7.2 mM	—	2 μM
$Fe_2(SO_4)_3$	—	1 μM	—	—	—
Na_2.EDTA	—	10 μM	—	—	—
Trace metal solution	—	—	P-II	'f/2'	TMS-I
$MnCl_2$	—	0.1 μM	—	—	—
Vitamin solution	—	—	—	—	S-3
Cyanocobalamin	—	1.0 μg	1.6 μg	0.5 μg	—
Biotin	—	—	0.8 μg	0.5 μg	—
Thiamine.HCl	—	—	20 μg	100 μg	—
Tris	—	—	0.66 mM	—	0–5 mM
Glycylglycine	—	—	—	—	5 mM
Soil extract	50 ml	—	—	—	50 ml
Liver extract	—	—	—	—	10 mg

TABLE 2-6. *Enriched media*

Additive	Concentration/liter				
	ASP-2	ASP-6	ASP-12	Müller	ASP-M
NaCl	308 mM	411 mM	479 mM	457 mM	400 mM
KCl	8.1 mM	9.4 mM	9.4 mM	9.8 mM	10 mM
$MgSO_4$	20.3 mM	32.5 mM	28.4 mM	26.6 mM	20 mM
$MgCl_2$	—	—	19.7 mM	23 mM	20 mM
$CaCl_2$	2.5 mM	3.75 mM	10 mM	13.5 mM	10 mM
$NaHCO_3$	—	—	—	2.4 mM	2 mM
Na_2CO_3	0.30 mM	—	—	—	—
$NaNO_3$	0.59 mM	3.53 mM	1.18 mM	1.18 mM	0.5–2.0 mM
NaH_2PO_4	—	—	—	—	50–100 μM
Na_2HPO_4	—	—	—	140 μM	—
K_3PO_4	—	—	47 μM	—	—
K_2HPO_4	29 μM	—	—	—	—
Na_2.glycerophosphate	—	317 μM	31.7 μM	—	—
Na_2SiO_3	528 μM	246 μM	528 μM	0.07 mM	200 μM
NTA	523 μM	—	523 μM	—	—
Fe.EDTA	—	—	—	—	2 μM
Fe (as Cl^-)	9 μM	—	—	—	—
Trace metal solution	P-II[a]	P-8	P-II + S-II	M-I + M-II	TMS-I + TMS-II
Vitamin solution	S-3	8A	—	—	S-3
Cyanocobalamin	—	—	0.2 μg	0.5 μg	—
Biotin	—	—	1.0 μg	0.5 μg	—
Thiamine.HCl	—	—	100 μg	100 μg	—
Tris	8.25 mM	8.25 mM	8.25 mM	—	0–5 mM
Glycylglycine	—	—	—	—	5–10 mM
pH to	7.6–7.8	7.4–7.6	7.8–8.0	—	7.5
H_2O to	1000 ml	1000 ml	1000 ml	1000 ml	1000 ml

[a] Added at three times the concentration indicated in Table 2-2.

3. ES (Provasoli 1968; J. West, personal communication). This is a good general purpose medium in which West has cultured numerous unicellular and benthic species. The medium is similar to Grund medium (2, preceding) but more complete with respect to trace metals and vitamins. There are several inconsistencies in Provasoli (1968) regarding preparation of the medium and micronutrient concentrations. The composition given in Table 2-5 is that used by West.

The seawater is sterilized by filtration or steaming. The enrichments are assembled into a single solution, sterile filtered and added aseptically to the medium.

4. 'f/2' (Guillard and Ryther 1962; R. R. L. Guillard, personal communication). Guillard has used this medium for isolation and cultivation of a wide spectrum of unicellular species. The medium is prepared with various dilutions of seawater (see Guillard Chp. 4). Although it has not been used to any extent for benthic species, there is no reason why it should not be equally effective. In the original reference (Guillard and Ryther 1962), the formulation for medium 'f' was presented. Formulation 'f/2' as given in Tables 2-2 and 2-5 is now recommended (R.R.L. Guillard, personal communication).

The nutrients and seawater are autoclaved together. The medium may be buffered with either 4.13 mM tris (f/2-t) or 3.8 mM glycylglycine (f/2-g) and the pH adjusted to 7.4.

Nitrogen and phosphorus can be lowered considerably, e.g., to 100 μM and 5 μM respectively, without affecting yields or rates of growth. Some species will assimilate reduced nitrogen only, and various quantities of NH_4Cl, to a maximum of 500 μM, are substituted for $NaNO_3$. The medium is then designated 'h/2'.

5. SWM (McLachlan 1964; Chen, Edelstein and McLachlan 1969). This medium has been used for growth and nutritional studies of benthic and unicellular species representative of various classes. Life histories of Chlorophyceae, Chrysophyceae, Phaeophyceae and Rhodophyceae species have been completed in this medium.

The nutrients are added to the seawater and autoclaved (SWM-1, SWM-2, SWM-3) or filter sterilized (SWM-4).

Glycylglycine (SWM-1, SWM-2) may be increased to 10 mM for mass cultures or when CO_2 gas is added to the medium. Tris (SWM-3, SWM-4) may be reduced to 1 mM or omitted; rapid growth will cause a marked increase in pH.

Maximal concentrations of nitrogen and phosphorus are added for mass cultures or conservation; otherwise minimum concentrations are employed and the N:P ratio is maintained between 10:1–20:1. For organisms which require reduced nitrogen, $NaNO_3$ is replaced by NH_4Cl (100–500 μM) or urea (0.5–1.0 mM) and designated SWM-2.

Soil extract and liver extract (SWM-3) may be omitted for most species, but a few do considerably better in the presence of these extracts.

B. *Synthetic* (Tables 2-2 to 2-4, 2-6)

There is considerable variability amongst synthetic media, especially with respect to salinity and concentrations of calcium and magnesium. These media are generally buffered, contain mixtures of vitamins and chelated trace metals and may be autoclaved or sterile filtered. Although they were designed primarily for unicellular species, many benthic species have been grown in these media. Certain modifications can be made to achieve better growth, such as: (1) the concentration of buffer may be lowered to 4–5 mM to prevent precipitation during autoclaving; (2) addition of 2 mM $NaHCO_3$; and (3) supplements of soil or other extracts or small volumes of natural seawater may be effective with concentrations determined by trial and error.

1. ASP-2 (Provasoli, McLaughlin and Droop 1957; Provasoli 1963*a*). This medium is satisfactory for growth of a variety of unicellular species. The salinity and calcium and magnesium content may be too low for some benthic species.

Axenic cultures of a Rhodophyceae and several species of Laminariales (Phaeophyceae) have been obtained in this medium. For the latter silicate was omitted and the medium supplemented with 5.2 μM Na_2MoO_4; 0.4 μM KI; 815 μM KBr; 2.3 mM $NaHCO_3$ and 0.57 mM IAA (Druehl and Hsiao 1969).

2. ASP-6 (Provasoli *et al.* 1957; Provasoli 1963*a*). This is a general purpose medium which has been used for both unicellular and benthic species including axenic cultures of several Rhodophyceae, Phaeophyceae and Chlorophyceae species. The salinity and calcium and magnesium concentrations are higher than in ASP-2 (1, preceding).

Fries' (1963) effective modifications consisted of substituting NTA for Versonal and supplementing the medium with 0.4 μM KI and 815 μM KBr. Pedersén (1968) added in addition 20–40 μM kinetin for axenic cultures of phaeophytes.

3. ASP-12 (Provasoli 1963*a*; Iwasaki 1967). This medium was formulated for Dinophyceae but has been used for other unicellular and Rhodophyceae species. It more closely approximates seawater with respect to salinity and the calcium and magnesium content than the other ASP media.

Axenic cultures of several red algal species have been obtained in this medium; the elements in S-II metal mix (Table 2-3) stimulated growth of the conchocelis-stage.

4. ASP-M (McLachlan 1964). A wide variety of unicellular and benthic species have been cultured in this medium. The salinity and calcium and magnesium content are intermediate between ASP-6 and ASP-12.

Nutritive additives are employed in the same manner as in SWM (A.5, preceding) and the medium is autoclaved or filter sterilized. The TMS-II metal mix may be omitted for most unicellular species.

5. Müller (Müller 1962; J. Z. Page, personal communication). Page has used this medium for growth of unialgal cultures, and has had more success with it than with Erdschreiber medium prepared from Long Island (New York) seawater.

Dinoflagellates, several red and brown algae and coenocytic green species have been grown successfully. The Chlorophyceae *Derbesia* (Codiales) and *Acetabularia* (Dasycladales) form reproductive structures (see Chp. 6).

The medium is unbuffered and some precipitation results from autoclaving, but the pH returns to normal about 36 h after sterilization.

V. ACKNOWLEDGMENTS

I am indebted to my colleagues at the Atlantic Regional Laboratory who have assisted me in many ways, and to Dr R. R. L. Guillard who has generously provided valuable unpublished information and suggestions based on his experiences. Drs M. Parke and L. Provasoli have, in innumerable ways, been continuously helpful, and to them I am especially grateful.

VI. REFERENCES

Behrens, E. W. 1965. Use of the Goldberg refractometer as a salinometer for biological and geological field work. *J. Mar. Res.* **23**, 165–71.

Berglund, H. 1969. On the cultivation of multicellular marine green algae in axenic culture. *Sv. Bot. Tidskr.* **63**, 251–64 (ASP-6 medium).

Bernhard, M. and Zattera, A. 1970. The importance of avoiding chemical contamination for successful cultivation of marine organisms. *Helgoländer wiss. Meeresunters.* **20**, 655–75 (presents a survey of 70 different materials for potential toxic or inhibitory effects towards 6 species of unicellular algae).

Boalch, G. T. 1961*a*. Studies on *Ectocarpus* in culture. I. Introduction and methods of obtaining uni-algal and bacteria-free cultures. *J. Mar. Biol. Ass. U.K.* **41**, 279–86 (enriched seawater medium).

Boalch, G. T. 1961*b*. Studies on *Ectocarpus* in culture. II. Growth and nutrition in a bacteria-free culture. *J. Mar. Biol. Ass. U.K.* **41**, 287–304 (comparison of enriched seawater with synthetic media).

Chen, L. C.-M., Edelstein, T. and McLachlan, J. 1969. *Bonnemaisonia hamifera* Hariot in nature and in culture. *J. Phycol.* **5**, 211–20 (SWM-3 medium).

Craigie, J. S. and McLachlan, J. 1964. Excretion of colored ultraviolet-absorbing substances by marine algae. *Can. J. Bot.* **42**, 23–33 (procedure for washing activated charcoal).

Droop, M. R. 1958*a*. Optimum, relative and actual ionic concentrations for growth of some euryhaline algae. *Verh. Internat. Ver. Limnol.* **13**, 722–30 (growth of unicellular species in various concentrations of conservative elements).

Droop, M. R. 1958*b*. Requirement for thiamine among some marine and supra-littoral Protista. *J. Mar. Biol. Ass. U.K.* **37**, 323–9 (introduction of glycyl-glycine as a pH buffer).

Droop, M. R. 1961. Some chemical considerations in the design of synthetic culture media for marine algae. *Bot. Marina* **2**, 231–46 (discussion of trace metals, O_2 and redox potentials and pH).

Droop, M. R. 1969. Algae. In Norris, J. R. and Ribbons, D. W., eds., *Methods in Microbiology* **3B**, 269–313. Academic Press, New York.

Druehl, L. D. and Hsiao, S. I. C. 1969. Axenic culture of Laminariales in defined media. *Phycologia* **8**, 47–9 (ASP-2 medium).

Føyn, B. 1934. Lebenszyklus, Cytologie und Sexualität der Chlorophycee *Cladophora suhriana* Kützing. *Arch. Protistenk.* **83**, 1–56 (introduction of Erdschreiber).

Fries, L. 1963. On the cultivation of axenic red algae. *Physiol. Plant.* **16**, 695–708 (6 species cultured in modified ASP-6 medium).

Fries, L. 1966. Influence of iodine and bromine on growth of some red algae in axenic culture. *Physiol. Plant.* **19**, 800–8.

George, E. G. 1966. University of Cambridge culture collection of algae and protozoa. List of strains. *The Botany School, Cambridge*, pp. 1–67 (media used for maintenance of the Cambridge collection).

Gross, F. 1937. Notes on the culture of some marine plankton organisms. *J. Mar. Biol. Ass. U.K.* **21**, 753–68 (Erdschreiber medium for planktonic species; the procedures are basically the same as those used presently at the Marine Biological Association Laboratory, Plymouth, England).

Guillard, R. R. L. 1963. Organic sources of nitrogen for marine centric diatoms. In Oppenheimer, C. H., ed., *Symposium on Marine Microbiology*, pp. 93–104. Charles C Thomas, Springfield, Illinois. ('f' medium).

Guillard, R. R. L., and Ryther, J. H. 1962. Studies of marine planktonic diatoms. I. *Cyclotella nana* Hustedt and *Detonula confervacea* (Cleve) Gran. *Can. J. Microbiol.* **8**, 229–39 ('f' medium).

Hamilton, R. D. and Carlucci, A. F. 1966. Use of ultra-violet-irradiated sea-water in the preparation of culture media. *Nature* **211**, 483–4.

Hutner, S. H., Cury, A. and Baker, H. 1958. Microbiological assays. *Anal. Chem.* **30**, 849–67 (numerous useful techniques on media, glassware, sterilization, etc.).

Hutner, S. H., Provasoli, L., Schatz, A. and Haskins, C. P. 1950. Some approaches to the study of the role of metals in the metabolism of micro-organisms. *Proc. Amer. Phil. Soc.* **94**, 152–70 (discussion of metal chelates in media).

Iwasaki, H. 1961. The life cycle of *Porphyra tenera* in vitro. *Biol. Bull.* **121**, 173–87 (synthetic and enriched seawater media).

Iwasaki, H. 1967. Nutritional studies of the edible seaweed *Porphyra tenera*. II. Nutrition of conchocelis. *J. Phycol.* **3**, 30–4 (ASP-12 medium and effects of micronutrients).

Johnston, R. 1963. Sea water, the natural medium of phytoplankton. I. General features. *J. Mar. Biol. Ass. U.K.* **43**, 427–56.

Johnston, R. 1964. Sea water, the natural medium of phytoplankton. II. Trace metals and chelation, and general discussion. *J. Mar. Biol. Ass. U.K.* **44**, 87–109.

Lewin, J. C. 1955. Silicon metabolism in diatoms. II. Sources of silicon for growth of *Navicula pelliculosa. Plant Physiol.* **30**, 129–34.

Lewin, J. C. 1966. Boron as a growth requirement for diatoms. *J. Phycol.* **2**, 160–3 (preparation of boron-free medium and synthetic medium for diatoms).

Lewin, J. C. and Busby, W. F. 1967. The sulfate requirements of some unicellular marine algae. *Phycologia* **6**, 211–17 (variation in sulphur in a synthetic medium).

McLachlan, J. 1959. The growth of unicellular algae in artificial and enriched sea water media. *Can. J. Microbiol.* **5**, 9–15.

McLachlan, J. 1960. The culture of *Dunaliella tertiolecta* Butcher – a euryhaline organism. *Can. J. Microbiol.* **6**, 367–406 (variation of conservative elements in ASP medium).

McLachlan, J. 1964. Some considerations of the growth of marine algae in artificial media. *Can. J. Microbiol.* **10**, 769–82 (comparison of growth in enriched seawater media with ASP media).

McLachlan, J. and McLeod, G. C. 1959. The use of conversion factors for the determination of the concentration of nutrients in culture media. *Limnol. Oceanogr.* **4**, 218–19.

Müller, D. 1962. Über jahres-und lunarperiodische Erschcinungen bei einigen Braunalgen. *Bot. Marina* **4**, 140–55.

Packer, E. L., Hutner, S. H., Cox, D., Mendelow, M. A., Baker, H., Frank, O. and Amsterdam, D. 1961. Use of amine buffers in protozoan nutrition. *Ann. New York Acad. Sci.* **92**, 486–90 (includes a discussion of tris).

Pedersén, M. 1968. *Ectocarpus fasciculatus*: marine brown alga requiring kinetin. *Nature* **218**, 776 (ASP-6 medium).

Pedersén, M. 1969*a*. The demand for iodine and bromine of three marine brown algae grown in bacteria-free cultures. *Physiol. Plant.* **22**, 680–5 (ASP-6 medium).

Pedersén, M. 1969*b*. Marine brown algae requiring vitamin B_{12}. *Physiol. Plant.* **22**, 977–83.

Pintner, I. J. and Provasoli, L. 1958. Artificial cultivation of a red-pigmented marine blue-green alga, *Phormidium persicinum*. *J. Gen. Microbiol.* **18**, 190–7 (synthetic and enriched seawater media).

Pintner, I. J. and Provasoli, L. 1968. Heterotrophy in subdued light of 3 *Chrysochromulina* species. *Bull. Misaki Mar. Biol. Inst., Kyoto Univ.* No. 12, pp. 25–31 (requirement for selenium).

Pringsheim, E. G. 1946. *Pure Cultures of Algae*. Cambridge Univ. Press, London. 119 pp. (contains many useful procedures and techniques).

Provasoli, L. 1958. Effect of plant hormones on *Ulva*. *Biol. Bull.* **114**, 375–84 (synthetic and enriched seawater media, kinetin and adenine).

Provasoli, L. 1963*a*. Growing marine seaweeds. In DeVirville, D. and Feldmann, J., eds., *Proc. Int. Seaweed Symp.* **4**, 9–17. Pergamon Press, Oxford (composition of marine media with an indication of suitability for various species).

Provasoli, L. 1963*b*. Organic regulation of phytoplankton fertility. In Hill, M. N., ed., *The Sea*, **2**, 165–219. Interscience Publishers, New York (review of some of the factors influencing growth of marine algae).

Provasoli, L. 1968. Media and prospects for the cultivation of marine algae. In Watanabe, A. and Hattori, A., eds., *Cultures and Collection of Algae*. Proc. U.S.–Japan Conf. Hakone, Sept. 1966. *Jap. Soc. Plant Physiol.* pp. 63–75.

Provasoli, L., McLaughlin, J. J. A. and Droop, M. R. 1957. The development of artificial media for marine algae. *Arch. Mikrobiol.* **25**, 392–428 (theory and comprehensive treatment of marine media).

Provasoli, L., McLaughlin, J. J. A. and Pintner, I. J. 1954. Relative and limiting concentrations of major mineral constituents for the growth of algal flagellates. *Trans. New York Acad. Sci.* **16**, 412–17.

Schreiber, E. 1927. Die Reinkultur von marinem Phytoplankton und deren Bedeutung für die Erforschung der Produktionsfähigkeit des Meerwassers. *Wiss. Meeresuntersuch.* N. F. **16**, 1–34 (also as National Research Council, Canada Tech. Transl. 1108, 1964).

Shapiro, J. 1961. Freezing-out, a safe technique for concentration of dilute solutions. *Science* **133**, 2063.

Spencer, C. P. 1958. The chemistry of ethylenediamine tetra-acetic acid in sea water. *J. Mar. Biol. Ass. U.K.* **37**, 127–44.

Starr, R. C. 1964. The culture collection of algae at Indiana University. *Amer. J. Bot.* **51**, 1013–44 (media for maintenance of cultures at Indiana University).

Tatewaki, M. and Provasoli, L. 1964. Vitamin requirements of three species of *Antithamnion*. *Bot. Marina* **6**, 193–203 (ASP media).

von Stosch, H. A. 1963. Wirkung von Jod und Arsenit auf Meeresalgen in Kultur. In DeVirville, D. and Feldmann, J., eds., *Proc. Int. Seaweed Symp.* **4**, 142–50. Pergamon Press, Oxford.

von Stosch, H. A. 1969. Observations on *Corallina*, *Jania* and other red algae in culture. In Margalef, R., ed. *Proc. Int. Seaweed Symp.* **6**, 389–99. Dirección General de Pesca Maritima, Madrid.

West, J. A. and Norris, R. E. 1966. Unusual phenomena in the life histories of Florideae in culture. *J. Phycol.* **2**, 54–7 ('f' and ES media).

NOTES ADDED IN PROOF

These recent papers provide either new information or techniques or update previous information.

Davey, E. W., Gentile, J. H., Erickson, S. J. and Betzer, P. 1970. Removal of trace metals from marine culture media. *Limnol. Oceanogr.* **15**, 486–8. (Trace metals can be removed from natural and synthetic seawater by use of a chelating resin. This procedure permitted the removal of toxic concentrations of heavy metals, and to reduce the concentration below the nutritional requirements. It should be possible by use of this technique to study the metabolism of individual metals.)

Eagle, H. 1971. Buffer combinations for mammalian cell culture. *Science* **174**, 500–3. (A number of organic buffers have been examined to stabilize the pH in the range 6.4–8.3. If these buffers prove to be non-toxic to algal material, they may be useful for some purposes. The buffers are obtainable commercially through ICN Nutritional Biochemicals Division.)

Lewin, J. and Chen. C.-H. 1971. Available iron: A limiting factor for marine phytoplankton. *Limnol. Oceanogr.* **16**, 670–5. (The availability of iron in natural seawater, and the use of EDTA in solubilizing iron in freshly collected and stored seawater has been examined.)

Rosowski, J. R. and Parker, B. C., eds. 1971. *Selected Papers in Phycology*. The Department of Botany, University of Nebraska, Lincoln, Nebraska. 876 pp. (A considerable amount of useful specific and general information is contained in this volume.)

Shephard, D. C. 1970. Axenic culture of *Acetabularia* in a synthetic medium. In Prescott, D. M., ed., *Methods in Cell Physiology* **4**, 49–69, Academic Press, New York. (The medium is probably generally useful and the techniques of culture applicable to many organisms.)

Siegelman, H. W. and Guillard, R. R. L. 1971. Large-scale culture of algae. In San Pietro, A., ed., *Methods in Enzymology* **23**, 110–15. Academic Press, New York. (Media and techniques for large-scale culture of unicellular species.)

Starr, R. C. 1971. The culture collection of algae at Indiana University – additions to the collection July 1966–July 1971. *J. Phycol.* **7**, 350–62. (Additions to the list of media used for maintenance and experimental purposes are given.)

3: Methods for microscopic algae

ROBERT W. HOSHAW
Department of Biological Sciences,
University of Arizona, Tucson, Arizona 85721; and

JAMES R. ROSOWSKI
Department of Botany,
University of Nebraska, Lincoln, Nebraska 68508

CONTENTS

I. OBJECTIVE

This chapter deals with selected methods for the isolation and purification of freshwater or marine microscopic algae. Microscopic algae as defined for this chapter are non-swimming, unicellular or filamentous eukaryotic algae from freshwater, marine, soil, and atmospheric environments. Allen (Chp. 7) discusses isolation of the prokaryotic Cyanophyceae. It is also possible to use methods in the present chapter to isolate bluegreen algae. Application of the following methods to natural collections of microscopic algae produces unialgal or axenic cultures. Unialgal cultures are considered as clonal cultures with bacteria. These methods are not difficult, although the preparation of axenic cultures (bacteria-free) may require patience and perseverance. It is intended that the investigator can use these methods with minimal reference to other reports. However, it may be profitable to read the discussions on the isolation and purification of algae by Bold (1942), R. Lewin (1959), and Pringsheim (1946, 1950, 1951).

Several other chapters in this section are related to the present one. For details on media used with the isolation and purification methods, refer to Chps. 1, 2. Success of purification can be tested with methods described in VII, this chapter. Methods described in Chps. 4–7 may offer advantages for certain organisms ordinarily included in the present chapter. Chapter 11 contains descriptions of the environmental conditions of light and temperature suitable for algal growth.

II. EQUIPMENT

The equipment named includes items used in the isolation or purification methods described. The list does not include many small items such as microscope slides, cover glasses, razor blades, forceps, etc. The methods of isolation include the following: (1) capillary pipette, (2) capillary pipette (simplified), (3) streak plating, (4) spray plating, (5) isolation on agar, and (6) isolation from the atmosphere. Four methods are described for purification: (a) washing, (b) ultrasonics, (c) antibiotics, and (d) potassium tellurite.

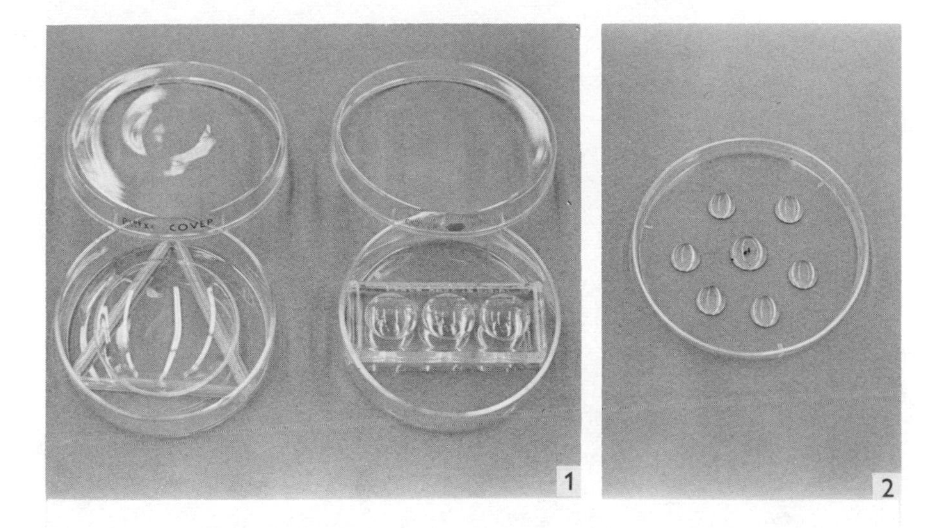

Figs. 3-1, 2. Isolation dishes. Fig. 3-1. Isolation dishes prepared with a chemical watch glass supported by triangle (left) and a spot plate (right). Fig. 3-2. Isolation dish from an inverted plastic petri dish top showing natural collection in center surrounded by six washing droplets.

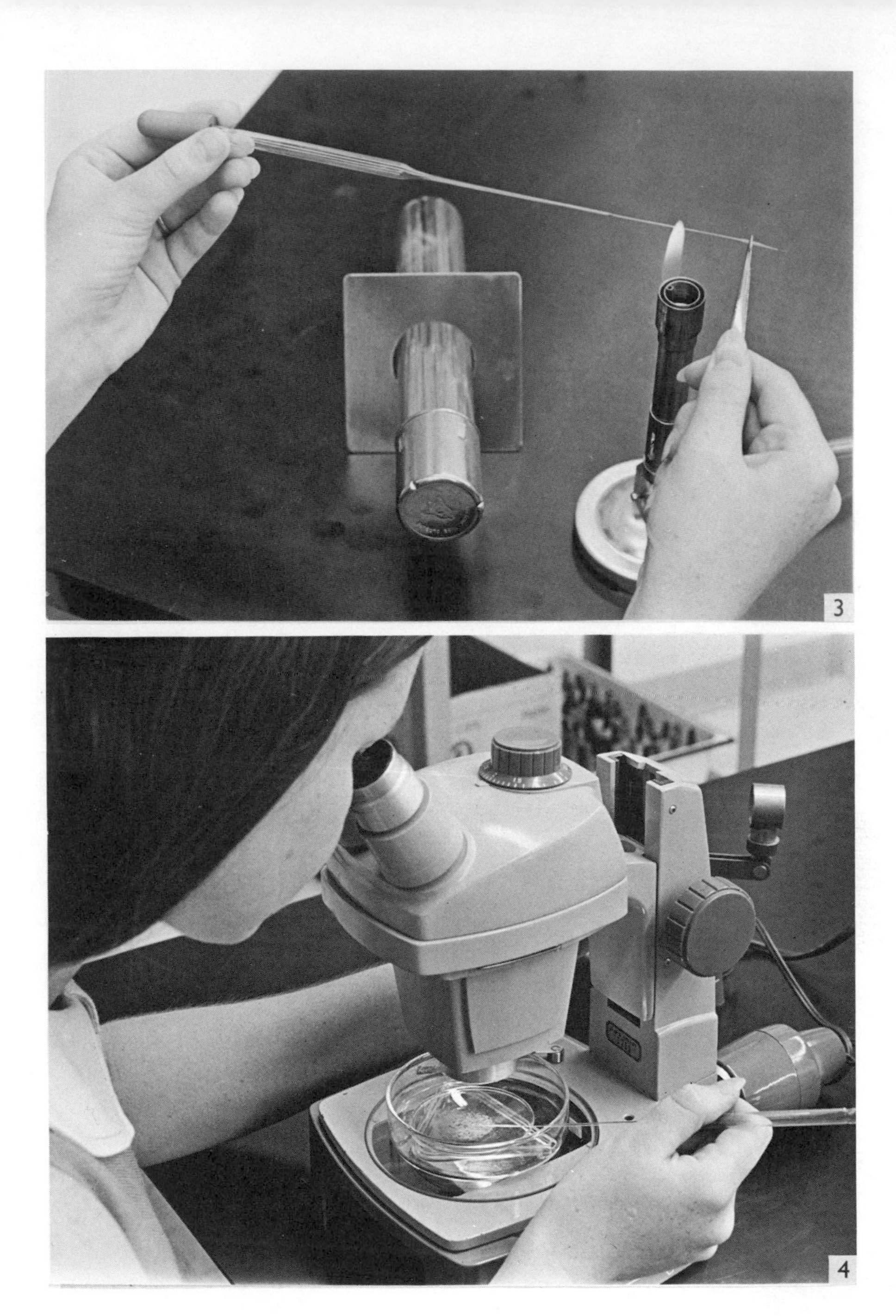

Fig. 3-3. Isolation by capillary pipette. Drawing a fine capillary pipette in the pilot flame of a Bunsen burner. In center is large pipette container placed in mounting plate and with closure.

Fig. 3-4. Isolation with a capillary pipette using the stereomicroscope.

The series of numbers and/or letters after each item of equipment refers to those methods named in the preceding two statements. Those items in which no number or letter is given are used in all methods of isolation and/or purification.

A. *Optical* (used for all isolation and purification methods)

1. Stereomicroscope with substage illumination, magnification 30–100×.
2. Compound microscope, magnification 43–1000×.

B. *Instruments, apparatus*

1. Burner (with pilot), see Fig. 3-3.
2. Incubator (lighted) (see Chp. 11, II and III).
3. Sterile transfer chamber (see Chp. 11, IV).
4. Nichrome wire inoculating needle with 4 mm loop, see Fig. 3-5, (3) (a, b, d) (p. 68).
5. Fine platinum or nichrome needle (5).
6. Microspatula (5).
7. Membrane filter sterilizer (Gelman; Millipore) or Seitz filter (c) (see Chp. 12, II.E).
8. Ultrasonic apparatus (b, d) (such as, Disontegrator System 80, Model G-80C1, Ultrasonics Industries Inc., 141 Albertson Ave., Albertson, Long Island, New York).
9. Centrifuge, to 4000 rpm (a, b, d).
10. Waste container for pieces of glass.
11. Used pipette container.

C. *Glass, plastic supplies*

1. Test tubes (18 × 150 mm or 20 × 150 mm) with closures (e.g., Bellco or cotton plugs for media; Bellco Glass Inc., 340 Edrudo Rd, Vineland, New Jersey 08360).
2. Pasteur-type pipette – 9 in, plugged at wide end with cotton (e.g., Bellco 12-1305).
3. Rubber bulbs (1 or 2 ml) to fit Pasteur-type pipette.
4. Cannister for sterilizing and storing Pasteur-type pipette (e.g., Bellco 1721, 38 × 300 mm with metal closing – Bellco B-38 or cotton plug), see Fig. 3-3 (or pipette can, such as Bellco 1715).
5. Mounting plate (e.g., Bellco 1722) for holding Pasteur-type pipette cannister, see Fig. 3-3.

6. Glass and plastic petri dishes, see Figs. 3-1, 2 – 100 × 20 mm, glass, for watch glasses (8, following) and spot plates (9, following), see Fig. 3-1 (a, b, d) – 100 × 15 mm – 65 × 15 mm.
7. Centrifuge tubes (a, b, d).
8. Chemical watch glasses, 65 mm diameter (e.g., Corning 9985), see Fig. 3-1, left (1).
9. Spot plates (1 × 3 in) with 3 depressions, 7 × 22 mm (e.g., Corning 7223), see Fig. 3-1, right (1).
10. Erlenmeyer flasks, 125 ml (c).
11. Glass rod, 4 × 220 mm, bent as 70° isosceles triangle, see Fig. 3-1, left (1).
12. Glass rod, 3 × 150 mm, bent at 50–60° angle at 50 mm (3) (a).

III. TEST ORGANISMS

The methods of isolation and purification described are general methods applicable to thousands of algal species. These methods are not restricted to microscopic, non-swimming eukaryotic algae from freshwater and marine environments although they are the organisms for which the methods have been designed. The sources of algae for isolation include plankton samples, mud or soil samples, air samples, etc., and are referred to as natural collections.

IV. METHODS

A. *Isolation*

The isolation of a single algal unit into medium suitable for growth is required to establish a clonal, unialgal culture. The term *unit* as used in this chapter denotes any unicell, colony, filament, thallus piece, or reproductive body. A reproductive body must produce only a single individual when it develops to be considered a *clonal culture*. Established unialgal cultures may be the source of material for preparing axenic cultures.

Six methods of isolation are described. The capillary pipette and streak plate methods are recommended for most isolations. Three additional methods are described to show variations of these basic methods.

The method of isolation depends on the thallus type and size. Algal units less than 10 μm in diameter are difficult to see, and, therefore to isolate with a capillary pipette. (Suitable methods are given in IV.A.3–5, following.) Larger thalli, such as filaments, can be isolated by the capillary pipette method. Sometimes it is easier to isolate into unialgal culture newly-released reproductive cells (zoospores, aplanospores, etc.) rather than the vegetative thallus. It is often possible to establish axenic clones from these reproductive cells. After suitable growth has occurred, axenic cultures may be established as in IV.B, following.

1. Sterile Pasteur-type pipette

a. Heat sterilize (Chp. 12, II.D.2.b) isolation dishes by placing watch glasses on glass triangles in 100 × 20 mm petri dishes (Fig. 3-1, left, facing p. 54). A spot plate with three depressions can be substituted for the watch glasses and triangles. Also heat sterilize the Pasteur-type pipette in the pipette cans or cannisters.

b. Place 6–8 drops of liquid culture medium in the watch glasses or 3–4 drops in the depressions of the spot plate portion of each isolation dish. Keep the isolation dishes covered except when in use.

c. Place 2–3 drops of a natural collection into the medium in watch glass, or 1 drop into a depression in the spot plate. The dilution of the natural collection may be modified, depending on the number of units of the desired species present.

d. Prepare a fine capillary pipette by taking aseptically a sterile Pasteur-type pipette and putting a rubber bulb on the wide end. Use forceps to hold the narrow end of the pipette so that part of the pipette is in a low or pilot flame of a burner (Fig. 3-3, facing p. 55). As the glass softens, and reddens, gently and with a smooth lengthwise pulling action, remove the pipette from the flame. (Sudden jerking and/or pulling while the pipette is in the flame does not result in a capillary pipette.) Break off the tip of the capillary pipette near the point where the glass bends under its own weight. The bore of the pipette should be several times (*ca.* 75–150 μm) the diameter of the algal units being isolated. (Also see Chp. 4, IV.C.3 for pipette pulling techniques.)

e. Locate the desired algal units for isolation while looking through the stereomicroscope.

f. Manipulate the tip of the fine capillary pipette to a position just out of the liquid and above the algal unit (Fig. 3-4, facing p. 55). Gently dip the pipette into the liquid and by capillary action the liquid with the desired cells should flow into the pipette. This flow of liquid may be controlled by slight pressure on the rubber bulb, although this is generally not necessary. Only one algal unit should be drawn into the pipette at each dip. Depending on size of algal unit, several may be picked up in one pipette.

g. Place the tip of the capillary pipette in the liquid of the second isolation dish. Gently squeeze the rubber bulb of the pipette until an air bubble is released. The pipette should be repulled between successive transfers.

h. Wash the units by transferring them singly with a capillary pipette to the liquid in the remaining two isolation dishes.

i. Transfer single algal units by a capillary pipette to tubes of liquid culture medium. As a precautionary step, the single algal unit in a droplet may be transferred first to a small piece of cover glass resting on a microscope slide; the droplet may then be examined with the compound microscope. The small piece of cover glass with the algal unit is then transferred to the tube of culture medium. This precaution ensures that a single algal unit is isolated and that it goes into the culture medium. Also, isolation dishes should be changed between successive transfers. Never use the same washing dish twice.

j. Place tubes of isolated algal units under conditions suitable for growth (Chp. 11).

2. *Capillary pipette (simplified)*. A single inverted plastic petri dish top is used as an isolation dish.

a. Place 10–15 drops of a natural collection in the center of an inverted plastic petri dish top (Fig. 3-2, facing p. 54).

b. Place 6–8 drops of suitable liquid medium in six positions encircling the natural collection (Fig. 3-2). Mark one position as number 1.

c. Using a newly pulled, sterile, capillary pipette, transfer the desired algal units by capillary pipette from the natural collection to one of the six drops of liquid medium as outlined in A.1.e,f, preceding.

d. Transfer a single algal unit from the first drop to a second drop of the liquid medium as preceding (moving clockwise).

e. Repeat the transfer process through the remaining drops of liquid medium until you are certain that only a single algal unit is present in a drop.

f. Transfer the single algal unit to tubes of liquid culture medium, as in 1.i,j, preceding.

3. *Streak plating*. When algal units are 10 μm or less in diameter, they are isolated more easily by streak plating or spray plating. Either of these methods serves to isolate an algal unit and may produce an axenic culture.

a. Prepare petri dishes containing growth medium solidified with 1–1.5% agar medium (Chp. 1, II.B). The agar should be $\frac{1}{2}$–$\frac{2}{3}$ the depth of the dish. Plastic or glass (100 × 15 mm) petri dishes are suitable.

b. Place 1–2 drops of natural collection near the periphery of the agar. Flame sterilize either a wire loop (Chp. 12, II.D.2.a), or a bent-glass rod (dip rod in 70% ethanol, flame and remove from flame; repeat several times, being careful flame is extinguished before dipping in alcohol each time). Using aseptic technique use either the sterile loop or glass rod to make parallel streaks of the suspension on the agar (see Fig. 3-5, p. 68). If desired, a wax pencil outline may be drawn on the outside bottom of the dish (Chp. 5, Fig. 5-2).

c. Cover, invert the plate and incubate 4–8 days under suitable growth conditions (Chp. 11, II or III).

d. Observe with a stereomicroscope and select the desired colonies that are free of other organisms for further isolation. Remove a sample using a fine capillary pipette or fine wire needle, and place in a drop of sterile culture medium on a cover glass. Observe with the high power objective of the compound microscope that the desired species has been isolated and is unialgal.

e. Repeat the streaking procedure with algal units from single colony and again allow colonies to develop. This second streaking reduces the possibility of bacterial contamination and of colonies originating from more than one algal unit.

f. Transfer algal units from a desired colony to liquid or agar medium.

4. Spray plating. This method is similar to streak plating except the algal units are sprayed onto the surface of the agar in a petri dish rather than streaked with a wire loop. A detailed report on one method of spray plating has been given by Wiedeman, Walne, and Trainor (1964).

a. Prepare petri dishes as in 3.a, preceding.

b. Draw a drop of the natural collection into a Pasteur-type pipette. Depending upon the spray-size desired, the pipette may be pulled in a manner similar to that of capillary pipette, 1.d, preceding.

c. Fasten the Pasteur-type pipette on a table so that the thin end is free (Fig. 3-6, p. 68). An air source (compressed air, pump, etc.) should be attached to the blunt end of the pipette so a jet of air may be used to create the spray.

d. Hold the petri dish with solidified agar *ca.* 200 mm perpendicular to the pipette (Fig. 3-6).

e. Place the tip of the Pasteur-type pipette with the algal units into a jet of air passing through a drawn pipette attached to an air source (see Fig. 3-6). The algal suspension will atomize and spray onto the agar surface. The degree of separation of algal units is controlled by varying the diameter of the Pasteur-type pipette or by varying the velocity of air flow.

f. Follow steps 3.c–f, preceding.

5. *Isolation on agar*. Certain algal units may be separated from contaminants more quickly and with greater certainty by manipulating them on an agar surface or by allowing the units to isolate themselves by creeping away from other contaminating algae.

a. Prepare agar petri dishes as in 3.a, preceding.

b. Remove the lids from the petri dishes to allow the agar surface to dry slightly, forming an invisible pellicle. (This pellicle varies with locality and season and must be determined by the investigator.)

c. Place one drop of a natural collection near the center of the agar surface.

d. Flame sterilize a fine wire needle. Using the fine needle and while making observations with the stereomicroscope, manipulate a single algal unit in a zigzag trail until it is isolated. When allowed enough time (several hours), some species such as diatoms (Bacillariophyceae) and desmids (Chlorophyceae), will move and slowly isolate themselves.

e. Draw the isolated algal unit into a fine capillary pipette, or remove the unit by cutting out a small agar block with a sterile microspatula. (Sterilize the microspatula by flaming in 70% ethanol as in 3.b, preceding.)

f. Transfer the isolated algal unit to a tube of liquid culture medium, as in 1.i,j, preceding.

6. *Isolation from atmosphere*. It is a simple task to isolate algal units from the atmosphere. These algae originate primarily from the soil. Although elaborate sampling devices have been used, a simple method was described by Brown, Larson and Bold (1964) for collecting and isolating airborne algae. Another successful method is that of Luty and Hoshaw (1967).

a. Prepare agar medium petri dishes as in 3.a, preceding.

b. Expose dishes in a vertical position directly to the atmosphere for 5–10 min from a moving automobile at 50–70 mph.

c. Cover, invert, and incubate the exposed dishes for 2–6 weeks under conditions suitable for algal growth (Chp. 11).

d. Follow steps for streak plating in 3.d–f, preceding.

B. *Purification*

Physiological and biochemical studies of microalgae require axenic cultures. Streak and spray plating may yield axenic cultures directly; however, algal units isolated with a capillary pipette may have associated contaminants. Repeated washing and transfer of algal units with a capillary pipette may yield axenic cultures. If contaminants adhere to the algal units, they may be physically separated by ultrasonic vibration. But for those algal units with tenaciously attached contaminants, it may be necessary to kill or inhibit the growth of contaminants *in situ* by chemical methods. Often it is easier to rid a newly-isolated unialgal culture of contaminants rather than work on xenic cultures that have been maintained for a period of time.

Four methods yielding axenic cultures are described here. They may be used singly or in combination. To test that an algal culture is bacteria-free, standard microbiological media may be employed. Examples of these are given in VII, following and in Chps. 4, IV.D; 5, III.D. In addition, putative axenic cultures should be examined using phase contrast microscopy or light microscopy at high magnification.

The methods of purification described omit reference to the use of sterile equipment. *It should be assumed that all equipment, media and other materials are STERILE* (see Chp. 12).

1. Centrifugation. Purification by washing is accomplished by repeated transfer of algal units through a liquid medium, and centrifuging.

a. Place algal units in centrifuge tubes half-filled with distilled water or culture medium. Filamentous fragments, or small thallus pieces may be used. The thallus may be cut on a glass surface with a scalpel or a series of razor blades bound 1.5 mm apart prior to use.

b. Centrifuge until algal units are loosely packed; decant supernatant. Time and speed of centrifugation have to be determined by trial and error.

c. Suspend algal units in fresh liquid and centrifuge. Repeat the washing process and centrifuge at least five times.

d. Using a Pasteur-type pipette, transfer single algal units to test tubes containing fresh culture medium as in A.1 or 2, preceding.

e. When the tubes show algal growth, test for bacterial contamination as outlined in VII, following.

f. Streak or spray plating methods (A.3 or 4, preceding) may be used after repeated washing and centrifuging if bacterial contamination is not eliminated. Follow steps a–c, then A.3 or 4, preceding.

2. Ultrasonic treatment. Securing axenic cultures by ultrasonic treatment has been described by Brown and Bischoff (1962). This method uses a low intensity (90 k cycles/s) ultrasonic water bath in which contaminants are physically separated from algal units. The algal units freed of contaminants yield axenic cultures with repeated washings and centrifugation.

a. Place algal units in a centrifuge tube half-filled with distilled water or culture medium as in 1.a,b, preceding.

b. Place centrifuge tube in an ultrasonic water bath for 5 s to 20 min, depending on algal units. Time is determined by viewing algal units microscopically for evidence of cell damage and by testing them for viability. Death may occur before visual evidence is available.

c. Centrifuge until algal units are loosely packed, decant supernatant and add fresh liquid.

d. Repeat sonication and centrifugation.

e. Wash algal units at least five times, centrifuging and resuspending them in fresh liquid between washes.

f. Streak or spray the material onto dishes of suitable agar medium as outlined in A.3,4, preceding.

g. Test mature cultures for bacterial contamination as outlined in VII, following.

3. Antibiotic treatment. It may be difficult to free algal units of certain contaminants by repeated washing and/or ultrasonic treatment. In these instances use of a chemical method rather than a physical method is preferred, but physical methods such as washing and ultrasonic treatment may precede. One such chemical method uses antibiotics singly or in combination to kill or inhibit the growth of tenaciously attached contaminants. The results of applying an antibiotic treatment are found in reports by Droop (1967), Machlis (1962), Provasoli (1958), Spencer (1952), and Tatewaki and Provasoli (1964). Further antibiotic treatments are outlined by Guillard (Chp. 4, IV.D), and Chapman (Chp. 5, III.D.2). These include placing the antibiotics in the agar as well as the liquid technique described here.

a. Prepare the antibiotic solution as follows: dissolve 100 mg penicillin G (K or Na salt) and 50 mg streptomycin.SO_4 together in 10 ml distilled water; add 10 mg chloramphenicol dissolved in 1 ml 95% ethanol to the penicillin–streptomycin solution; mix well.

b. Filter the triple antibiotic solution quickly, using a membrane, or Seitz filter apparatus.

c. Place 1 ml of algal suspension to be purified in each of six 125 ml Erlenmeyer flasks, each containing 50 ml culture medium.

d. Add one of the following volumes of antibiotic solution to each of the flasks: 3.0, 2.0, 1.0, 0.5, 0.25, 0.125 ml. This provides penicillin levels ranging from approximately 20–500 mg/l and corresponding levels of the other two antibiotics.

e. Place the flask cultures under conditions suitable for growth (see Chp. 11).

f. After 24 and 48 h, aseptically transfer some algal units from each flask to tubes of sterile, antibiotic-free culture medium. Prepare tubes in triplicate at each time interval.

g. Place tube cultures under conditions suitable for growth (see Chp. 11).

h. Check tube cultures for bacterial contamination after 2–3 weeks as outlined in VII, following.

4. Potassium tellurite treatment, sonication. In addition to antibiotics, a chemical method using potassium tellurite has been described by Ducker and Willoughby (1964) and by Rosowski and Hoshaw (1970). This compound acts as a bacteriostatic agent, allowing algal units to grow away from regions of bacteria. The clean algal units are used to establish axenic cultures.

a. Prepare 1–1.5% agar medium petri dishes as in A.3.a containing 10 mg/l potassium tellurite (K_2TeO_3).

b. Place the algal units in centrifuge tube half-filled with distilled water or culture medium, as described in 2.a–c, prior to use.

c. Streak or spray a suspension of the algal units to be purified onto the agar medium plates.

d. Cover, invert, incubate dishes under conditions suitable for growth (Chp. 11).

e. Using the stereomicroscope examine the dishes after 4–8 days for bacteria-free algal units. If bacteria-free units are present isolate them into liquid or onto agar medium as outlined in A.3,4.

f. Test mature cultures for bacterial contamination as outlined in VII, following.

5. Potassium tellurite (liquid) (K_2TeO_3). In addition to the method described in 4, preceding, the potassium tellurite may be used in liquid medium. Use the method outlined for antibiotic treatment (3, preceding), substituting 10 mg/l potassium tellurite for the antibiotic.

V. PROBLEM AREAS

A. *Algal contaminants*

Sometimes unialgal cultures are difficult to achieve due to contamination by other algae as the Cyanophyceae and Bacillariophyceae (diatoms). Some success has been achieved in controlling diatom growth with 1–10 mg/l of germanium dioxide (J. Lewin 1966). This is described by Chapman in Chp. 5, II.D.1 on culturing of marine algae. Bluegreen algae can be reduced or removed by using antibiotics. Media containing *ca.* 25 mg/l streptomycin is effective. Follow the general directions given in IV.B.3, preceding. Other methods of ridding cultures of alga contaminants are given by Guillard in Chp. 4, IV, V.

B. *Failure of purification treatment*

The success of a purification treatment is dependent upon the critical factor of timing and, in the case of chemical treatments, also upon the concentration of the chemical agent. When purification fails, isolates are either killed or the cultures produced are not axenic. In the ultrasonic treatment, too long an exposure in the water bath results in damage to cells. If exposure is too short, contaminants are not removed from the algal units. The use of antibiotics or potassium tellurite requires careful attention to the concentration of these chemical agents and to the length of time the algal units are exposed. Concentrations of chemical agents and times of exposure to them have been suggested, but experimentation is usually required for each alga. If a method does not work the first time, it is appropriate to say – try, try again. Do not hesitate to try modifications of a method! Conditions are never identical even on successive trials.

VI. ALTERNATIVE TECHNIQUES

A. *Isolation*

1. Glass hook. This method is identical to the capillary pipette method except a glass hook is used instead of a fine capillary pipette to isolate the algal units. The method is especially useful for isolating simple and branched filaments. The glass hook is made by passing a fine capillary pipette through a flame. The capillary bends to form a hook with the end sealed in the form of a glass ball. This ball prevents algal units from slipping off the glass hook during the steps of isolation.

2. Cover glass attachment. In this method algal units from a sample of a natural collection become attached to pieces of cover glass placed in petri dishes or other suitable containers of liquid culture medium. Algal units may become established on pieces of cover glass from structures such as motile and nonmotile reproductive bodies. These units should then be isolated with a capillary pipette or suitable instrument.

B. *Purification*

In addition to the chemical agents mentioned in this chapter, other chemicals that have been used successfully include caffeine (Browne 1964), sulfonamids (Reich and Kahn 1954), alcoholic iodine (Soli 1963), and 'Jodopax' (see Chapman Chp. 5, III.D.2).

A very weak agar medium (0.9% w/v) has been used by Guillard (personal communication). The agar at this concentration may be poured at temperatures as low as 30–35 °C. The algal suspension is added just before pouring the plates. Incubation is as given in V.A, preceding. In some instances it may be necessary to autoclave the agar and medium separately (see Allen Chp. 7, IV.B.1).

VII. STERILITY TESTS FOR AXENIC CULTURES

(J. Stein, personal communication)

A. *Microbiological media*

General media should be tested at different temperatures as well as in the light and dark. Depending on the algae being studied, media may be made with either filtered seawater or glass-distilled water. All media should be sterilized by heat (steam or autoclave) or membrane filter, depending upon the treatment used for the algae under study. Other methods than those listed here are presented in Chp. 5, III.D.1.a.

1. Nutrient broth – use as directed on package or 1.0% (w/v).
2. SST

glucose	1 g
tryptone	1
yeast extract	0.5
water	100 ml

3. Peptone – glucose

glucose	1 g
peptone	1
water	1 l

4. Any nutrient medium in which the alga grows plus any organic additive, singly or in combination, at 0.1% or 1–2% (w/v) (also see Chp. 1, IV.B)

acetate
glucose
sucrose
malt extract
peptone
tryptone
yeast extract

5. 1–2% (w/v) agar may be added to any of the previous (1–4) media.

Cultures should be incubated at least 2 weeks, but observations may be made daily after 24 h. Liquid cultures supporting bacterial growth may appear cloudy within 24 h. Macroscopic bacterial colonies are apparent on agar media after approximately the same time. Marine media may take as much as 12 days to support visible bacterial growth (Fries 1963). The appearance of fungal contaminants is usually indicated within 7–14 days.

B. *Microscopic observation*

Use of phase contrast microscopy can be employed to indicate the presence of contaminants.

1. Aseptically remove some algal material from a culture.
2. Place on clean (preferably sterile) slide. If necessary add sterile nutrient solution to the preparation.
3. Cover with clean (preferably sterile) cover glass.
4. Place on stage of microscope equipped with phase contrast illumination and objectives.
5. Using the highest power available, look for presence of microorganisms in addition to the algae.

C. *Procedure*

Tests for axenic cultures should include both growth in microbiological media and direct, microscopic observation. Because many algae have a great deal of associated gelatinous material, it is often easier to establish axenic cultures from new isolates rather than those maintained in culture over a period of time.

VIII. ACKNOWLEDGMENTS

We wish to thank H. P. Hostetter, A. E. Dennis, and Page Lindsey for their comments on the operational aspects of the methods described in this chapter. We thank Ruth S. Hoshaw for assistance with the illustrations, and Page Lindsey for demonstrating the techniques in them.

IX. REFERENCES

Bold, H. C. 1942. The cultivation of algae. *Bot. Rev.* **8**, 69–138.

Brown, R. M., Jr and Bischoff, H. W. 1962. A new and useful method for obtaining axenic cultures of algae. *Phycol. Soc. Amer. News Bull.* **15**, 43–4.

Brown, R. M., Jr, Larson, D. A. and Bold, H. C. 1964. Airborne algae: their abundance and heterogeneity. *Science* **143**, 583–5.

Browne, S. W. 1964. Purification of algal cultures with caffeine. *Nature* **204**, 801.

Droop, M. R. 1967. A procedure for routine purification of algal cultures with antibiotics. *Brit. Phycol. Bull.* **3**, 295–7 (describes an excellent procedure with results for purifying algae with antibiotics).

Ducker, S. C. and Willoughby, L. G. 1964. Potassium tellurite as a bacteriostatic agent in isolating algae. *Nature* **202**, 210.

Fries, L. 1963. On the cultivation of axenic red algae. *Physiol. Plant.* **16**, 695–708.

Lewin, J. 1966. Silicon metabolism in diatoms. v. Germanium dioxide, a specific inhibitor of diatom growth. *Phycologia* **6**, 1–12.

Lewin, R. A. 1959. The isolation of algae. *Rev. Algol.* n.s. **3**, 181–97 (an excellent discussion on the isolation and purification of algae).

Luty, E. T. and Hoshaw, R. W. 1967. Airborne algae of the Tucson and Santa Catalina Mountain areas. *J. Arizona Acad. Sci.* **4**, 179–82.

Machlis, L. 1962. The nutrition of certain species of the green alga *Oedogonium*. *Amer. J. Bot.* **49**, 171–7.

Pringsheim, E. G. 1946. *Pure Cultures of Algae*. Cambridge Univ. Press, London. 119 pp. (the classic work on the isolation and purification of algae).

Pringsheim, E. G. 1950. The soil-water culture technique for growing algae. In Brunel, J., Prescott, G. W. and Tiffany, L. H., eds., *The Culturing of Algae*, pp. 19–26. Antioch Press, Yellow Springs, Ohio.

Pringsheim, E. G. 1951. Methods for the cultivation of algae. In Smith, G. M., ed., *Manual of Phycology*, pp. 347–57. Chronica Botanica Co., Waltham, Massachusetts.

Provasoli, L. 1958. Effect of plant hormones on *Ulva*. *Biol. Bull.* **114**, 375–84.

Reich, K. and Kahn, J. 1954. A bacteria-free culture of *Prymnesium parvum* (Chrysomonadina). *Bull. Res. Council Israel* **4**, 144–9.

Rosowski, J. R. and Hoshaw, R. W. 1970. Cultivation of a filamentous alga in quantity on agar plates. *J. Phycol.* **6**, 220–2.

Soli, G. 1963. Axenic culture of a pelagic diatom. In Oppenheimer, C. H., ed., *Symposium on Marine Microbiology*, pp. 122–6. Charles C Thomas Publisher, Springfield, Illinois.

Spencer, C. P. 1952. On the use of antibiotics for isolating bacteria-free cultures of marine phytoplankton organisms. *J. Mar. Biol. Ass. U.K.* **31**, 97–106.

Tatewaki, M. and Provasoli, L. 1964. Vitamin requirements of three species of *Antithamnion*. *Bot. Marina* **6**, 193–203.

Wiedeman, V. E., Walne, P. L. and Trainor, F. R. 1964. A new technique for obtaining axenic cultures of algae. *Can. J. Bot.* **42**, 958–9.

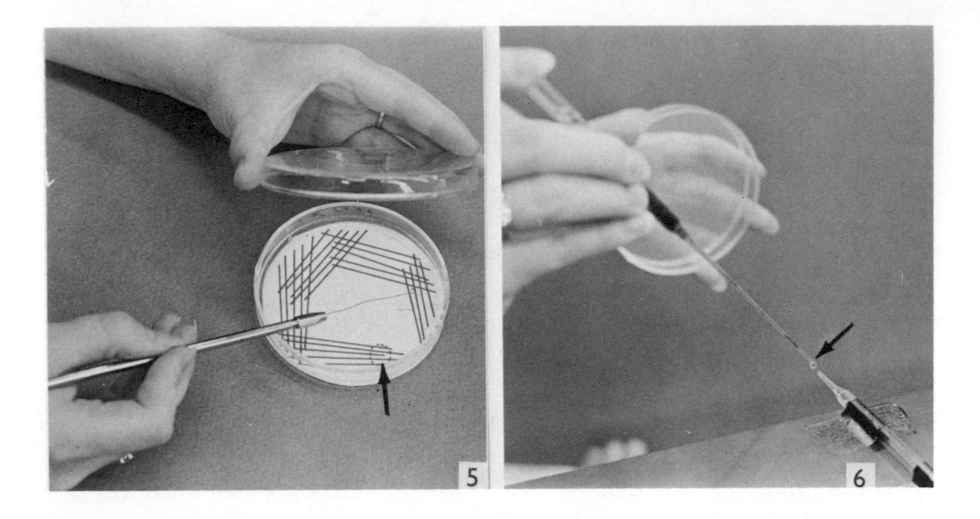

Figs. 3-5, 6. Streak and spray plating. Fig. 3-5. Diagrammatic representation showing pattern of streaks on agar surface. Position of natural collection shown by dotted lines (arrow). Fig. 3-6. Tip of capillary pipette (arrow) with air source producing a spray of the algal suspension onto the agar surface.

4: Methods for microflagellates and nannoplankton*

ROBERT R. L. GUILLARD†
Woods Hole Oceanographic Institution,
Woods Hole, Massachusetts 02543

CONTENTS

* Contribution Number 2501 from the Woods Hole Oceanographic Institution. This work was supported by National Science Foundation GB 7682.

† The methods for purification by phototaxis were contributed by Dr E. Paasche, Institute for Marine Biology B, University of Oslo, P.B. 1069, Blindern, Oslo 3, Norway.

I. OBJECTIVE

This chapter describes special methods for obtaining unialgal or axenic cultures of microalgae. The agar plate technique (Chp. 3, IV.A.3–5) is generally the method of choice for algae that will grow on agar; however, many species will not, whereas others cannot be free of algal contaminants on agar. Sometimes a desired species is so rare in the material from which it is to be isolated that individual specimens must be sought and freed of other organisms. Under any of these circumstances pipette isolation is a matter of necessity, but in other instances it may be the method of choice for reasons of logistics or experimental design.

The procedure outlined here involves four possible steps between collection of the natural sample and the establishment of a clonal axenic culture; not all steps are applicable in every instance, and more than one step may be achieved in one operation. The separate stages are: (1) raw enrichment culture; (2) dilution culture from natural collections or raw enrichments; (3) pipette isolation of individual cells or colonies to achieve unialgal clonal cultures, not necessarily axenic; and (4) achievement of axenic culture by either (a) antibiotic treatment, or (b) phototactic migration. The logic of each step and indication for application are discussed in the appropriate methods section.

General isolation techniques are described by Pringsheim (1946, 1950, 1951) and R. Lewin (1959). Droop (1954) gives methods especially suited to isolation of small marine flagellates.

II. EQUIPMENT

A. *Glassware* (also see Chp. 3, II.C)

1. Glass petri dishes 100 × 15 mm or 100 × 20 mm.
2. Test tubes with closures, or cotton plugs.
3. Pasteur-type pipettes with rubber bulbs.
4. Chemical watch glass, 65 mm diameter.
5. Erlenmeyer flasks, 1 or 2 l; or Fernbach flasks, 2.6 l with beakers for covers.

6. Spot plates of borosilicate glass 3 or 9 depression, 22 × 7 mm (Corning 7220, 7223) or 12 depression of varying diameter (Corning 7220). Depression slides, 75 × 25 mm, with 1 or 2 depressions, 3 mm deep, occasionally are needed (see Chp. 5, II.C).

7. Triangle, made of 6–8 mm glass tubing, 30–60 mm on a side, to support and steady watch glasses and depression microscope slides.

8. Device to expel cells from Pasteur-type pipettes, such as 50 mm length of heavy-walled rubber or plastic tubing plugged with glass rod at one end. At times it is desirable to watch cells emerge from the capillary of the drawn Pasteur-type pipette which is done by gentle pressure on the tubing. Alternatively, use a plastic mouthpiece and length of small rubber tubing.

9. Capillary pipette for purification of flagellates by phototactic migration (IV.E, following; Paasche 1971) shown in Fig. 4-1 (see p. 82). This can be made of soft glass tubing 200 mm in length. The tubing can be presterilized, or pipettes sterilized after manufacture. The lower part of the pipette comprises four bulbs retaining the original 4 mm diameter of the tubing, separated by sections *ca.* 0.1 mm diameter bore. This design restricts mixing of water in adjacent sections of the pipette without reducing the total volume unduly. The upper part is narrowed to *ca.* 1.5 mm internal diameter and provided with a double bend in the form of a sharp 'S'. Midway between the uppermost bulb and the first bend the pipette is scored lightly with a diamond or file (Fig. 4-1, see arrow).

10. Conical flasks capped with glass beakers (optional for specific procedures – see IV.B.1).

B. *Optical*

1. Stereomicroscope 90 × magnification minimum, with clear stage and substage mirror preferably rotating in all directions. The best optical resolution obtainable can be used profitably (some workers have had low power condensers fitted to dissecting microscopes to increase resolution). The illuminator of choice is the twin 4 watt fluorescent tube type with optional diffusing cover.

2. Compound binocular microscope, 35–100 × magnification. If isolation is to be done with the compound microscope, a reversing prism incorporated into the instrument is helpful. Thus the image and object movement are in the same direction, as with a dissecting microscope. (Reichert microscopes already contain reversing prisms: sold by American Optical Corp., Eggert and Sugar Rds, Buffalo, New York 14215.)

C. *Bench facilities*

Local atmospheric dust content determines the type of bench facility that can be used. Purification can be achieved at an open bench in clean locations; in dustier ones, a hood of some kind is necessary. Plywood will do for a hood of a simple design, having a transparent plastic top with a louvre to exhaust burner heat. Fluorescent and ultraviolet light fixtures for sterilization can be installed if desired (see Chp. 12, II.F). In very dusty areas a standard transfer cabinet is needed (Chp. 11, IV). An arrangement to darken the isolation area partially or totally is necessary to get illumination contrast best suited to the organisms being isolated. Experience has shown the importance of arranging the bench area so that it can be converted rapidly from right-handed to left-handed use, and of having all accessories conveniently arranged at all times.

1. Burner, preferably with pilot (e.g., 'Touch-o-matic' of Bellco) or alcohol burner.
2. Metal waste can (on floor below the burner, for waste glass from drawn pipettes – some care is needed to protect the wrists from injury by glass needles).
3. Wire pipette holder (to hold drawn sterile pipettes, sterile forceps, etc.).
4. Used pipette holder; a cut-off plastic quart-size bottle will do – fill with water to keep pipettes cleanable.
5. Dropping bottle, *ca.* 50 ml, for 70% ethanol.
6. Fine and medium point forceps.
7. A cart, or second best, a table for carrying glassware, media, and trays of warm or cold water for holding cultures at the correct temperature. This cart is placed beside the isolator where all items can be reached from the microscope seat (which should have wheels).

D. *Media*

Media for algae are treated by Nichols (Chp. 1) and McLachlan (Chp. 2). Remarks are made on specific points in the methods sections that follow. Also see IV, B.1 on precautions.

III. TEST ORGANISMS

Any culture or source available may be used to practice isolation techniques. Skill in picking up larger organisms from a unialgal culture should be attained before isolating algae of different sizes from mixed samples. The ones suggested are graded roughly by size; the largest (given first) being over 50 μm in diameter, the smallest *ca.* 5 μm. These taxa are available from most culture collections or biological supply companies (see introductory chapter).

A. *Marine flagellates*

Gymnodinium nelsoni (*splendens*); *Peridinium trochoideum*; *Amphidinium carteri*; *Platymonas* (*Tetraselmis*) sp.; *Rhodomonas* (*Cryptomonas*) sp.; *Dunaliella tertiolecta*; *Isochrysis galbana*; *Monochrysis lutheri*.

B. *Marine Bacillariophyceae* (diatoms)

Ditylum brightwelli; *Thalassiosira fluviatilis*; *Thalassiosira pseudonana* (*Cyclotella nana*.)

C. *Freshwater flagellates*

Synura sp.; *Pandorina morum*; *Euglena gracilis*; *Cryptomonas ovata*; *Chlamydomonas eugametos* (*moewusii*).

D. *Freshwater Bacillariophyceae* (diatoms)

Asterionella formosa; *Fragilaria crotonensis*; *Cyclotella meneghiniana*; *Navicula pelliculosa*; *Thalassiosira pseudonana* (estuarine form).

E. *Freshwater nonmotile Chlorophyceae* (green algae)

Closterium sp.; *Scenedesmus* sp.; *Chlorella* sp.; *Nannochloris* sp.

IV. METHODS

A. *Raw enrichment*

These may be made for any of several reasons: merely to maintain or increase the number of the desired species; to determine what physical or chemical treatments permit survival of the species desired, so that isolation attempts can succeed; to favor growth of the species desired (selective enrichment culture) or to depress growth of competitors to facilitate isolation chances.

1. Sampling, dilution. Using non-toxic samplers and containers (e.g., glass, polycarbonate, teflon), collect enough water so that each enrichment culture has a volume of 500 ml or more, if possible. Protect the water samples from heat, cold, bright light, and violent agitation. Transfer portions to sterile culture Erlenmeyer or Fernbach flasks. It is usually best to have a 40–50 mm liquid layer in enrichment cultures, but some flagellates are favored by conditions in deep narrow-mouthed vessels. Sometimes it is best to dilute a dense natural population 2–10-fold with membrane filtered (sterile) water from the collection site.

2. *General enrichment.* The nutrient enrichment should be dilute and should approximate natural conditions of the water being enriched. For sea and estuarine waters to not exceed the following: nitrate 35 μM; phosphate 1.5 μM; silicate 25 μM; trace metals $\frac{1}{5}$ the levels in medium 'f/2' (Chp. 2, IV.A.4); vitamins $\frac{1}{25}$ the levels in 'f/2'. A general enrichment as 'f/50' is usually adequate.

For freshwater, no additives should exceed the levels corresponding to $\frac{1}{5}$ of a modified Chu no. 10 (Chp. 1, IV.A.5); or the nitrate, phosphate, silicate, trace metal, and vitamin levels should not exceed the values given for marine media preceding.

The addition of *ca.* 5 μM NH_4Cl and 0.5% soil extract (see Chp. 1, IV.B.4) is usually favorable and seldom harmful.

3. *Selective growth.* Techniques of selective growth are discussed further in Chapters 3, 5, 6, 7 by the use of antibiotics, growth substances, specific chemicals (e.g., germanium dioxide, potassium tellurite, etc.).

Some algae cannot use nitrate, hence must be supplied ammonia. These species commonly tolerate higher levels of ammonia than do nitrate users and may be supplied *ca.* 1 mM NH_4Cl. Ammonia requirers often grow in the first enrichment culture without added ammonia, giving the false impression they do not need it.

Diatoms and perhaps other silica depositing forms can be suppressed by omitting silicate and using non-toxic plastic vessels or paraffin-coated glass. One to 10 mg/l germanium dioxide inhibits diatom growth (J. Lewin 1966) (see Chp. 5, II.D.1).

Actidione (cycloheximide) selects for Cyanophyceae as other algae are generally inhibited by *ca.* 50 mg/l (Zehnder and Hughes 1958). This antibiotic can be heat sterilized (Chp. 12, II.C). It is easy to select against blue-green algae as they are sensitive to most antibiotics; such as, low levels of streptomycin (*ca.* 25 mg/l) (see Chp. 3, IV.B.3).

A change of salinity can be a drastic selective agency for marine populations; 2–3 dilutions of the natural water by a factor of two each time are suggested. Increasing the total solid content of freshwaters can have the same effect (*not* using nitrate, phosphate, or trace metal enrichment). Changing the ion ratios of freshwaters is less strongly selective (Provasoli and Pintner 1960).

4. *Incubation*. Some enrichment cultures should be kept under conditions of temperature and light duration approximating those of the collection site. However, the intensity of full daylight should be approached only with the greatest caution. One-tenth of this (*ca*. 10000 lux) is an approximate safe upper limit, and half of this is usually ample. Fluorescent or filtered incandescent light is commonly used (see Chp. 11, II, III), but reduced natural light may be best – the traditional north window, of the Northern Hemisphere.

Exposing replicate enrichment cultures to different temperatures is an obvious and effective selective treatment; varying light intensity and duration is less strikingly effective.

5. *Processing*. The enrichment cultures should be examined periodically as soon as 48 h after establishment; do *not* wait for obvious growth to appear. A hand lens (4–10×) aids in examining the unshaken culture vessels. Sample separately the surface-dwelling, bottom-dwelling, and phototactically aggregated populations, using a Pasteur-type pipette.

A relatively large volume of water (*ca*. 1 ml) placed in a depression slide or Sedgwick–Rafter cell (Chp. 19, III.B.3.a) may be used. Begin isolation efforts as soon as indications of survival are apparent. The results of the enrichment treatments are a guide for incubation of the isolation tubes.

Sub-cultures may be made from raw enrichment cultures as suggested by population developments in the latter. If it is desired to maintain the mixed native population, keep the nutrient levels low and transfer often, at 2–3 day intervals. In addition, keep the old cultures, but in dim light (500–1000 lux).

From the enrichment or sub-cultures, the next stage is to establish unialgal and possibly axenic cultures by any isolation technique.

B. *Dilution cultures*

The basic principle is estimation of the algal population in a sample and dilution of a portion to reduce the density to one or a few cells per specified volume (e.g., 1 ml). One ml portions are then dispensed into each of many tubes of enriched medium. (This is a method of single cell isolation without handling individual specimens.) Successful incubation yields many unialgal cultures. The method has been used both to obtain cultures and to estimate algal numbers in natural waters (Knight-Jones 1950; Ballantine 1953).

The dilution technique is not used for rare species, but is of maximal value when several relatively abundant species are desired from a sample.

Dilution cultures can be made at sea under conditions too rough for other isolation methods.

The number of tubes inoculated and extent of dilution depend upon several factors. If the alga desired is a dominant, few tubes are needed, but if it is less abundant, more are necessary. As not all specimens survive, it is best to use two or three dilutions, allowing several algae per tube in the most concentrated one. Media and glassware can be prepared in advance; counting, dilution, and inoculation are quick. Some of the sample can be kept for raw enrichment cultures and later pipette isolation.

1. Precautions. The considerations given here apply to all the techniques for isolating single cells.

Media to be used for isolating single cells should receive minimum exposure to materials or treatments that might impart toxicity (see Chp. 14). One or a few cells in mildly toxic medium will not grow usually, though a large inoculum can often dilute the unfavorable influence and yield a culture. Medium autoclaved in proximity to certain plastic materials may become toxic (Chp. 14, IV.A.1.a). Plastic screw caps for test tubes or plastic tubes themselves, may be toxic. Stainless steel closures cause no difficulty unless they come into contact with seawater. Certain plastic closures are non-toxic, but this should not be assumed. Foam plastic plugs should be tested, as should cotton and cheesecloth used to make plugs. (Test by using a small inoculum of a known sensitive species.) It is usually necessary to wash cotton and cheesecloth with hot water, then autoclave, using the 'fast exhaust and dry' cycle if available. 'Water conditioners' added to steam supplies for autoclaves may contaminate media (Kordan 1965). Impurities leach from plastic filter units (Simpson 1966), while detergents as well as other organic materials leach from membrane filters (Parker 1967, and references therein). Chapter 14 gives more details concerning toxic and inhibitory materials.

For utmost certainty, dry sterilize glass capped tubes, or autoclave them in a small autoclave that uses its own distilled water reservoir (e.g., a pressure cooker). The screw caps or other closures can be autoclaved separately and exchanged for the glass caps later.

Media may be autoclaved in tubes, or, to reduce precipitation, autoclave some 50 or 100 ml of media (especially marine ones) in glass capped conical flasks. As with media sterilized by filtration, these must be aseptically poured or pipetted later into sterile tubes.

The depth of liquid in the tube into which a cell is isolated is sometimes critical to its survival, for reasons unknown but apparently connected with the redox potential (Lees and Tootill 1952). Isolates may survive better in one depth or another (although all media are dispensed at the same time after autoclaving) (Wall, Guillard and Dale 1967). A range 30–90 mm depth is suggested.

2. *Estimation of population.* Methods for using counting chambers are given in Chp. 19. Preliminary tenfold concentration of some samples by centrifugation in 15 ml conical tubes is helpful. The Coulter Counter supplemented by microscopic examination may be used (see Chp. 22). Only a rapid order of magnitude count is necessary. Attention should be paid to organisms 5 μm or less in size, such as cells of *Nannochloris* (Chorophyceae), *Micromonas* (Prasinophyceae), or *Synechococcus* (Cyanophyceae), or spores of *Chaetoceros* (Bacillariophyceae) which may be by far the most abundant organisms present.

3. *Medium preparation*

a. The best water for dilution and culture is membrane filter sterilized water from the sample site; failing this, autoclaved water from the sample site or any sterile medium matching the natural water in important properties such as salinity, total solids, etc., should be used. The temperature of the dilution water should be that of the sample.

b. Nutrient addition to the final growth medium should be dilute, as recommended in A.2, preceding. It may be convenient to use enriched water for the dilution series because of the sequence of pipetting steps involved in dilution and inoculation of the final isolation tubes.

4. *Dilution.* The range of dilution is from 10^3 to 10^6 depending on the population density, and may be achieved by any sequence of steps, using flasks or tubes. The final addition to empty or partially filled tubes is by ordinary or automatic pipette. Liquid in the dilution series must be mixed thoroughly, but without violent agitation.

In the field, it is often more convenient to carry empty sterile tubes. Dilution should then be carried to the point that there is *ca.* one specimen per 8–10 ml. This amount is dispensed to the empty tubes. In addition, the dilution series should be done using enriched water as described in 3.b, preceding.

5. *Incubation.* As discussed in A.4, previously.

6. *Processing.* Examine tubes with a hand lens or low power of the dissecting microscope after *ca.* 1 week and at intervals of a few days thereafter. Transfer successful cultures and test for the presence of algal or

other microbial contaminants. Tests for the latter are given in Chp. 3, VII. Contaminating algae and fungi can be detected microscopically, especially in old cultures.

While the dilution technique may be repeated if the culture is not unialgal, isolation with a pipette is generally preferable. This is especially true when the culture is unialgal and a clonal strain is desired.

C. *Pipette isolation*

Single cells or colonies (*algal units*) can be recognized under relatively low magnification (*ca.* 15–100 ×), picked up in a Pasteur-type capillary pipette, and ejected into sterile medium as described in Chp. 3, IV.A.1. The isolates can be washed in drops of sterile medium or transferred to a slide for microscopic examination at high magnification for contaminants or for identification. Some isolates will even survive recovery from the observation slide and successfully establish a culture.

Cells must be protected from adverse conditions during the isolation procedure; notably, abrupt changes of temperature, salinity, pH, light intensity, and excessive agitation of the medium. Rapid and gentle handling of the cultures is essential for success. Speed and skill in drawing capillary pipettes should be acquired by practice, and familiarity gained in use of the optical equipment available.

In any specific isolation attempt, the isolator should first arrange the work area and materials, then experiment with some of the algae wanted in order to select the illumination, magnification, types of slides, quantity of liquid per well, etc., that favor recognition and capture. With preliminaries done, a skilful isolator can sample material held in a growth chamber and return the first tubes of isolates to the same chamber in less than ten minutes – this is the timing needed for fragile organisms.

1. Media, washing fluid. Media to receive isolates are like those used for the dilution cultures described in B.3, preceding.

The fluid used to 'wash' isolates and to dilute raw enrichments must be particle-free or small algae are difficult to discern. Seawater or freshwater media should be membrane filtered and stored in clean beaker-covered flasks (see Chp. 12, II.E). If the seawater or medium precipitates on autoclaving, it must be sterilized by membrane filtration.

2. Glassware. Observe precautions outlined in B.1, preceding, regarding sterilization of tubes, etc. Pasteur-type pipettes, which are used both for transfer and isolation, are prepared and sterilized as described in Chp. 3,

II.C.2 and 4; Chp. 12, II.D.2.b. Slides and watch glasses must be washed free of particles. Trial may show that sterilization by dry heat (in baking dishes) yields glassware with less particulate matter.

3. Isolation

a. Arrangement – arrange work area (for right-handed person) with microscope directly ahead, lamp beyond it. The tube of isolating pipettes and burner are to the left and the metal waste can below the bench. To the right place the ethanol bottle, and ethanol-sterilized wire pipette holder (for loops, forceps, drawn pipettes). A cart or table at isolator's left holds trays of ice or water baths, if necessary, and tubes of transfer and isolating pipettes, dishes of depression slides and watch glasses previously brought to suitable temperature. A compound microscope should be nearby.

b. Use a sterile Pasteur-type pipette to add dilution medium to depression slides or watch glasses. Fill all wells of a multi-depression slide at once; fill a few watch glasses at a time and keep covered.

c. Add one to a few drops of the natural material (or enrichment culture, etc.) to the first well or watch glass; dilute to the extent previously found appropriate.

d. Draw a capillary on a Pasteur-type pipette as follows (also see Chp. 3, IV.A.1). If right-handed, hold the pipette in the left hand by the mouthpiece, open tip to the right. Grasp the tip with sterile forceps, and with one or both wrists supported by the bench edge, move the portion of the pipette 10–15 mm from the tip into the edge or tip of the flame. Rotate the pipette (about the long axis) with fingers of the left hand, supporting tip loosely with the forceps. Flame heat should be moderate or use the pilot light. In 3–4 s, when the flame is the sodium yellow color, the glass is ready to draw. Grasp the tip with the forceps, simultaneously draw the pipette several mm from the flame and with continuous *but not sudden motion*, draw out a capillary 50–150 mm in length. This takes about a half-second! Hot glass pulled too quickly breaks; too slowly, sags. By varying draw speed and distance, capillaries of perhaps 10–500 μm can be made. Break the capillary at a convenient length, or where the bore of the tube is right, by grasping at the point desired with sterile forceps and pulling sharply with a slight twist. The sterile pipette may be set on the wire holder.

e. Use the microscope to find a specimen desired. Change to lowest practical magnification. Organisms can often be recognized under surprisingly low magnification. Flagellates in particular may have a motion identifiable by speed, change of direction, tumbling, or rotation; some migrate to a particular spot in the well. Flashing of the chloroplast is often characteristic. Some flagellates will swim to the side or bottom of the depression and assume a position such that the chloroplast does not glow; tilting the mirror can reveal such cells if the medium is free of particles. (Background haze sometimes can be reduced by putting a little distilled water between the stage and a depression slide, which helps when hunting small cells.)

f. To capture, put the tip of the capillary in view just above the liquid surface. (Support for the wrist aids steadiness and maintaining the correct angle.) Approach the specimen with a single dipping motion of the capillary pipette. Remove the capillary from the liquid as soon as the specimen is in the bore. The dilution of the algal culture and the diameter of the bore should be such that one or only a few specimens move into the capillary at each attempted capture. To check the number of cells isolated, gently blow out the pipette (by mouth, tube, or bulb, according to the technique in use) into a small drop of medium on a flat slide. Observe with either the dissecting or compound microscope. It is also possible to observe the organisms in the capillary bore, by gently placing it on the microscope stage.

g. The specimen may be recaptured and washed more than once. One dilution plus one wash will often yield a unialgal culture; it does not pay to wash more than six times even if trying to establish axenic culture by this means. Repeated capture stops some flagellates, so use of the devices to permit watching the cell emerge from the pipette is suggested. A pipette may be redrawn and used until too short; the flaming preserves sterility.

h. After the last wash, the specimen is expelled into a tube of medium. Some species grow well if the isolate is left in the bore of the capillary, which is gently broken in the tube of medium.

4. Incubation. As described in A.4 and B.5, previously.

D. *Antibiotics*

Spencer (1952) noted that when phytoplankton subcultures were started using a small inoculum taken from cultures growing with antibiotics, some of the subcultures were bacteria-free. The method recommended here combines this observation with Oppenheimer's (1955) that the number of viable bacteria in seawater containing antibiotics reaches a minimum some hours (*ca.* 1 day) after the antibiotics are added, then increases rapidly. The idea is to expose a healthy, rapidly dividing population of algae to antibiotics for 10–36 h. During this interval make several (1–3 times) pipette isolations of one to a few algal cells into tubes of sterile medium.

The antibiotic treatment can be combined with washing the algae by repeated transfer of single cells to fresh sterile medium, but much washing is unnecessary if growth of the algae and depression of the bacteria by antibiotics have produced a favorable ratio of algae to bacteria.

It is important that the algae be growing well and that contaminants be relatively few in species and in numbers before attempting the isolation. Dilution techniques and preliminary pipette isolation should be used.

1. Antibiotic solutions. The suggested basic antibiotic solution contains penicillin G and streptomycin.SO_4; chloramphenicol is often added (see Chp. 3, IV.B.3). Experience has shown this basic solution to be tolerated by all but a few algae. More elaborate mixtures are given by McLaughlin and Zahl (1957), Provasoli (1958), and Tatewaki and Provasoli (1964).

Antibiotic stock solutions are best prepared just before use unless frozen immediately after sterilization. Unless known sterile antibiotic solutions or dry powders are available, it is best to sterilize the antibiotic mixes.

Dissolve 100 mg (about 162 500 units) of penicillin G (Na or K salt with no procaine or other additions) and 50 mg of streptomycin (or dihydro-streptomycin) sulfate in 10 ml distilled water, and sterilize with membrane filter. When chloramphenicol is to be used, separately dissolve up to 20 mg chloramphenicol in 0.5–1.0 ml 95% ethanol. Add this to the penicillin–streptomycin solution. Membrane filter the resulting solution quickly. (Chloramphenicol is soluble to 2.5 mg/ml in hot water, but penicillin should not be heated.)

2. Procedure (also see Chp. 3, IV.D.3)

a. First grow the algae in a known satisfactory medium and obtain a culture that is moderately dense for the particular species or assemblage, but still growing rapidly.

b. Transfer 1 ml of the above algal culture into each of five 125 ml flasks containing 50 ml of culture medium. To the series of flasks add antibiotic mix as follows: 0, 0.25, 0.5, 1.0, 2.0 ml each. (This yields penicillin levels of 0, 50, 100, 200, and 400 mg/l and corresponding levels of the other antibiotic(s). Experience may suggest that some of the levels can be omitted in the series, for some kinds of algae.) Incubate the flasks in suitable conditions (see Chp. 11) favorable for the algae, near the end of a working day.

About 15 h later (next morning) compare the appearance and motility of cells in the flasks with and without antibiotics. This will determine if the antibiotics are toxic at any level. In general, make isolations from the highest non-toxic level.

At several intervals between 15 and 48 h from the time of placing the

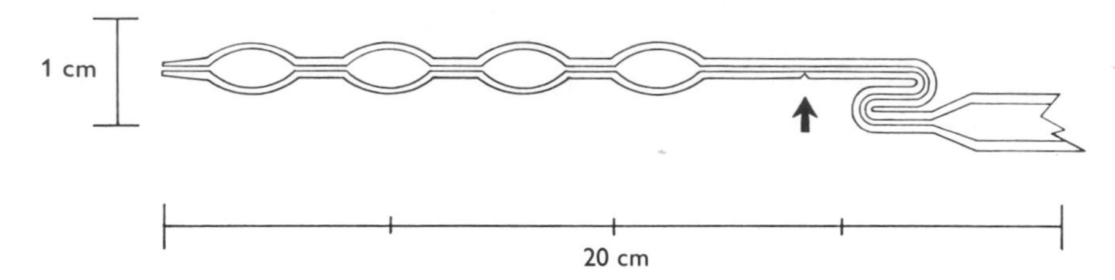

Fig. 4-1. Pipette for purification of flagellates by phototaxis. Arrow indicates score mark at which the pipette is eventually broken. The width of the pipette has been exaggerated for greater clarity.

algae in the antibiotics (up to 72 h for species living at temperatures near freezing), isolate algal cells individually or in small numbers into tubes of sterile medium (see C, preceding). Five to fifteen tubes each time is sufficient. A few isolates should go into bacteria test media to indicate success or failure of the treatment (see Chp. 3, VII).

Allow 5–20 days for growth to appear in the isolation tubes; follow as described in B.6. When growth is visible, transfer to tubes or flasks of maintenance medium and test critically for microbial contaminants (see Chp. 3, VII). Some test media should be low in total organics (less than 1 g/l).

E. *Phototactic migration* (after Paasche 1971)

With phototactic flagellates, pipetting can be combined with migration to achieve pure unialgal and/or axenic cultures. Cells are released from a Pasteur-type pipette near one edge of a small volume of sterile medium in a dish. Illumination from one side induces swimming to the opposite edge, where they can be picked up anew. The process can be repeated. If the flagellates are allowed to swim in a pipette rather than in a series of dishes, the need for various preparations and handling is reduced to a minimum, thus eliminating potential damage.

The special pipette described in II.A.9, preceding, is used here (hereafter referred to as 'pipette'), see Fig. 4-1.

1. Draw sterile medium the length of the pipette to the first bend (beyond the score mark).

2. Draw in sufficient flagellate suspension to replace *ca.* half the medium in the bulb nearest the tip. The air–water interface is now moved beyond the second bend.

3. Place the pipette horizontally near a window or in front of a lamp (be careful that there is no excess heat). Point the tip toward or away from the light, depending on whether the flagellates are negatively or positively phototactic.

4. Cover the pipette with a transparent lid to protect from contamination. After 30–120 min (depending on the organism and the number of cells originally introduced), ten to many hundreds of cells accumulate in the first bend. They remain trapped here with no risk of their being absorbed onto the air–water interface. Thus the time of interruption of the experiment is not critical.

5. Break the pipette at the score mark, and expel the cells into a tube of sterile medium.

6. The tubes are incubated and followed as described in A.4 and B.6, preceding.

V. ALTERNATIVE TECHNIQUES

A. A micromanipulator may be used to isolate cells; this is described in Chp. 8, by Throndsen.

B. Droop (1967) has described a procedure using antibiotics, but not involving pipette isolation, that can be used to obtain axenic cultures from unialgal ones.

C. Substances other than antibiotics, as well as various processes, have been used to assist in obtaining axenic cultures. These include: caffeine (Browne 1964); tellurite as in Chp. 3, IV.B.4, 5 (Ducket and Willoughby 1964); sulfonamids (Reich and Kahn 1954); alcoholic iodine (Soli 1963); ultrasonic treatment as in Chp. 3, IV.B.2 (Brown and Bischoff 1962); ultraviolet light (Gerloff, Fitzgerald and Skoog 1950); filtration (Soli 1964).

VI. REFERENCES

Ballantine, D. 1953. Comparison of the different methods of estimating phytoplankton. *J. Mar. Biol. Ass. U.K.* **32**, 129–48.

Brown, R. M., Jr and Bischoff, H. W. 1962. A new and useful method for obtaining axenic cultures of algae. *Phycol. Soc. Amer. News Bull.* **15**, 43–4 (ultrasonic treatment).

Browne, S. W. 1964. Purification of algal cultures with caffeine. *Nature* **204**, 801.

Droop, M. R. 1954. A note on the isolation of small marine algae and flagellates for pure culture. *J. Mar. Biol. Ass. U.K.* **33**, 511–41.

Droop, M. R. 1967. A procedure for routine purification of algal cultures with antibiotics. *Brit. Phycol. Bull.* **3**, 295–7.

Ducker, S. C. and Willoughby, L. G. 1964. Potassium tellurite as a bacteriostatic agent in isolating algae. *Nature* **202**, 210.

Gerloff, G. C., Fitzgerald, G. P. and Skoog, F. 1950. The isolation, purification, and nutrient solution requirements of blue-green algae. In Brunel, J., Prescott, G. W. and Tiffany, L. H., eds., *The Culturing of Algae*, pp. 27–44. Antioch Press, Yellow Springs, Ohio (UV treatment).

Knight-Jones, E. W. 1950. Preliminary studies of nannoplankton and ultra-plankton systematics and abundance by a quantitative culture method. *J. Cons. Intern. Expl. Mer.* **17**, 140–55.

Kordan, H. A. 1965. Fluorescent contaminants from plastic and rubber laboratory equipment. *Science* **149**, 1382–3.

Lees, K. A. and Tootill, J. P. R. 1952. The assay of vitamin B_{12}. Relationship between growth response of *Lactobacillus leichmanii* 313 in tubes and diffusion of oxygen into the medium. *Biochem. J.* **50**, 455–9 (redox potential).

Lewin, J. C. 1966. Silicon metabolism in diatoms. v. Germanium dioxide, a specific inhibitor of diatom growth. *Phycologia* **6**, 1–12.

Lewin, R. A. 1959. The isolation of algae. *Rev. Algol.* n.s. **3**, 181–97.

McLaughlin, J. J. A. and Zahl, P. A. 1957. Studies in marine biology. II. *In vitro* culture of zooxanthellae. *Proc. Soc. Exp. Biol. Med.* **95**, 115–19 (antibiotic mix).

Oppenheimer, C. H. 1955. The effect of bacteria on the development and hatching of pelagic fish eggs, and the control of such bacteria by antibiotics. *Copeia* **1955**, 43–9.

Paasche, E. 1971. A simple method for establishing bacteria-free cultures of phototactic flagellates. *J. Cons. Int. Explor. Mer.* **33**, 509–11.

Parker, B. C. 1967. Influence of method for removal of seston on the dissolved organic matter. *J. Phycol.* **3**, 166–73.

Pringsheim, E. G. 1946. *Pure Cultures of Algae.* Cambridge Univ. Press, London. 119 pp.

Pringsheim, E. G. 1950. The soil-water culture technique for growing algae. In Brunel, J., Prescott, G. W. and Tiffany, L. H., eds., *The Culturing of Algae*, pp. 19–26. Antioch Press, Yellow Springs, Ohio.

Pringsheim, E. G. 1951. Methods for the cultivation of algae. In Smith, G. M., ed., *Manual of Phycology*, pp. 347–57. Chronica Botanica, Waltham, Massachusetts.

Provasoli, L. 1958. Effect of plant hormones on *Ulva. Biol. Bull.* **114**, 375–84 (antibiotic mix).

Provasoli, L. and Pintner, I. J. 1953. Ecological implications of *in vitro* nutritional requirements of algal flagellates. *Ann. New York Acad. Sci.* **56**, 839–51.

Provasoli, L. and Pintner, I. J. 1960. Artificial media for freshwater algae: problems and suggestions. In Tryon, C. A., Jr and Hartman, R. T., eds., *The Ecology of Algae*, pp. 84–96. Special Publ. No. 2, Pymatuning Laboratory of Field Biology, Univ. Pittsburgh.

Reich, K. and Kahn, J. 1954. A bacteria-free culture of *Prymnesium parvum* (Chrysomonadina). *Bull. Res. Council Israel* **4**, 144–9.

Simpson, L. 1966. Toxic impurities in Nalgene filter units. *Science* **153**, 548.

Soli, G. 1963. Axenic cultivation of a pelagic diatom. In Oppenheimer, C., ed., *Marine Microbiology*, pp. 121–6. Charles C Thomas, Springfield, Illinois.

Soli, G. 1964. A system for isolating phytoplankton organisms in unialgal and bacteria-free culture. *Limnol. Oceanogr.* **9**, 265–8 (filtration).

Spencer, C. P. 1952. On the use of antibiotics for isolating bacteria-free cultures of marine phytoplankton organisms. *J. Mar. Biol. Ass. U.K.* **31**, 97–106.

Tatewaki, M. and Provasoli, L. 1964. Vitamin requirements of *Antithamnion*. *Bot. Marina* **6**, 193–203 (antibiotic mix).

Wall, D., Guillard, R. R. L. and Dale, B. 1967. Marine dinoflagellate cultures from resting spores. *Phycologia* **6**, 83–6 (influence of volume of culture fluid on success of isolation procedure).

Zehnder, A. and Hughes, E. O. 1958. The anti-algal activity of acti-dione. *Can. J. Microbiol.* **4**, 399–408.

5: Methods for macroscopic algae

A. R. O. CHAPMAN
Biology Department,
Dalhousie University, Halifax, Nova Scotia, Canada

CONTENTS

I. INTRODUCTION

Methods of obtaining crude cultures of microscopic stages of macroscopic (benthic) algae have long been known. However, it is only in recent years, with the development of refined media and handling techniques, that growth and reproduction of larger stages of some species in culture has become feasible. Many of the problems involved in inducing growth and reproduction of the larger seaweeds have yet to be solved and this account must in many ways be regarded as preliminary.

The methods used for setting up cultures are largely determined by the objectives of the study in hand. Thus multialgal cultures with bacteria, which are in many cases suitable for taxonomic work, are obviously inappropriate for nutritional studies. For the major marine classes (Chlorophyceae, Phaeophyceae and Rhodophyceae) methods are given for setting up; (a) crude cultures which may be multialgal and contain bacteria; (b) unialgal cultures with bacteria (xenic); and (c) unialgal cultures without bacteria (axenic). Crude cultures are the easiest to establish and often give the best results in terms of the growth rate. It may well be that species in a multialgal culture interact to their mutual benefit in much the same way that de Wit (1960) recorded for higher plants. It is fairly obvious, however, that crude cultures are unsuitable for many purposes, particularly in life history studies. In these instances the establishment of unialgal cultures is necessary. Axenic cultures are used almost exclusively for nutritional, physiological and biochemical studies where synthetic media are required. In only a limited number of marine algae has there been an attempt to establish axenic cultures in synthetic media; and in most, growth of the alga has been relatively poor.

II. EQUIPMENT

A. *Temperature control*

Successful cultivation of large seaweeds often depends on the use of large volumes of medium. Maintenance of controlled conditions in such large volumes obviously requires fairly extensive culture facilities. In many laboratories use is made of expensive walk-in plant growth chambers which have built-in lights, temperature control systems and, in some instances, humidity controls. A number of scientific equipment companies manufacture these chambers (see Chp. 11, III), but they all suffer from a relatively high rate of breakdown. It is very rare not to have a cooling failure at least once in six months. The cultivation of large seaweeds is often a long term project, sometimes involving 12 months where plants are long-lived. A single overnight breakdown with a rise in temperature may ruin the work of months.

An alternative to specially manufactured plant growth chambers is the commercial refrigerated meat box used by butchers. The advantages with such a unit are low cost and long term reliability, which results from the relative simplicity of the system which also gives it its disadvantages. Chief among these is the coarse temperature control ($\pm$ 1–2 °C). In addition there is no control over humidity which may be important in moist coastal areas where fungal infections are common in the summer months.

Undoubtedly one of the most successful methods of long term culturing of larger seaweeds is through the use of refrigerated water baths. These provide very accurate temperature control and reliability. At the present time little use is made of large refrigerated baths and consequently they are not available commercially. They are, however, relatively easy to construct in the laboratory using immersion coolers (available from Frigid Units Inc., 3214 Sylvania Ave., Toledo, Ohio 43613). A one horse-power unit is necessary for a tank of *ca.* 400 gal (US) capacity. The tank should be rectangular, of fiberglass or galvanized steel, and insulated with polystyrene or polyurethane foam. A drain valve is needed on the underside and the whole tank should be supported on a metal frame. Culture vessels are supported on a perforated zinc sheet suspended in the cooling water to a depth appropriate to the size of the culture vessels.

Smaller refrigerated baths may be constructed in the same manner using portable cooling units (available from Frigid Units Inc.: Blue M Electric Co., 138th and Chatham St., Blue Island, Illinois 60406: Grant Instruments, Barrington, Cambridge, England). By adding immersion heating devices available from scientific supply companies very accurate temperature control both above and below ambient may be attained.

For some seaweeds, such as the crustose corallines (Rhodophyceae), maintenance in an 'open' system of running seawater is the best method of cultivation (Adey 1970). This obviously has the inherent difficulty of supplying running seawater to the laboratory and there are other, less obvious, problems. Foremost among these is the risk of contamination in

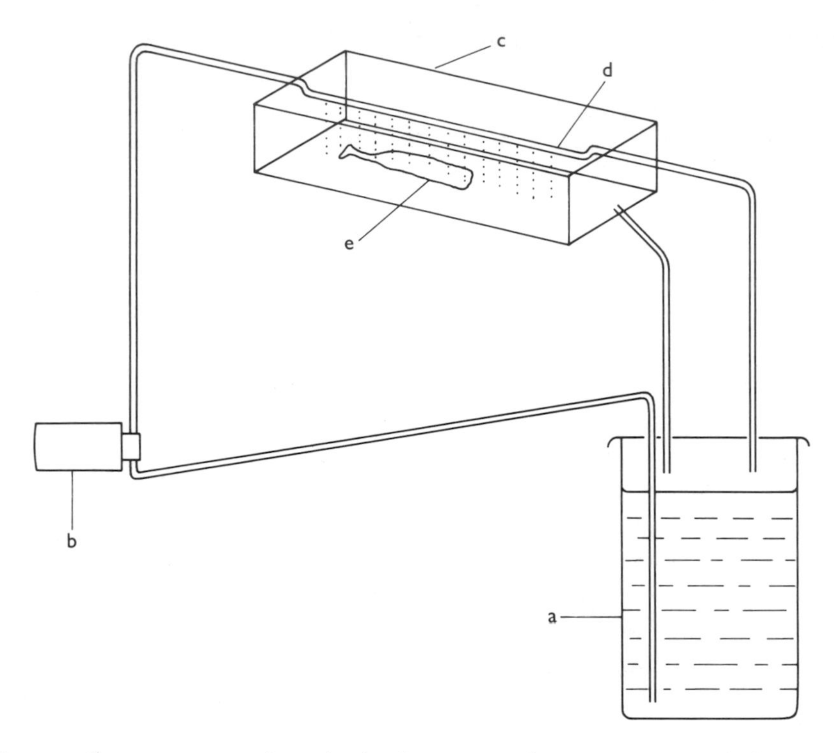

Fig. 5-1. Spray apparatus for culturing large seaweeds. a – reservoir tank; b – pump; c – plexiglass aquarium; d – perforated hose; e – alga.

untreated water. It is not really feasible to filter sterilize large volumes of water in such a system on a continuous basis and perhaps the best method of treatment is ultraviolet irradiation. Even this latter approach is not without risk and some diatoms may survive. Thus, generally speaking, cultures in running seawater will contain contaminants.

A very successful approach to the culturing of macroscopic seaweeds using a spray apparatus has recently been developed by L. Hanic (personal communication). Basically the system involves subjecting the algae to a continuous shower of medium recirculated to and from a large capacity reservoir tank. If the apparatus is used in a constant temperature room, it may be constructed as shown in Fig. 5-1. The polyethylene reservoir tank may be 50–200 gal (US) capacity (Cole-Parmer no. 6310 or 6317, 7425 North Oak Park Ave., Chicago, Illinois 60648). The piping, $\frac{1}{2}$ in inside diameter, black polyethylene, is usually obtained from building suppliers. The length of piping passing through the culture tank is perforated repeatedly to provide the spray. In the absence of a constant temperature room, the reservoir tank is cooled with a titanium immersion cooler (Frigid

Units, Inc.). Algal growth in this system is relatively rapid and it holds great promise for large forms.

Except for the refrigerated meat units, all the systems mentioned are suitable for tropical algae provided a heating system is available to hold the temperature *ca.* 25 °C or warmer.

B. *Illumination*

A light source is essential for autotrophic organisms and, in general, fluorescent tubes are the most useful. A very wide range of spectra is covered by the large number of different types of tubes now available. For general purposes many laboratories use cool-white tubes as these most closely resemble immediate sub-surface illumination in the sea. However, at least for the Phaeophyceae, the spectral emission curve of daylight tubes fits more closely to the absorption spectrum of the plants and, on this basis, would be expected to give better growth (Edwards 1969).

Many algae show considerable absorption in the red end of the visible spectrum, and it is usual to supplement fluorescent lamps (deficient in these wavelengths) with incandescent bulbs in the ratio of 10 input watts of incandescent to 100 watts of fluorescent light. The exclusive use of incandescent bulbs is not recommended because of high heat output and short bulb life.

In constant temperature rooms it is usual to fix light tubes to the undersides of the shelves bearing culture vessels. With water baths a light bank is set up above the bath. The number of tubes and their distance from the plants will determine the amount of light energy available. Many of the smaller algae are saturated at low illuminance levels (Edwards 1969) and this is true of microscopic stages of the larger seaweeds (Chapman and Burrows 1970; Kain 1966). However, large plants tend to have higher saturation levels and an illuminance of about 4000 lux may be required (Kain 1966). Generally, it is advisable to install a large number of lamps and then reduce illuminance as necessary by switching some off or by shading with black nylon, chiffon or net. High output fluorescent lamps are now available from Sylvania (VHO) and General Electric (Power Groove) and these help providue sufficient light for macroscopic seaweeds (also see Chp. 11, II.C).

The wiring for fluorescent lamps is complicated and reference should be made to General Electric Bulletin (e.g., GCE-983-12) on this. It is preferable with fluorescent tubes to wire them so that the ballasts (which preheat electrodes for the electron arc) are outside the cooling chamber. Where time clocks are needed, the watt rating of the unit should be checked. Those clocks available from hardware stores are usually not suitable for more than one or two lamps. Time clocks from Tork Time Controls Inc., 1 Grove St., Mt Vernon, New York 10550 are recommended.

Most laboratories have a simple lux meter or photometer for measuring the illuminance of light sources. However, with this equipment and the conversion tables produced by ISCO (4700 Superior, Lincoln, Nebraska 68504) it is possible to convert illuminance readings into approximate radiometric terms (e.g., μg. cal/cm^2 s) which are more meaningful in comparative studies.

C. *Glassware*

For microscope stages use of Pyrex 100 × 80 mm storage dishes is recommended. Only two microscope slides are fitted in each dish, which holds *ca.* 300 ml medium. The glass lids may be replaced with 100 mm plastic petri dish tops which are easily drilled for insertion of aeration tubes. When plants are *ca.* 10 mm long, the rate of growth usually decreases in such small vessels and transfer to larger tanks should be made.

Borosilicate glass should be used where possible, but tanks of 5 l or more are extremely expensive. For routine culturing it is, therefore, necessary to resort to flint glass such as 6 quart rectangular tanks (Carolina Biological Supply Co., Burlington, North Carolina 27215). Once again the lids should be drilled for insertion of aeration tubes. About 30 plants up to 50 mm in length may be cultured in a vessel of this size.

Plants over 300 mm in length are best kept in aquaria of approximately 900 × 300 × 300 mm. About 25 l of medium are required. These tanks are most conveniently fabricated from Plexiglass of $\frac{3}{8}$ inch thickness. These tanks should be supported in a water bath before filling, as the seams may collapse.

Other glassware required is:

1. Petri dishes 60 × 15 mm, 100 × 15 mm.
2. Microscope slides, 25 × 75 mm.
3. Shallow depression slides 1 × 3 in, depression 3 mm deep (no. A-1475, Clay-Adams, Division of Becton, Dickinson and Co., 299 Webro Rd, Parsippany, New Jersey 07054; 2464 S. Sheridan Way, Clarkson, Ontario).
4. Pasteur-type pipettes (see Chp. 3, II.C).

Glassware for media preparation and methods of cleaning are given in Chps. 1 and 2).

D. *Media*

Full details of widely used media are given in Chp. 2. The thalloid Chlorophyceae and filamentous Phaeophyceae grow well in Erdschreiber medium which is easily prepared. The larger Phaeophyceae grow best in heavily enriched media of which the ES and SWM types are suitable (Chp. 2, IV.A.3, 5).

1. Contaminants. In crude cultures there is a very high risk of diatom contamination and subsequent overgrowth. These may be eliminated by the addition of germanium dioxide (6 mg/l) to the medium (J. Lewin 1966). Germanium dioxide is not very soluble in water and Lewin used concentrated sodium hydroxide to dissolve it. However, in a stock solution of 250 mg/l distilled water, a sufficient quantity is in solution to retard diatoms when 2 ml stock are added to 1 l culture medium. When diatoms have been eliminated, GeO_2 should be discontinued.

Bacterial infection and overgrowth of crude cultures in heavily enriched media is also a risk. Should such an infection occur, addition of 0.01% (w/v) potassium tellurite (K_4TeO_3) will usually have a bacteriostatic effect (Ducker and Willoughby 1964). Dissolve the K_4TeO_3 in a small volume of distilled water before adding to the medium.

2. Seawater sterilization. Seawater for cultures may be rid of contaminants by tyndallization, or steaming. The water is heated to 73 °C on three consecutive days with intermittent cooling at room temperature (see Chp. 12, II.C.2.b). This usually kills all the bacteria. To remove algal contaminants, a single heating to *ca.* 75 °C usually suffices, presumably because marine algae do not have resistant resting stages.

3. Aeration. Vigorous aeration of medium is recommended for large seaweeds as this circulates the nutrients. The air should always be passed through a wash bottle of water, particularly if it is supplied from a central compressor which may discharge oil into the gas. Aeration often results in plants floating to the surface of the medium and sublittoral algal tissue damage may occur. This can be avoided by using a net of the largest possible mesh to keep plants below the surface.

Alternatives to aeration are the use of a shaker or submersible pump which will keep the medium in constant movement.

III. METHODS

A. *Collection*

With all macroscopic seaweeds, excepting *Desmarestia* (Phaeophyceae), the plants are scraped from the rock surface, shaken to remove as much water as possible, and placed in polyethylene bags. *Desmarestia*, however, has a very high cellular acid content which is released when plants are removed from water and causes extensive tissue damage. Therefore, specimens should be collected under water and placed in large volume bags or boxes of water for transport. At no time should the thallus be exposed to air.

Plants should be kept cool during transport, either in a portable refrigerator or in an ice box.

B. *Crude cultures from spore or gamete suspensions*

1. Establishment

a. Presence of reproductive structures – depending on the species, flagellated spores may remain motile for a considerable period (e.g., 24 h in *Laminaria*, Phaeophyceae) and use may be made of this characteristic in establishing cultures. With *Laminaria* the spores will remain active and swimming at the surface while most contaminating diatom cells sink rapidly in undisturbed medium.

i. Clean thalli carefully with cheesecloth or muslin to remove as many potential contaminants as possible.

ii. Immerse small pieces of reproductive tissue in 100 × 15 mm petri dish of sterilized seawater. With very bulky material use *ca.* 300 ml of seawater in a 100 × 80 mm dish.

iii. Using 40 × magnification with stereomicroscope observe release of flagellated spores (in Chlorophyceae, Phaeophyceae).

iv. Allow the cell suspension to stand several hours, preferably at a temperature similar to that of the collection. Using a Pasteur-type capillary pipette (Chp. 3, IV.A.1 or Chp. 4, IV.C.3), remove the actively motile cells from the top mm of the water. In some Phaeophyceae, such as *Desmarestia*, this procedure is not applicable. In the Rhodophyceae the nonmotile spores will sink to the bottom of the container and should be pipetted off soon after release, before attachment occurs.

v. Treat the suspension as outlined in B.2; C, D, following.

b. Induction of reproductive structures – plant material that is not reproductive when collected from the field, may sometimes be induced to form reproductive structures by environmental shock. In *Enteromorpha* (Chlorophyceae) this involves transferring field plants to enriched medium. Formation of sporangia or gametangia occurs in 3–5 days. Other methods of inducing reproduction are temperature and salinity shock, where in each the material is subjected to a sudden change. This method is successful in the morphologically simple seaweeds whereas in the structurally complex Phaeophyceae and Rhodophyceae reproductive plants from the field are required.

2. *Treatment.* For establishing crude cultures, one of two procedures should be followed depending on the intensity of spore release.

a. Heavy suspension (visibly turbid)

i. Pour sterilized, cooled seawater to a depth of about 30 mm in a 5 l glass tank.

ii. Space five or six clean, sterile microscope slides on the bottom of the tank.

iii. With a bulb Pasteur-type pipette, pick up *ca.* 3–5 ml of cell suspension from just below the surface of the medium; expel over the slides. Repeat four or five times.

iv. Cover tank with a glass lid and place in a constant temperature room or chamber with low illumination (400–500 lux). The temperature of the chamber should be near that of field conditions.

v. Leave for 24 h; then flood with fresh medium. Light aeration is given and illumination may be raised to *ca.* 3000 lux.

With very thick suspensions, inoculation with an automatic dispenser is possible. In replicate cultures this method may be used to ensure even density of spores.

b. Light suspension (use stereomicroscope)

i. Cut double pieces of cheesecloth to cover the bottom of a 100 × 80 mm culture dish.

ii. Pour in sterile seawater to soak the cheesecloth. Pour out excess water.

iii. Using the dissecting microscope at 30–100 ×, pipette a suspension of motile cells. A capillary Pasteur-type pipette (as in Chp. 3, IV.A.1; Chp. 4, IV.C.3) may be used.

iv. Expel two or three drops onto clean microscope slides.

v. Draw rings around the spots with a diamond pencil.

vi. Place slides on wet cheesecloth in culture dish and cover.

vii. Keep dishes in constant temperature room with low illumination for 24 h.

viii. Remove cheesecloth and flood slides with fresh medium.
ix. Place dishes under higher illumination, if desired.

3. Examination of germlings. Germlings are best examined under a compound microscope fitted with a water immersion objective (Vickers Instruments Inc., 15 Waite Ct, Malden, Massachusetts 02148). Using this type of objective, germlings need not be subjected to excessive drying out.

As soon as plants are large enough to be handled they should be thinned. Eventually plants will become dislodged from glass slides which do not form a suitable surface for permanent attachment. Most algae will grow very well lying loose.

A regular change of medium is recommended, with intervals of one week usual. Some species, however, react adversely to this procedure. For instance, young sporophytes of *Desmarestia* will shed the meristems. Similarly, organized plants of *Codium* (Chlorophyceae) will degenerate to the filamentous condition. In these instances, maintaining plants in unchanged medium seems most satisfactory. It is possible that a build-up of extracellular products is necessary for growth and development and that changing medium prevents this. Unfortunately, nutrients are finally exhausted from the medium.

When algae are to be grown in a spray system or are to be placed in the field, glass slide cultures are inappropriate as the plants do not attach well. Instead, spores should be pipetted onto a suitable surface such as stone, brick or ceramic clay plugs for this purpose (L. Hanic, personal communication).

C. *Unialgal cultures*

1. Washing. It is often possible to establish unialgal cultures of macroscopic algae using methods employed with unicellular algae. Washing of unicellular reproductive bodies is a common procedure, and similar to techniques in Chps. 3, 4.

a. Prepare Pasteur-type pipettes, as outlined in Chp. 3, IV.A.1; Chp. 4, IV.C.3. The inside diameter of the pipette should be *ca.* 100 μm. A rubber bulb is fitted over the wide end of the pipette.

b. Place a drop of fresh sterile medium in each of the cavities of a depression slide.

c. Observe the release of spores using a stereomicroscope at 40× magnification. Quickly dip the pipette into the suspension. Capillary action will draw up spores.

d. Expel contents into medium of one cavity of the depression slide and examine for contaminants using 100× magnification.

e. If there are algal contaminants, repeat the procedure with a fresh pipette and expel into the next slide depression.

f. When the desired cells have been isolated, the droplet containing them may be pipetted onto a flat slide which is then placed in a moist chamber for settlement of the cells.

About 10–20 cells at a time are conveniently washed in this manner. It is not easy to isolate single cells and excessive manipulation may result in damage.

2. *Hanging drop.* This method, developed by M. Wynne (personal communication), is suitable for algae with motile reproductive cells. The procedure depends on the gravitational fall of the inoculating material to the bottom of a hanging drop and the capacity of released swarmers to swim up and settle on a cover slip.

a. Place a drop of spore suspension or small pieces of fertile plant fragments on a cover glass (no. 1, 22 mm^2).

b. Invert the cover glass quickly leaving the drop intact.

c. Place the inverted drop over a glass or plastic ring, that is attached by Vaseline to a slide. A depression slide (3–7 mm deep) may also be used.

d. Observe release and settling of swarmers using a compound microscope.

e. After settlement, remove parent plant and contaminants by gently flicking the drop off the cover glass.

f. Wash cover glass two or three times with sterile medium.

g. Place separate cover glasses in separate dishes to prevent cross contamination.

3. *Agar streak.* Where the handling of spores with pipettes is not possible and the isolation of single plants is desired, streaking on agar seems to be the best procedure.

a. Prepare petri plates containing growth medium solidified with 1–1.5% agar (see Chp. 1, II.B; Chp. 3, IV.A.3). The petri dishes may be inverted for storage to prevent condensate running onto the agar.

b. Heat a platinum or nickel microbiological loop to red heat (Chp. 3, II.B.5; Chp. 12, II.D.2.a). Allow the loop to cool, then pick up a loop-full of spore suspension.

c. If the spore suspension is very dense streak out on the surface of the

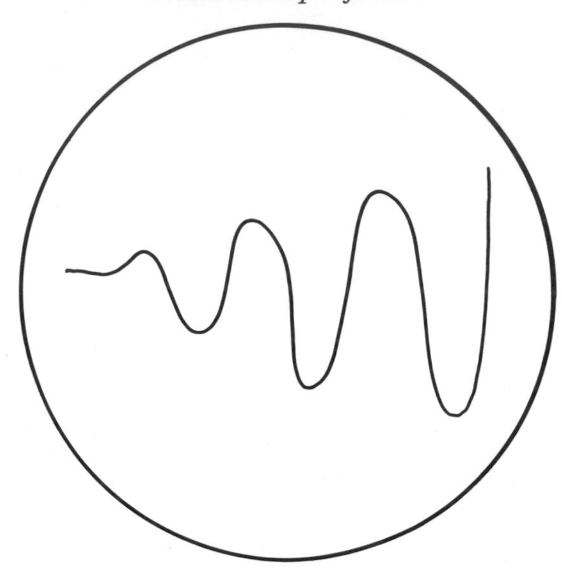

Fig. 5-2. Pattern for streaking on agar in 100 mm petri dish.

agar as shown in Fig. 5-2. It is helpful to have marked out the path of the streak on the petri dish bottom with a wax pencil beforehand.

d. If the suspension is dilute, simply smear a loop-full in a small spot. Several spots may be made on each plate.

e. Incubate at suitable temperature and illumination conditions for growth.

f. When germlings are visible they should be isolated with aid of a stereomicroscope. With a glass needle or sterile microspatula (Chp. 3, II.B.6) cut out the block of agar on which the plant stands. Transfer this to fresh medium in a depression slide. Examine for contaminants and, if free, transfer to a suitable vessel.

D. *Axenic cultures*

Axenic cultures are unialgal and bacteria-free. Their establishment and maintenance requires the normal sterile techniques used for microbiological investigations (see Thomas and Prier 1956).

1. Mechanical isolation. Mechanical methods of establishing axenic cultures are not in wide use since the widespread availability of antibiotics. However, mechanical methods have much to recommend them. In particular, they avoid any possible physiological disturbance of the organisms under study which may be induced by antibiotics.

Thalli of macroscopic algae are almost invariably superficially contaminated with bacteria, fungi and other algae. Thus the easiest mechanical method of establishing axenic cultures is to use reproductive bodies which are generally formed under bacteria-free conditions in sporangia or gametangia. The methods used here are simply refinements of those used for setting up unialgal cultures (see C, preceding).

a. Washing – the methods for washing spores are given previously in C.1. For axenic cultures it is necessary to use sterile Pasteur-type pipettes as described in Chp. 3, IV.A.1 and C, preceding.

i. 10–20 cells are washed 4–5 times before inoculation into sterile medium in sterile dishes.

ii. Incubate the cultures for growth at the necessary conditions of light and temperature.

iii. Test for axenic cultures by use of sterility media as outlined in Chp. 3, VII, using seawater media plus 0.1% glucose and 0.1% bactopeptone (Druehl and Hsiao 1969).

iv. Incubate cultures for 12 days at the growth temperature, as marine bacteria grow slowly (Fries 1963). Contaminated cultures become cloudy.

v. The axenic cultures should also be examined using phase contrast microscopy and 1000× magnification.

b. Agar streak – a suspension of spores is streaked onto agar made up in culture medium and containing 0.02% Bacto-tryptone (see Chp. 3, VII.A, using seawater). With the additional organic nutrient, bacterial colonies are evident after a few days. Isolated germlings are cut out of the agar with a flattened sterile spatula and transferred to fresh sterile medium. Once again the use of a sterility test medium is required (see preceding, 1.a).

2. *Surface sterilization.* A wide range of sterilants with powerful bacteriocidal action is now available. It is necessary to start from unialgal cultures when using these bacteriocides to establish axenic cultures. Surface sterilization has been used on vegetative thalli (Provasoli 1958; Boalch 1961*a,b*; Fries 1963) and on reproductive thalli to prevent contamination of bacteria-free endogenous spores following release (Druehl and Hsiao 1969). The following method was used successfully by Provasoli (1958) to obtain bacteria-free cultures of thalli of *Ulva* (Chlorophyceae).

a. Provasoli method

i. Make up antibiotic solution so that 1 ml contains: K penicillin G 12000 IU; chloramphenical 50 μg; polymyxin B 50 μg; neomycin 60 μg. Membrane filter sterilize the mixture.

ii. Pour 0.3 ml aliquots of antibiotic mixture into the bottom of sterile petri dishes (20 × 100 mm).

iii. Add 20 ml of sterile 1.4% agar media (liquid) at 45 °C to each dish. Thoroughly mix antibiotics by twirling. Cool to solidify.

iv. Place pieces of alga (*Ulva*) thallus on the agar surface. Incubate 7 days at proper temperature and light conditions.

v. Transfer alga aseptically to sterile liquid medium.

Boalch (1961*a*) obtained bacteria-free cultures of *Ectocarpus* (Phaeophyceae) sporelings in liquid medium where Na penicillin G was at a concentration of 4000 iu/ml. The algae were removed after 9 days to fresh medium.

Fries (1963) has successfully employed molecular iodine (in the form of 'Jodopax') as a surface disinfectant to obtain axenic red algal cultures. The iodine in the commercial product is 5% and is diluted with sterile medium to 0.01% iodine. Thalli are treated for 2 min in the disinfectant; then transferred to sterile medium.

b. Druehl and Hsiao (1969) used the following procedure to establish axenic cultures of selected Laminariales (Phaeophyceae) from meiospores.

i. Cut mature sori from the thallus. Wipe with paper towelling, rinse several times in sterile seawater.

ii. Place the sorus for 3 h in an antibiotic mixture of: 1 l sterile seawater; 622.5 mg penicillin G; 250.0 mg streptomycin SO_4; 100.0 mg chloramphenicol.

iii. Transfer sorus to 1% sodium hypochlorite solution in seawater for 20–30 min.

iv. Trim sorus edge aseptically.

v. Wash sorus in five rinses of culture medium.

vi. Place sorus in culture medium. Meiospores are released within 2 h.

vii. Inoculate spores into fresh medium (see preceding, B.2).

For all of these procedures a sterility test is required as previously outlined (D.1.a).

IV. GROWTH MEASUREMENTS

The usual criteria of algal performance in culture are growth and reproduction. Detection of reproduction is obvious, but measurement of growth may not be as easy. Growth measurements are dependent on construction and mode of growth; and the purpose of the measurements. Seaweeds grow in a variety of ways, largely dependent on the presence or absence of a discrete meristem and its position in the thallus. The method of determining growth depends on the particular study; e.g., dry weight measurements are inappropriate in monitoring continuous cultures. A few simple growth measurement procedures and their relative advantages and disadvantages are discussed here.

A. *Filamentous types*

1. Filament length. Russell (1963) used filament length for estimating growth of *Pilayella littoralis* (Phaeophyceae) in culture. Using a calibrated ocular micrometer in a compound microscope, changes in filament length were measured. These are the result of three separate processes: (a) cell elongation; (b) cell division; (c) cell loss. A fairly high cell loss due perhaps to swarmer production will obscure actual growth of filaments through cell elongation and division.

Filament length measures are probably most appropriate in filaments with apical meristems.

2. Cell number. Boney (1960) used cell numbers, or counts, as a measure of growth in early stages of filamentous red algae. However, with larger filaments it is not feasible to count the high numbers involved.

3. Filament volume. Dring (1967) used cell volume for estimations of growth of filaments of *Porphyra* (conchocelis-phase) (Rhodophyceae). Volume was measured by centrifugation of filaments in liquid medium at 4000 rpm. In the calibrated blood sedimentation tubes used, the volume of plant material was simply read from the graduations.

4. Dry weight. Measures of dry weight are terminal, Boalch (1961*b*) used the following procedure for *Ectocarpus*.

a. With a vacuum pump remove the algal material to a sintered glass crucible of porosity 40–90 μm.

b. Wash quickly with distilled water and dry for 25 h in an oven at 100 °C.

c. Cool for 1 h in a desiccator and weigh.

B. *Thalloid types*

1. Thallus length. Measurements of thallus length are useful in seaweeds where most of the growth is linear and there is a relatively small increase in thallus width (e.g., *Desmarestia aculeata*). In branched, thalloid types with this mode of growth an outline of the plant may be obtained by placing it on a piece of photographic paper under a safelight, switching on an ordinary incandescent bulb over the paper for a few seconds. This is followed by developing and fixing of the paper. The total length of plant material may then be recorded by using an accurate map mileage recorder calibrated in centimeters or inches. Length measurements for each outline should be made at least three times to ensure accuracy.

Thallus length measurements are inappropriate where there is a relatively large increase in thallus width or where the meristem is in an intercalary position and considerable loss of the thallus takes place distally as in *Laminaria*.

2. *Surface area.* Measures of surface area have been used extensively by Burrows (1964*a,b*) for growth measurements of *Fucus* and *Laminaria*. Undoubtedly the simplest method of recording surface area is to make a pencil or photographic outline of the plant and then make measurements with a planimeter (Gelman Instrument Co., 600 South Wagner Rd, Ann Arbor, Michigan 48106). Alternatively, the outline may be traced on squared paper and the number of square mm counted, or the tracing may be cut out and weighed.

Surface area, of course, does not give a measure of increasing thallus thickness. Burrows (1964*b*) recorded marked differences in the thickness of *Laminaria saccharina* thalli grown at 5 and 10 °C. With monostromatic or distromatic thalli this problem does not arise.

3. *Punched holes.* A widely used method of recording growth in laminarians is to punch holes in the blade above the meristem and to follow their distal movement (Parke 1948; Sundene 1964; Norton and Burrows 1969).

4. *Killed thallus measurements.* The normal method of recording growth in terminal cultures is to measure dry weight. For many thalloid seaweeds this is extremely difficult. With mucilaginous laminarians and fucoids obtaining a constant dry weight may not be possible because the mucilage appears to retain some moisture. Fresh weight measures are equally difficult for the same reason. The mucilage produced by a thallus can vary widely with relation to tidal immersion and emmersion, temperature, rainfall and sun.

Lipid, carbohydrate, and protein concentrations may be used as an expression of biomass. The methods described by Strickland and Parsons (1968) can be employed.

V. ACKNOWLEDGMENTS

I would like to thank Dr E. M. Burrows for the early training I received in the culturing of seaweeds. Many of the methods mentioned here were developed by her. The spray technique was developed by Dr L. A. Hanic for whose advice I am grateful.

VI. REFERENCES

Adey, W. H. 1970. The effect of light and temperature on growth rates in boreal–subarctic crustose corallines. *J. Phycol.* **6**, 269–76.

Boalch, G. T. 1961*a*. Studies on *Ectocarpus* in culture I. Introduction and methods of obtaining uni-algal and bacteria-free cultures. *J. Mar. Biol. Ass. U.K.* **41**, 279–86.

Boalch, G. T. 1961*b*. Studies on *Ectocarpus* in culture II. Growth and nutrition of a bacteria-free culture. *J. Mar. Biol. Ass. U.K.* **41**, 287–304.

Boney, A. D. 1960. Nurture of a fruiting *Antithamnion* tuft and the physiological condition of the liberated spores. *Brit. Phycol. Bull.* **2**, 38–9.

Burrows, E. M. 1964*a*. Ecological experiments with species of *Fucus*. In DeVirville, D. and Feldmann, J., eds., *Proc. Int. Seaweed Symp.* **4**, 166–70. MacMillan Co., New York.

Burrows, E. M. 1964*b*. An experimental assessment of some characters used for specific delimitation in the genus *Laminaria*. *J. Mar. Biol. Ass. U.K.* **44**, 137–43.

Chapman, A. R. O. and Burrows, E. M. 1970. Experimental investigations into the controlling effects of light conditions on the development and growth of *Desmarestia aculeata* (L.) Lamour. *Phycologia* **9**, 103–8.

de Wit, C. T. 1960. On competition. *Versl. Landbouwk, Onderze. Ned.* **66**, 1–82.

Dring, M. J. 1967. Effects of daylength on growth and reproduction of the *Conchocelis*-phase of *Porphyra tenera*. *J. Mar. Biol. Ass. U.K.* **47**, 501–10.

Druehl, L. D. and Hsiao, S. I. C. 1969. Axenic cultures of Laminariales in defined media. *Phycologia* **8**, 47–9.

Ducker, S. C. and Willoughby, L. G. 1964. Potassium tellurite as a bacteriostatic agent in isolating algae. *Nature* **202**, 210.

Edwards, P. 1969. Field and cultural studies in the seasonal periodicity of growth and reproduction of selected Texas benthic marine algae. *Contr. Mar. Sci.* **14**, 59–114.

Fries, L. 1963. On the cultivation of axenic red algae. *Physiol. Plant.* **16**, 695–708.

Kain, J. M. 1966. The role of light in the ecology of *Laminaria hyperborea*. In Bainbridge, R., Evans, C. G. and Rackman, O., eds., *Light as an Ecological Factor*, pp. 109–30. Oxford Univ. Press, London.

Lewin, J. 1966. Silicon metabolism in diatoms V. Germanium dioxide, a specific inhibitor of diatom growth. *Phycologia* **6**, 1–12.

Norton, T. A. and Burrows, E. M. 1969. Studies on marine algae of the British Isles. 7. *Saccorhiza polyschides* (Lightf.) Batt. *Brit. Phycol. J.* **4**, 19–53.
Parke, M. 1948. Studies on British Laminariaceae I. Growth in *Laminaria saccharina* (L.) Lamour. *J. Mar. Biol. Ass. U.K.* **27**, 651–709.
Provasoli, L. 1958. Effect of plant hormones on *Ulva*. *Biol. Bull.* **114**, 375–84.
Russell, G. 1963. A study in populations of *Pilaiella littoralis*. *J. Mar. Biol. Ass. U.K.* **43**, 469–83.
Strickland, J. D. H. and Parsons, T. R. 1968. A Practical Handbook of Sea-water Analysis. *Fish. Res. Bd., Canada*, Bull. 167.
Sundene, O. 1964. The ecology of *Laminaria digitata* in Norway in view of transplant experiments. *Nytt Mag. Bot.* **11**, 83–107.
Thomas, R. M. and Prier, J. E. 1956. *An Illustrated Guide for Introductory Bacteriology*. Burgess Press, Minneapolis, Minnesota. 94 pp.

NOTES ADDED IN PROOF

McLachlan, Chen and Edelstein (1971) have recently reported that germanium dioxide (used for suppressing diatoms) at concentrations of 5–10 mg/l induced apical necrosis in the embryos and mature thalli of *Fucus* spp. They also reported toxic effects of germanium dioxide in other brown algae. Rhodophycean species tested were not similarly affected.

McLachlan, J., Chen, L. C.-M., and Edelstein, T. 1971. The culture of four species of *Fucus* under laboratory conditions. *Can. J. Bot.* **49**, 1463–9.

6: Methods for coenocytic algae

JOANNA ZIEGLER PAGE*
Systematics and Environmental Biology,
University of Connecticut, Storrs, Connecticut 06268

CONTENTS

* Present address: Blue Meadow Road, Belchertown, Massachusetts 01007.

I. OBJECTIVES

For purpose of this discussion, the term 'coenocytic algae' will be used to include marine members of the Siphonocladales, Dasycladales, and Codiales (Chlorophyceae). As with many other algae, the degree of freedom from other organisms and the degree of definition of culture conditions desirable depends largely on the purpose for which the organism is being cultured. Some taxonomic, life history, and morphological studies can be accomplished perfectly well with collections of plants which contain many other organisms and which will simply survive for a period in the lab, or which will grow well under undefined conditions. On the other hand, some investigations of these problems and most physiological studies require unialgal or axenic cultures which will thrive, show a fairly normal morphology, and possibly reproduce in well-defined media under controlled conditions. This chapter will attempt to describe various methods used for chlorophyceaean coenocytes.

II. EQUIPMENT, MEDIA

Specific culture conditions used successfully with particular organisms are given in Table 6-1. Formulae for the media listed are given in Table 6-2, or by McLachlan in Chp. 2. In general, nutritional requirements for coenocytes have not been investigated, but many grow more luxuriantly and with a more normal morphology in a medium which contains vitamins (e.g., Erdschreiber or seawater with the vitamins from Guillard's 'f/2' medium, see Chp. 2, IV.A.4). With *Derbesia* spp. (Codiales) (Page, unpublished) increasing freedom from other organisms seems to be accompanied by an additional need for vitamins. Gibor (B. M. Sweeney, personal communication) found that axenic *Acetabularia* (Dasycladales) grows better with soil extract and thiamine and cyanocobalamin. Coenocytes usually will grow and develop normally in fluorescent light alone. Optimal temperature for growth and/or reproduction is species-specific, but the tropical and subtropical forms (including all the Dasycladales, Siphonocladales, and most of the Codiales) thrive at a temperature of 20–28 °C. Certain codialian species of *Bryopsis*, *Codium*, *Derbesia*, and

Halicystis may require lower temperatures in line with their natural habitats.

No unique equipment or culture vessels are required (see Chp. 5). The size of the vessel depends on the size of the organism and the length of the growth period desired. Anything from large aquaria to small test tubes or petri dishes may be used. With small, shallow vessels, however, evaporation of the medium resulting in increased salt concentration may indicate the kind of culture vessel (e.g., flasks or tubes *vs* petri dishes) or the frequency of medium change or transfer. Growth of forms kept in large containers may be enhanced by bubbling air through the medium (see Chp. 15, IV.B; Chp. 16, II.E). Implements for handling the organisms again depend on the organisms in question; stainless steel implements are desirable to avoid rust. Some of the smaller filamentous forms have a pronounced tendency to stick to metal and may be best handled with glass needles drawn from glass rod or tubing, or with glass droppers or pipettes. Special techniques may require special equipment and/or chemicals (e.g., see Shephard's method for *Acetabularia*, V.B.3, following).

III. ORGANISMS

Table 6-1 lists some members of the Siphonocladales, Dasycladales, and Codiales which have been grown successfully (as opposed to maintenance only of collected plants). This table does not cover all studies but is representative of organisms that have been grown. Numbers in the second column beginning with LB are those from the Indiana University Culture Collection of Algae (Starr 1964, 1966, 1971). This IUCC number is also given when applicable in the text.

It should be emphasized that many coenocytes may be collected and kept for at least a short period of time in seawater in aquaria supplied with natural or artificial light, in which the temperature is kept below *ca.* 23 °C and in which the seawater is changed fairly frequently. The rest of this discussion assumes that the investigator wishes to obtain plants actively growing in cultures which can be maintained indefinitely. The general steps to be followed in culturing an organism collected from nature are: (1) elimination of other organisms, starting with the largest first (IV, following); (2) finding a suitable medium and other culture conditions; and (3) obtaining normal morphology and reproduction.

IV. ELIMINATION OF CONTAMINANTS

A. *Mechanical methods*

Collections of marine algae usually contain an assortment of invertebrates, bacteria, and fungi as well as several algae. Attempts to remove unwanted organisms should be made as soon as possible after a collection. Depending on the purpose for which the plants are to be used, they may be placed in crude culture at this point or further treatments may be used to isolate them.

1. Rinse vigorously with a syringe and sterile seawater to remove larger animals and planktonic organisms. Use sterile vessels to avoid introducing new contaminants. Remove large epiphytes with forceps. Use of a soft, sterile brush (moist heat sterilization, Chp. 12, II.C.2.c) helps remove microscopic epiphytes.

2. Thalli such as *Bryopsis* and *Derbesia*, which regenerate readily from small fragments may be grown; then the new, cleaner parts excised several times with a sterile microspatula thus eliminating many epiphytes.

3. Filamentous forms such as *Derbesia* may grow well on or into agar ahead of contaminants (both bacteria and other algae); the clean filament ends can then be excised. Lay the plant or a portion of it on the surface of a plate of medium solidified with 1.5% agar (see Chp. 1, II.B). Use a medium with few organic constituents (e.g., Müller's artificial seawater, Chp. 2, IV.B.5; Erdschreiber without soil extract, Chp. 2, IV.A.1; or Müller's artificial seawater with half the concentration of vitamins from Guillard's 'f/2' medium added, Chp. 2, IV.A.4).

4. Isolation of reproductive cells, such as zoospores, zygotes, cysts (Dasycladales) and aplanospores (Dasycladales, Siphonocladales) may be used as outlined by Chapman (Chp. 5, III.B). Isolate the cells with a sterile Pasteur-type capillary pipette using a stereomicroscope (Chp. 3, IV.A.1). Wash the cells several times. Place the cells in individual culture vessels of appropriate medium. Unlike many unicellular bacteria and algae, however, these reproductive cells lack cell walls and are fragile or do not divide well on agar, thus spraying and agar plating techniques (as in Chp. 3, IV.A.3, 4) are not successful. Phototactic swarmers like those of some Dasycladales can be partially purified (as given in Chp. 4, IV.E) by allowing them to swim through several changes of sterile medium; but usually treatment of cysts or fertile branches with antibiotics gives better results more easily.

B. *Use of growth inhibitors*

To date growth inhibitors such as antibiotics and detergents have been used successfully only with *Acetabularia mediterranea* (Dasycladales) and *Codium fragile* (Codiales), though there is no reason to believe that the methods employed would not work with other coenocytes also.

1. Gibor and Izawa (1963) obtained axenic cultures of *A. mediterranea* by combining a growth inhibitor and antibiotic treatments.

a. Place mature intact plant in 10% (w/v) 'Argyrol' seawater solution (Argyrol is manufactured by Crooks-Barnes Laboratory, Inc., Wayne, New Jersey) for 1–2 h. (Also, young plants may be used.)

b. Wash thoroughly in sterile seawater.

c. Place the plants in a membrane filtered sterile antibiotic-seawater (100 ml seawater; 200 mg streptomycin.SO_4; 100 mg penicillin; 20000 units mycostatin; 20 mg chloramphenicol; 20 mg neomycin) for 5 days. (See Guillard, Chp. 4, IV.D.)

d. Remove axenic cysts and induce germination.

2. Lateur (1963) reported a slightly different method.

a. Place mature caps with cysts in detergent solution 1% (v/v) (Teepol) in sterile distilled water in dark at 20 °C for 25 h to remove protista.

b. Mix sterile seawater gradually into solution to avoid bursting cysts.

c. Place caps in membrane filtered sterile antibiotic-seawater solution (100 ml seawater; 1000 units mycostatin; 1 g neomycin) in dark at 12 °C for 2–5 days to remove bacteria and fungi.

Green *et al.* (1967) used a solution of 100 mg penicillin, 20 mg neomycin, and 20000 units mycostatin per 100 ml to obtain axenic plants of *A. mediterranea*.

Shephard (1970) has compiled an extensive list of antibiotics and bacteriocides for use with *A. mediterranea*; among the most useful and least harmful were the bacterial wall inhibitors ampicillin, penicillin G, carbenicillin, vancomycin (used at a concentration of 1 mg/ml for any desired time period), cephalothin and cycloserine (100 μg/ml, for any desired time). Other antibiotics, including neomycin (100 μg/ml for 4 days) and streptomycin (1 mg/ml for any desired time) were found to be unpredictable in their effects. Of the bacteriocides, the most useful were 2% formaldehyde for 5 min and 1% sodium lauryl sulfate for 12 h, both used on the cysts. Aureomycin, mycostatin, novobiocin, polymixin, argyrol, copper sulfate, domiphen bromide, hexyl resorcinol, hydrogen peroxide, phenol, sodium hypochlorite, and zephrin chloride were all found to be excessively toxic to plants and cysts, despite apparently satisfactory use of some of these by other workers. Shephard's complete schedule for obtaining and growing axenic plants is given in V.B, following.

Diatoms are frequently among the most difficult of other algae to eliminate from cultures. They may be checked by the use of 5–10 μg/ml germanium dioxide in the culture medium (Provasoli 1968; Borden and Stein 1969).

Puiseux-Dao (1962) eliminated epiphytes, especially diatoms by putting cysts in 70% ethanol for a few minutes.

The growth of bluegreen algae may be inhibited by the use of massive doses of penicillin (100 units/ml) and dihydrostreptomycin, but the latter may be toxic to other algae and may induce bleaching (Provasoli 1968).

C. *Manipulation of culture media*

Little is known of the nutritional requirements of the coenocytes, particularly with respect to a need for vitamins. However, crude cultures containing other organisms may grow in natural or artificial seawater without vitamins or other organic substances. It may be best to avoid organic materials until the plant has been freed at least from most contaminants, as the additives may favor the growth of other algae (particularly Chrysophyta), protozoa, and bacteria over the growth of the coenocytic algae. Once plants have been isolated, however, organic additives may be necessary for growth or reproduction.

V. MORPHOLOGY, REPRODUCTION IN CULTURE

Reproductive structures are frequently obtained from collections of plants; this discussion will be limited to work done with those organisms which will form reproductive entities in culture and will give a few methods for maintaining cultures and obtaining normal morphology not discussed fully in Table 6-1.

A. *Siphonocladales*

Members of this order reproduce by means of flagellated zoospores and gametes or by aplanospores.

Little work has been done with this order. Jonsson and Puiseux-Dao (1959) obtained biflagellated cells from *Siphonocladus pusillus* plants which developed in culture from plants collected as juveniles. See Table 6-1 for details of the culture conditions.

B. *Dasycladales*

These plants are characterized by the formation of special reproductive branches which contain numerous cysts; each cyst usually forms biflagellate swarmers which are gametes or zoospores.

Acetabularia has been a favorite organism for morphogenetic studies, including studies of the influence of the nucleus and the influence of environmental factors on morphogenesis. A discussion of most of these would be out of place here; readers are referred to the work of Terborgh and Thimann (1964, 1965); to studies cited by them; and to the books by Puiseux-Dao (1970) and Brachet and Bonotto (1970). Usually a white light source is satisfactory; however, Terborgh (1965) reported blue light (as opposed to red or white) was necessary to form caps having normal morphology, and this light had the greatest elongation of the axis.

1. *Acetabularia* will readily form caps and cysts in culture under the conditions used by various authors (see Table 6-1). Lateur (1963) reported that caps and cysts (following detergent and antibiotic treatment, see IV.B, preceding) kept in complete darkness at 12°C remained viable indefinitely if exposed to light for 12 h/month, but optimum swarmer formation occurred after the cysts had been darkened for 30 days, with a minimum of *ca.* 6 days. Swarmer release was initiated by placing the cysts under normal culture conditions (see Table 6-1) in seawater for 5 days; then changing the medium; placing the cysts in darkness for 2–3 days; and alternating these conditions 4–5 times. Swarmer formation occurred beginning with the second light period. Heating the flask with cysts to 30–32°C enhanced emission. With this schedule, no distilled water treatment was necessary. Swarmer release was not as good in enriched seawater (phosphate, nitrate, bicarbonate, and soil extract). Aggregation and clumping of zygotes was prevented by transferring swarmers into new flasks and placing these in the dark for at least 24 h. Lateur also gave detailed instructions on methods for keeping stocks of juvenile plants for long periods. Essentially, these are kept in the dark in flasks at 16°C, except for a 12–48 h exposure to light each month.

2. Gibor and Izawa (1963) used a simpler procedure to obtain gametes. Following treatment with antibiotics, the caps were washed, cut open, washed further with sterile seawater and placed for *ca.* 3 min in sterile distilled water. Cysts were then placed in seawater in culture flasks at 5380 lux for 16 h/day (16:$\overline{8}$). Gametes were formed within a week. Further handling of zygotes was similar to Lateur's process. The culture medium used for plants (Table 6-2) enhanced the growth of any contaminants and hence contamination was easily detectable as turbidity.

3. Shephard (1970) has given a very complete schedule for growing *A. mediterranea* through many generations. As the cysts are more resistant to handling and chemical treatment than plants, contaminants can be eliminated from cysts. Handling the resulting axenic plants should be avoided as much as possible.

a. Cysts

i. Grind the mature caps in a loose-fitting tissue grinder (such as Ten Broeck with 0.01 in clearance).

ii. Filter resulting slurry through 74 mesh bolting silk to remove cap fragments and large debris.

iii. Wash filtrate several times in Provasoli ASP-6 medium plus thiamine.HCl and cyanocobalamin (see Table 6-2; Chp. 2, IV.B.2) by shaking in test tubes. (Foam produced and debris can be poured off in supernatant while the cysts settle.)

iv. Place the cysts in a sterile washing funnel fitted with a filter of 270 mesh bolting silk. Drain off liquid.

v. Aseptically add fresh sterile media and a sterile magnetic stirring bar.

vi. The cysts are washed by constant agitation with the stirring bar and repeated draining and addition of sterile medium.

vii. After several washings, add a 1% sodium lauryl sulfate solution made up in Provasoli ASP-6 medium.

viii. Stir and drain. Repeat steps vii and viii several times in a 10 min period.

ix. Wash out the sodium lauryl sulfate with several changes of sterile medium.

x. Fill funnel to overflowing with 2% formaldehyde (5% formalin) in sterile medium.

xi. Stir the cysts for 5 min.

xii. Wash out formaldehyde with at least six full-funnel rinses with sterile medium over a 30 min period.

xiii. Suspend cysts in 10 ml sterile medium.

xiv. Plate 0.5 ml of cyst suspension on glucose–tryptone–yeast (1%, w/v) extract agar to test for sterility. Incubate plates at 22–24 °C for a week. Check each day for contamination.

xv. For remaining suspension add 0.05% of glucose and store in dark (room temperature).

xvi. If mold colonies appear on sterility-test plates, repeat steps x–xiv. Success is usually achieved if washing is undertaken before the fungi sporulate. If bacteria are present, each different colony should be tested for antibiotic sensitivity. (For further details see Shephard 1970, pp. 54–5, 61–2.)

xvii. Axenic cysts should be stored in the dark.

b. Gamete release

i. With 270 mesh bolting silk make a long cylinder with cone-shaped closed bottom to fit a 250 ml, wide mouth, screw cap bottle. Attach cylinder to lid that has a hole covered with cotton and paper.

ii. Put 50 ml of sterile medium in 250 ml, wide mouth jar.

iii. Put decontaminated cysts from *ca.* 50 caps in the cylinder and suspend in bottle so that the cysts are covered by the medium.

iv. Place bottle in incubator at 22–24 °C, 2690 lux, 12:$\overline{12}$.

v. Carefully check the bottle daily *ca.* 5 h after light period begins for gamete release. The phototactic gametes will collect as a thin green line at the water surface nearest the light source.

vi. Prepare a second jar with 50 ml sterile medium. Place cylinder from first jar in this and repeat steps iii–v.

vii. Cover the first jar with a sterile lid. Cover the bottom $\frac{2}{3}$ of jar with black paper and return to incubator. The negatively phototactic zygotes will tend not to adhere to one another.

viii. After 3 weeks growth, gently use a magnetic stirrer or pipette to loosen the plants.

ix. Remove enough plants so that *ca.* 10000 are in the jar.

x. Add sterile medium to bring volume to 100 ml.

When the plants reach 2–3 mm the bottles can be stored in dark cabinets at room temperature. The plants will keep indefinitely in this condition if exposed to several days of light every 3 months. When further growth is desired, young plants from dark storage are placed in 200 ml sterile medium in 1 l Roux flasks. The number of cells inoculated depends on the final plant size desired. One thousand cells will grow to 10 mm length with no additional medium; 100 will grow to cap initiation; 50 will form large caps and 25 will grow to full maturity and complete the cycle. It is wise to check the young, dark storage plants for sterility before inoculation. If they are contaminated, test the contaminants for antibiotic sensitivity (as for cysts) and give appropriate treatment (IV.B, preceding). The plants will not stand sodium lauryl sulfate or formaldehyde treatments, however. Large cells may be freed from accidental contamination during manipulation by treatment for 5 min in 0.1% formaldehyde or overnight in 5 μg/ml sodium omadine.

4. The other Dasycladales studied is *Batophora* (Puiseux-Dao 1964). Cysts form swarmers after a 2–3 month dark period and the swarmers, unlike

those of *Acetabularia*, appear to develop without fusing and may be considered zoospores. The cysts can also germinate directly to produce new plants without swarmer formation.

c. *Codiales*

Members of the Codiales usually reproduce by means of biflagellate anisogametes. Other types of reproduction include production of zoospores and propagules (small branching structures formed on vegetative branch apices of *Derbesia* which will form new sporophytes), 'budding' (as in the gametophyte or *Halicystis*-stage of *Derbesia*), and fragmentation of filamentous and possibly rhizomatous forms.

1. Derbesia. Derbesia species (branched, coenocytic sporophytes) may have as gametophytes either spherical coenocytic plants (*Halicystis*-stages) or dissimilar branched coenocytes (*Bryopsis*-stages). In certain other species of *Bryopsis* (gametophytes), the zygotes may develop into creeping filaments whose contents cleave into stephanokontous zoospores (multiflagellate) which grow into *Bryopsis* plants (*B. plumosa*, *B. hypnoides*, *B. monoica*), or the creeping filament may grow directly into a *Bryopsis* plant (*B. plumosa*, *B. hypnoides*; Rietema 1969, 1970, 1971 *a,b*).

Derbesia sp. (IUCC LB 1260) forms sporangia in either Erdschreiber or in Müller's artificial SW plus half the concentration of vitamins used in Guillard's 'f/2' medium (Table 6-1; also see Chp. 2, IV.A.4; B.5). In order to maintain zoospore production, however, the cultures are transferred *ca.* every 2 weeks as zoospore production decreases and young gametophytes develop.

As noted in Table 6-1, different isolates produce sporangia at different temperatures (e.g. *D. osterhoutii*, *D. marina*, *Derbesia* sp. LB 1260). Manipulation of light intensity controls production of *D. neglecta* sporangia and propagules (Hustede 1964: see Table 6-1). Species of *Derbesia* which do not form propagules must be maintained by transferring filament fragments, since zoospores develop into the dissimilar gametophyte generation. An exception is one isolate of *D. marina* (Sears and Wilce 1970) which reproduces the sporophyte by zoospores.

Gametogenesis in the *Halicystis* stage of *Derbesia* sp. (LB 1260) is controlled by a 3–5 day endogenous rhythm (Table 6-1). These plants are always heterothallic and asexual 'budding' permits the development of male and female clones (Köhler 1958; Ziegler and Kingsbury 1964). Female clones are harder to maintain than males, however, because of parthenogenic development of the sporophyte generation in female clones following gamete formation. These new sporophytes are capable of producing viable zoospores which develop into both male and female gametophytes.

The *Halicystis*-stages of other *Derbesia* species form gametes periodically if collected and kept under favorable conditions in the lab (for more information, see Abèlard and Feldmann 1958; Hollenberg 1936).

Bryopsis hypnoides (Neumann 1969*b*; Burr and West 1970) generally forms gametes in the laboratory as noted in Table 6-1. Jacobs (1951) has shown that auxin can control the morphology and differentiation of *B. plumosa*. Like *Derbesia*, *Bryopsis* plants can be maintained indefinitely in culture by transferring fragments (see Table 6-1).

2. Codium. Codium fragile has been grown successfully in the juvenile condition but the normal adult morphology and gametangia have not been obtained in these cultures (Borden and Stein 1969; Ott 1966). *Codium* is frequently collected with gametangia, however; the gametes are then released in the laboratory, yielding viable zygotes (Borden and Stein 1969).

3. Caulerpa. Caulerpa spp. have been grown in crude cultures for various morphological and growth rate studies (Chen and Jacobs 1966, 1968; Mishra and Kefford 1969) but no sexual reproduction has been reported in these cultures. *Caulerpa sertularioides* showed a decrease in number of erect branches and an accelerated rhizome elongation rate in seawater without substrate or with a sand substrate 10 mm thick; a more normal growth habit occurred if a 20–30 mm sand layer in which a black band of microflora developed was used, or if 2–10% *Caulerpa* sap (by volume) was added to the growth medium (Mishra and Kefford 1969). *Caulerpa* sp. with apparently normal morphology has been grown with other plants and animals in seawater aquaria with sand substrate in diffuse sunlight supplemented with incandescent light, and a diurnal temperature 20–24 °C (H. Spring, personal communication). *Caulerpa* does not usually regenerate from small fragments.

4. Halimeda. Halimeda monile, *H. discoidea*, and *Penicillus capitatus* (which do not regenerate from fragments) developed normally and became calcified in seawater aquaria (Colinvaux, Wilbur and Watabe 1965; Table 6-1). Several other genera belonging to the three orders discussed here also grew in such aquaria. *Halimeda* formed clusters of gametangia in this study (see Table 6-1 for culture conditions).

TABLE 6-1. *Culture data for coenocytic Chlorophyceae (IUCC = Indiana University Culture Collection, Starr 1964, 1966, 1971)*

Organism	Reference (IUCC no.)	Seawater medium[a]	Light (lux)	Temp. (°C)	Presence of other organisms	Morphology	Reproduction
SIPHONOCLADALES							
Anadyomene stellata	Ott 1966 (LB 1421)	Erdschreiber; artificial (Table 6-2, A)	—	—	—	—	—
Cladophoropsis membranacea	Ott 1966 (LB 1417)	Erdschreiber; artificial (Table 6-2, A)	—	—	—	—	—
Siphonocladus pusillus	Jonsson & Puiseux-Dao 1959	$+NO_3$, PO_4, soil-extract	4000 L:D 12:$\overline{12}$	30	—	Normal	2 sizes biflagellate swarmers
Valonia ventricosa	Ott 1966	Erdschreiber; artificial (Table 6-2, A)	—	—	—	—	—
DASYCLADALES							
Acetabularia crenulata	Terborgh & Thimann 1964	Erdschreiber	3230–5380 L:D 8:$\overline{16}$	23	Unialgal	—	—
A. crenulata	Hämmerling 1944	Erdschreiber	Natural + artificial	21–24	—	—	—
A. mediterranea	Gibor & Izawa 1963	(see Table 6-2, B)	5380 L:D 16:$\overline{8}$	23	Axenic	—	—
A. mediterranea	Green *et al.* 1967	Lateur's (Table 6-2, C)	1800 L:D 12:$\overline{12}$	21	Unialgal during growth	—	Caps, cysts

A. mediterranea	Lateur 1963	Lateur's (Table 6-2, C)	1000 L:D 12:12	Diurnal cycle, L:D 24:20	Axenic	Normal	Caps, cysts; directions for obtaining gametes
A. mediterranea	Puiseux-Dao 1962	Seawater; Erdschreiber; $+NH_4$ or NO_3; +amino acids	Natural+fluorescent+incandescent 4000	*ca.* 26		—	Caps, cysts; directions for obtaining gametes
A. mediterranea	Hämmerling 1944	Erdschreiber	Natural+artificial	21–24	—	—	Discusses cyst formation, germination
A. mediterranea	Shephard 1970 (LB 1764)	Mod. ASP-6 (Table 6-2, E)	Daylight, warm white fluorescent 2960	22–24	Axenic	—	Discusses cyst germination
A. polyphysoides	Hämmerling 1944	Erdschreiber	Natural; artificial	21–24	—	—	—
A. wettsteinii	Hämmerling 1944	Erdschreiber	Natural; artificial	21–24	—	—	Discusses cyst formation, germination to form gametes
Acicularia schenckii	Hämmerling 1944	Erdschreiber	Natural; artificial	21–24	—	—	—
Batophora oerstedii	Hämmerling 1944	Erdschreiber	Natural; artificial	21–24	—	—	Discusses gametes
B. oerstedii	Puiseux-Dao 1964	Erdschreiber; +amino acid (10 g/l)	2000–2500	*ca.* 25	—	Normal	Gametangia, cysts after 10 months
B. oerstedii	Puiseux-Dao 1962	Erdschreiber; +amino acid; $+NH_4$ or NO_3; seawater	Natural+incandescent+fluorescent 4000	*ca.* 26	—	Normal	Gametangia, cysts
B. oerstedii	Ott 1966 (LB 1515)	Erdschreiber; artificial (Table 6-2, A)	—	—	—	—	—
Cymopolia barbata	Hämmerling 1944	Erdschreiber	Natural; artificial	21–24	—	—	Direct growth of gametangium (no gametes)

[a] If name of base medium is not given in a list, it is seawater. Thus 'Erdschreiber; $+NH_4$ or NO_3' means Erdschreiber; *seawater* $+NH_4$ or NO_3.

TABLE 6-1. (*cont.*)

Organism	Reference (IUCC no.)	Seawater medium[a]	Light (lux)	Temp. (°C)	Presence of other organisms	Morphology	Reproduction
Dasycladus clavaeformis	Hämmerling 1944	Erdschreiber	Natural; artificial	21–24	—	—	—
CODIALES							
Blastophysa rhizopus	(LB 1029)	—	—	—	—	—	—
B. rhizopus	Sears 1967	Guillard's 'f/2' medium	L:D 16:$\overline{8}$	22	—	—	—
Bryopsis hypnoides	Neumann 1969*b*	Modified Schreiber-soln. (Table 6-2, D)	1000–5000	6 and higher	Unialgal	Normal, has protonemal stage	6° – gametes, better at 5000 lux than 1000
B. hypnoides	Burr & West 1970	20% Provasoli's ES	1200–2400 L:D 16:$\overline{8}$	18	Unialgal	—	Gametes induced by transferring 1 month axes to fresh medium (7–12 days to gametangia)
Bryopsis sp. (possibly *hypnoides*)	Page unpubl.	Erdschreiber; Müller's artificial (± ½ Guillard's 'f/2' vitamin mix)	Fluorescent 2700 L:D 12:$\overline{12}$	20–26	Unialgal	Fragments regenerate normal small plants; quickly becomes abnormal	—
B. hypnoides	Rietema 1971*b*	Mod. Erdschreiber (Table 6-2, F)	1600 L:D 16:$\overline{8}$	16 ± 1	—	Normal	♂, ♀ gametes; life cycles vary for different isolates (see text); zoospores
B. monoica	Rietema 1971*a*	Mod. Erdschreiber (Table 6-2, F)	1600 L:D 16:$\overline{8}$; 8:$\overline{16}$	16 ± 1	—	Initially abnormal; zoospores develop plants	♂, ♀ gametes with L:D 16:$\overline{8}$; alternate stage produced stephanokontous zoospores

B. plumosa	Rietema 1969, 1970	Mod. Erdschreiber (Table 6-2, F)	1600 L:D $16:\overline{8}$; $8:\overline{16}$	15 ± 1; 16 ± 1	—	Normal	♂, ♀ gametes; life cycles vary for different isolates (see text); zoospores
B. plumosa	Jacobs 1951	Seawater	Natural + incandescent	21.5 ± 1	Xenic	Auxin induces rhizoid formation	—
B. plumosa	Ott 1966 (LB 1409)	Erdschreiber; artificial (Table 6-2, A)	—	—	—	Normal thalli only with aeration	—
B. halymeniae (gametophyte of *Derbesia neglecta*)	Hustede 1960, 1964	+ soil extract	Incandescent 1200, 2500 L:D $16:\overline{8}$	20	Free from diatoms, bluegreens	Sexual dimorphism	Gametes; β-indole carboxylic acid, tryptophane hasten formation
Derbesia neglecta (sporophyte stage of *B. halymeniae*)	Hustede 1960, 1964	+ soil extract	Incandescent 1200, 2000, 2500 L:D $16:\overline{8}$ (see morphology)	20	Free from diatoms, bluegreens	Normal, propagules at 1200 lux (no sporangia); sporangia (no propagules) at 2500 lux; trytophane caused propagules at 2000 lux with daily medium change	Sporangia at 2500 lux; sporangia at 2000 lux with indole-carboxylic acid, tryptophane if medium not changed
Derbesia sp. (possibly *tenuissima*)	Ziegler & Kingsbury 1964; Page unpubl. (LB 1260)	Erdschreiber; Müller's artificial + ½ Guillard's 'f/2' vitamin mix	Fluorescent 215–6420 L:D $12:\overline{12}$ or $16:\overline{8}$	14–26	Unialgal; axenic	—	Sporangia at all light intensities, temperatures

[a]If name of base medium is not given in a list, it is seawater. Thus 'Erdschreiber; + NH_4 or NO_3' means Erdschreiber; *seawater* + NH_4 or NO_3.

TABLE 6-1. (*cont.*)

Organism	Reference (IUCC no.)	Seawater medium [a]	Light (lux)	Temp. (°C)	Presence of other organisms	Morphology	Reproduction
Halicystis sp. (possibly *parvula*; gametophyte of *Derbesia* sp. preceding)	Page & Kingsbury 1968; Page & Sweeney 1968; Ziegler & Kingsbury 1964 (LB 1260)	Erdschreiber; Müller's artificial and modifications; seawater	Fluorescent 215–4845 L:D $12:\overline{12}$; $8:\overline{8}$; $16:\overline{8}$	15–25	Unialgal	Normal	Gamete formation controlled by *ca.* 4 day endogenous rhythm at all light intensities; asexual reproduction by 'budding'
D. tenuissima (sporophyte stage of *H. parvula*)	Köhler 1958	Erdschreiber	—	—	—	Development of gametophyte (*H. parvula*)	Sporangia, zoospores; gametophyte stage (*H. parvula*) formed gametes
H. parvula (gametophyte of *D. tenuissima*)	Abèlard & Feldmann 1958	Seawater	Artificial L:D $12:\overline{12}$	(See reproduction)	—	—	18–20 °C – gametes every 2 weeks; 11–14 °C – no gametes
D. marina (sporophyte of *H. ovalis*)	Kornmann 1938	Erdschreiber	—	—	Unialgal	Normal, gametophyte (*H. ovalis*)	Sporangia, zoospores; gametophyte stage formed gametes
D. marina (sporophyte of *H. ovalis*)	Neumann 1969*a*	Mod. Schreiber-soln (Table 6-2, D)	*ca.* 3000	6–15	Unialgal	Normal	15 °C – sporangia in 2–3 weeks; 6 °C – sterile, formed many short lateral branches
D. marina (from Cape Cod. Massachusetts; appears in floras as *D. vaucheriaeformis*)	Sears & Wilce 1970	Guillard's 'f/2'	*ca.* 1615–2230 L:D $16:\overline{8}$	5–20	Unialgal	—	15–20 °C – sporangia, zoospores; 5–10 °C – no reproduction; life cycles vary for different isolates

Derbesia stage (sporophyte of *H. osterhoutii*)	Page 1970; unpubl.	Erdschreiber; Müller's artificial ± with ½ Guillard's 'f/2' vitamin mix	Fluorescent 215–6460 L:D 12:$\overline{12}$; natural light	20–28 (no growth at 14)	Unialgal	Germlings from sporophyte stage developed poorly	Sporangia, zoospores; small gametophytes (*H. osterhoutii*) forms gametes
Caulerpa prolifera	Chen & Jacobs 1966, 1968	Erdschreiber	1075–2150 L:D 12:$\overline{12}$	24.5 ± 1	Plants cleaned periodically	Normal	—
C. sertularioides	Mishra & Kefford 1969	Seawater	Fluorescent 4305 L:D 12:$\overline{12}$	26 ± 1	Other microorganisms	(See discussion in text)	—
Codium fragile	Borden & Stein 1969	Provasoli's ES medium; seawater in aquaria	1075–1615 L:D 16:$\overline{8}$	8 ± 2	No diatoms; not unialgal	Juvenile stage; no adult plants	Gametes not formed
C. fragile	Ott 1966	Erdschreiber; artificial (Table 6-2, A)	—	—	—	—	—
Halimeda spp.	Colinvaux *et al.* 1965	Filtered seawater in aerated aquaria	Fluorescent + incandescent 10765–16145 L:D 14:$\overline{10}$; fluorescent 2585–4415 for 15 h and 83–235 for 9 h	25 ± 0.5	Other algae present	Normal	Some gametangia
Penicillus capitatus	Colinvaux *et al.* 1965	Filtered seawater in aerated aquaria	Fluorescent + incandescent 10765–16145 L:D 14:$\overline{10}$; fluorescent 2585–4415 for 15 h and 83–235 for 9 h	25 ± 0.5	Other algae present	Normal	—

Colinvaux *et al.* 1965 also grew *Acetabularia*, *Batophora*, *Cymopolia*, *Neomeris*, *Dictyosphaeria*, *Valonia*, *Caulerpa*, and *Udotea* under preceding conditions but more complete discussion of development not given.

[a]If name of base medium is not given in a list, it is seawater. Thus 'Erdschreiber; + NH_4 or NO_3' means Erdschreiber; *seawater* + NH_4 or NO_3.

TABLE 6-2. *Formulae for media*

A. *Artificial seawater* (Ott 1966)	
Distilled water	500 ml
NaCl	21 g
$MgSO_4.7H_2O$	6
$MgCl_2.6H_2O$	5
$CaCl_2.2H_2O$	1
KCl	0.8
NaBr	0.1
$NaHCO_3$	0.2
H_3BO_3	0.06
$Na_2SiO_3.9H_2O$	0.01
$Sr(NO_3)_2$	0.03
B. *Enriched seawater* (Gibor and Izawa 1963)	
1. Autoclave; cool	
Seawater	900 ml
Distilled water	100
Soil extract	10
KNO_3	75 mg
2. Add sterile	
K_2HPO_4	15 mg
Glucose	500
Tryptose	500
3. Adjust pH to 7.5 with sterile K_2CO_3	
C. *Enriched seawater* (Lateur 1963)	
Seawater	1 l
Soil extract	0.1 ml
$NaHCO_3$	0.025 g
$NaNO_3$	0.1
Na_2HPO_4	0.02
D. *Modified Schreiber-soln* (Neumann 1969*a,b*)	
Seawater	1 l
$NaNO_3$	42.5 mg
$Na_2HPO_4.12H_2O$	10.75
$FeSO_4.7H_2O$	0.278
$McCl_2.4H_2O$	0.0198
E. *Modified ASP-6* (Shephard 1970)	
(final volume is 1 l)	
1. Add as solids	
NaCl	24 g
$MgSO_4.7H_2O$	12
$CaCl_2.2H_2O$	1

TABLE 6-2. (*cont.*)

Tris	1
KCl	0.75
$NaNO_3$	0.04
K_2HPO_4	0.001 (add last)
Adjust pH to 7.8 with 1N HCl (*ca.* 5 ml/l)	
2. To soln 1, add second soln with:	
$Na_2.EDTA$	12 mg
$ZnSO_4.7H_2O$	2
$Na_2MoO_4.2H_2O$	1
$FeCl_3.6H_2O$	0.5
$MnCl_2.4H_2O$	0.2
$CoCl_2.6H_2O$	2.0 μg
$CuSO_4.5H_2O$	2.0 μg
Autoclave soln 1 plus 2	
3. Make as separate soln; sterilize by filter membrane; add to sterile salt soln 1+2)	
$NaHCO_3$	0.1 g
Thiamine.HCl	300 μg
p-aminobenzoate	20
Ca.pantothenate	10
Cyanocobalamin	4
F. *Modified Erdschreiber* (Rietema 1970)	
Seawater (North Sea)	1 l
Soil extract	25 ml
EDTA	3.72 mg
$FeSO_4$	0.278
$Na_2HPO_4.12H_2O$	0.269
$NaNO_3$	42.5
$MnCl_2$	0.02
GeO_2	5.0

VI. REFERENCES

Abèlard, C. and Feldmann, J. 1958. Influence de la temperature sur la formation des gametes d'*Halicystis parvula* Schmitz. *Bull. Soc. Phycol. France* **4**, 10–11.

Borden, C. A. and Stein, J. R. 1969. Reproduction and early development in *Codium fragile* (Suringar) Hariot: Chlorophyceae. *Phycologia* **8**, 91–9.

Brachet, J. and Bonotto, S., eds. 1970. *Biology of Acetabularia*. Proc. 1st Int. Symp. *Acetabularia*. Academic Press, New York. 300 pp.

Burr, F. A. and West, J. A. 1970. Light and electron microscope observations on the vegetative and reproductive structures of *Bryopsis hypnoides*. *Phycologia* **9**, 17–37.

Chen, J. C. W. and Jacobs, W. P. 1966. Quantitative study of development of the giant coenocyte, *Caulerpa prolifera*. *Amer. J. Bot.* **53**, 413–23.

Chen, J. C. W. and Jacobs, W. P. 1968. The initiation and elongation of rhizoid clusters in *Caulerpa prolifera*. *Amer. J. Bot.* **55**, 12–19.

Colinvaux, L. H., Wilbur, K. M. and Watabe, N. 1965. Tropical marine algae: growth in laboratory culture. *J. Phycol.* **1**, 69–78.

Gibor, A. and Izawa, M. 1963. The DNA content of the chloroplasts of *Acetabularia*. *Proc. Nat. Acad. Sci.* **50**, 1164–9.

Green, B., Heilporn, V., Limbosch, S., Boloukhere, M. and Brachet, J. 1967. The cytoplasmic DNAs of *Acetabularia mediterranea*. *Proc. Nat. Acad. Sci.* **58**, 1351–8.

Hämmerling, J. 1944. Zur Lebensweise, Fortpflanzung und Entwicklung verschiedenes Dasycladaceen. *Arch. Protist.* **97**, 7–56.

Hollenberg, G. J. 1936. A study of *Halicystis ovalis*. II. Periodicity in the formation of gametes. *Amer. J. Bot.* **23**, 1–3.

Hustede, H. 1960. Über den Generationswechsel von *Derbesia neglecta* Berth. und *Bryopsis Halymeniae* Berth. *Naturwissenschaften* **47**, 19.

Hustede, H. 1964. Entwicklungsphysiologische Untersuchungen über den Generationswechsel zwischen *Derbesia neglecta* Berth. und *Bryopsis Halymeniae* Berth. *Bot. Marina* **6**, 134–42.

Jacobs, W. P. 1951. Studies on cell-differentiation: the role of auxin in algae, with particular reference to *Bryopsis*. *Biol. Bull.* **101**, 300–6.

Jonsson, S. and Puiseux-Dao, S. 1959. Observations morphologiques et caryologiques relatives a la reproduction chez le *Siphonocladus pusillus* (Kütz.) Hauck, Siphonocladacées, en culture. *C. R. Acad. Sci., Paris* **249**, 1383–5.

Köhler, K. 1958. Über den Generationswechsel von *Halicystis–Derbesia* im Golf von Neapel. *Pubbl. Staz. Zool., Napoli* **30**, 342–6.

Kornmann, P. 1938. Zur Entwicklungsgeschichte von *Derbesia* und *Halicystis*. *Planta* **28**, 464–70.

Lateur, L. 1963. Une technique de culture pour l'*Acetabularia mediterranea*. *Rev. Algol.* n.s. **7**, 26–37.

Mishra, A. K. and Kefford, N. P. 1969. Developmental studies on the coenocytic alga, *Caulerpa sertularioides*. *J. Phycol.* **5**, 103–9.

Neumann, K. 1969*a*. Beitrage zur Cytologie und Entwicklung der siphonalen Grünalge *Derbesia marina*. *Helgoländer wiss. Meeresunters.* **19**, 355–75.

Neumann, K. 1969*b*. Protonema mit Riesenkern bei der siphonalen Grünalge *Bryopsis hypnoides* und weitere cytologische Befunde. *Helgoländer wiss. Meeresunters.* **19**, 45–7.

Ott, F. D. 1966. A selected listing of xenic algal cultures. *Systematics–Ecology Program*, Mar. Biol. Lab., Woods Hole, Massachusetts. Contrib. 72, 1–45 (Mimeo.).

Page, J. Z. 1970. Existence of a *Derbesia* phase in the life history of *Halicystis osterhoutii* Blinks and Blinks. *J. Phycol.* **6**, 375–80.

Page, J. Z. and Kingsbury, J. M. 1968. Culture studies on the marine green alga *Halicystis parvula–Derbesia tenuissima*. II. Synchrony and periodicity in gamete formation and release. *Amer. J. Bot.* **55**, 1–11.

Page, J. Z. and Sweeney, B. M. 1968. Culture studies on the marine green alga *Halicystis parvula–Derbesia tenuissima*. III. Control of gamete formation by an endogenous rhythm. *J. Phycol.* **4**, 253–60.

Provasoli, L. 1968. Media and prospects for the cultivation of marine algae. In Watanabe, A. and Hattori, A., eds., *Cultures and Collections of Algae*. Proc. U.S.–Japan Conference HaKone, Sept. 1966. *Japan Soc. Plant Physiol.* 63–75.

Puiseux-Dao, S. 1962. Recherches biologiques et physiologiques sur quelques Dasycladacées. *Rev. Gen. Bot.* **69**, 409–503.

Puiseux-Dao, S. 1964. Le *Batophora oerstedii* J. Ag. (Dasycladacées) en culture au laboratoire. In DeVirville, D. and Feldmann, J., eds., *Proc. Int. Seaweed Symp.* **4**, 135–40. Macmillan Co., New York.

Puiseux-Dao, S. 1970. *Acetabularia and Cell Biology*. Springer-Verlag, New York, Inc. xii + 162 pp.

Rietema, H. 1969. A new type of life history in *Bryopsis* (Chlorophyceae, Caulerpales). *Acta Bot. Neerl.* **18**, 615–19.

Rietema, H. 1970. Life-histories of *Bryopsis plumosa* (Chlorophyceae, Caulerpales) from European coasts. *Acta Bot. Neerl.* **19**, 859–66.

Rietema, H. 1971*a*. Life-history studies in the genus *Bryopsis* (Chlorophyceae) III. The life-history of *Bryopsis monoica* Funk. *Acta Bot. Neerl.* **20**, 205–10.

Rietema, H. 1971*b*. Life-history studies in the genus *Bryopsis* (Chlorophyceae) IV. Life-histories in *Bryopsis hypnoides* Lamx. from different points along the European coasts. *Acta Bot. Neerl.* **20**, 291–8.

Sears, J. R. 1967. Mitotic waves in the green alga *Blastophysa rhizopus* as related to coenocyte from. *J. Phycol.* **3**, 136–9.

Sears, J. R. and Wilce, R. T. 1970. Reproduction and systematics of the marine alga *Derbesia* (Chlorophyceae) in New England. *J. Phycol.* **6**, 381–93.

Shephard, D. 1970. Axenic culture of *Acetabularia* in a synthetic medium. In Prescott, D., ed., *Methods in Cell Physiology*, **4**, 46–69. Academic Press, New York.

Starr, R. C. 1964. The culture collection of algae at Indiana University. *Amer. J. Bot.* **51**, 1013–44.

Starr, R. C. 1966. *Culture collection of algae. Additions to the collection July 1, 1963–July, 1, 1966*, pp. 1–8. Published by Indiana University, Bloomington, Indiana 47401 and available from the Culture Collection.

Starr, R. C. 1971. The culture collection of algae at Indiana University – additions to the collection July 1966–July 1971. *J. Phycol.* **7**, 350–62.

Terborgh, J. 1965. Effects of red and blue light on the growth and morphogenesis of *Acetabularia crenulata*. *Nature* **207**, 1360–3.

Terborgh, J. and Thimann, K. V. 1964. Interactions between day length and light intensity in the growth and chlorophyll content of *Acetabularia crenulata*. *Planta* **63**, 83–98.

Terborgh, J. and Thimann, K. V. 1965. The control of development of *Acetabularia* by light. *Planta* **64**, 241–53.
Ziegler, J. R. and Kingsbury, J. M. 1964. Cultural studies on the marine green alga *Halicystis parvula – Derbesia tenuissima*. I. Normal and abnormal sexual and asexual reproduction. *Phycologia* **4**, 105–16.

7: Methods for Cyanophyceae

MARY MENNES ALLEN
Department of Biological Sciences,
Wellesley College, Wellesley, Massachusetts 02181

CONTENTS

I. OBJECTIVES

The objectives of this chapter are to establish enrichment procedures specific to the prokaryotic Cyanophyceae which allow their enrichment, isolation, and purification from natural materials containing other algae and bacteria. Apart from those which fix nitrogen the bluegreen algae appear to have nutritional requirements similar to other algae, making unpromising the devising of specific enrichment methods for isolation based on nutritional selection. Studies by Kratz and Myers (1955) and Fogg (1956) suggest that bluegreen algae generally have a temperature range significantly higher than that characteristic of most other algae. Allen and Stanier (1968) showed that 35 axenic strains of Chroococcaceae, Oscillatoriaceae, and Nostocaceae could grow at 35 °C. A nutritionally, non-selective mineral medium incubated at 35 °C can be used to advantage in enrichment of many bluegreen algae from soil and water, essentially unaccompanied by eukaryotic algae. Such primary enrichments provide excellent material for further purification. The same medium, without a combined nitrogen source, can be used for the selection of Cyanophyceae able to fix atmospheric nitrogen.

II. EQUIPMENT

Temperature-controlled incubators set at 35 °C with cool-white fluorescent light sources are the only non-standard pieces of equipment required (see Chp. 11). Standard equipment for isolation (glassware, needle loops, etc., are listed by Hoshaw and Rosowski in Chp. 3, II).

III. ORGANISMS

The method outlined for enrichment and isolation has proven satisfactory both for unicellular strains, with and without sheath, and for filamentous strains from soil and freshwater sources. Samples may be taken from ponds, rivers, mud or soil whether or not bluegreen algae are initially detectable by microscopic examination.

TABLE 7-1. *Mineral medium no. 11 (Hughes, Gorham and Zehnder 1958)*

$NaNO_3$	1.5 g
K_2HPO_4	0.369
$MgSO_4.7H_2O$	0.075
$CaCl_2.2H_2O$	0.036
Na_2CO_3	0.020
$Na_2SiO_3.9H_2O$	0.058
Ferric citrate	0.006
Citric acid	0.006
EDTA	0.001
Minor element soln.[a]	0.08 ml
Distilled water	to 1 liter

[a] Minor element soln. in g/l: H_3BO_3 3.1: $MnSO_4.4H_2O$ 2.23: $ZnSO_4.7H_2O$ 0.287: $(NH_4)_6Mo_7O_{24}.4H_2O$ 0.088: $Co(NO_3)_2.4H_2O$ 0.146: $Na_2WO_4.2H_2O$ 0.33: KBr 0.119: KI 0.083: $Cd(NO_3)_2.4H_2O$ 0.154: $NiSO_4(NH_4)SO_4.6H_2O$ 0.198: $VOSO_4.2H_2O$ 0.02: $Al_2(SO_4)_3–K_2SO_4.24H_2O$ 0.474.

Initial pH 8.0–8.5; final pH 9–10.
Autoclave iron separately; add aseptically.

IV. METHOD

A. *Enrichment*

1. Medium. Medium no. 11 of Hughes, Gorham and Zehnder (1958) (see Table 7-1), modified to include 1.5 g/l $NaNO_3$, supports the growth of a wide spectrum of bluegreen algae as well as other algae. Fourteen different cyanophytes were obtained from one tropical soil sample (Allen 1966). A variety of media including ones supplemented with vitamins, as well as variation in other parameters such as pH and light intensity, might increase the utility of the enrichment technique. Medium no. 11 with no combined nitrogen source allows the selection of nitrogen-fixing bluegreen algae.

2. Sample preparation. Five ml of medium no. 11, either nitrogen-free or supplemented with $NaNO_3$, are dispensed into test tubes, which are plugged and sterilized. Liquid inocula should be homogenized by agitation on a rotary shaker for 30 min before sampling. Soil specimens should be mixed with medium and also agitated for 30 min.

Serial tenfold dilutions are prepared from each water or soil-suspension sample, and used to inoculate test tubes containing the medium.

3. Incubation. Tubes should be incubated at 35 °C without aeration, gas or agitation for 2 weeks. The light intensity should be maintained at 300–600 ft-c (3240–6480 lux) with cool-white fluorescent lights (see Chp. 11, II.C).

B. *Isolation, purification*

1. Media

a. Liquid medium (as given in Table 7-1).

b. Solid medium (Allen 1968).

i. Autoclave 3% (w/v) agar in distilled water (see Chp. 1, II.B).

ii. Autoclave double-strength mineral medium (Table 7-1).

iii. Cool both agar and mineral medium to 50 °C.

iv. Mix agar and mineral medium; pour immediately into sterile petri dishes (final concentration 1.5% agar).

v. By using 2% (w/v) agar in distilled water (instead of 3% as in step i), the final concentration will be 1% agar.

2. Procedure

a. Unicellular bluegreen algae

i. Transfer samples from primary enrichment tubes to fresh liquid medium; incubate as in A.3, preceding.

ii. From these secondary tubes, streak the algae on mineral agar plates (1.b.iv, preceding); incubate as in A.3, preceding.

iii. After incubation isolate colonies with a sterile loop or needle into new media; incubate as previously mentioned.

iv. Successive transfers from liquid to solid media usually yield unialgal and/or axenic cultures.

b. Filamentous bluegreen algae – unialgal cultures of filamentous forms usually can be obtained simply by repeated liquid transfer of small amounts of material. In many instances the phototactic response of the algae is utilized to separate them from other algae.

i. Place a mass of cells on a slide in liquid medium.

ii. Cover with cover glass: seal with Vasoline, paraffin, or similar compound.

iii. Incubate slide in unidirectional light for 2–10 h. The filaments move from the mass so that single filaments are observed at the outer edge.

iv. Isolate with a sterile Pasteur-type capillary pipette (see Chp. 3, IV.A.1). Transfer to a tube of liquid medium or the surface of solidified medium in a petri dish. A low agar concentration, less than 1.5%, is important for obtaining good growth, and the filament rapidly migrates through 1% agar whereas motile bacteria do not.

v. Material from liquid cultures can be placed on or stabbed into the center of 1% agar plates, then incubated in unidirectional light.

vi. Isolate filaments that are visibly free of bacterial contamination from the rapidly moving front on the illuminated side of the plate. Place in liquid medium or on fresh agar plates.

vii. Three to ten passages through agar established 18 axenic filamentous strains (Allen 1966).

3. Purity. All strains should be tested for contamination by methods such as those described in Chp. 3, VII.

V. SAMPLE DATA

A. *Enrichment*

1. Scoring. Samples from freshwater ponds and lakes, fresh or dried temperate mud and soil, and tropical soil are inoculated in dilution series into the nutritionally non-selective mineral medium, with and without $NaNO_3$, at 25 °C. After 2 weeks incubation, each dilution series is examined macroscopically for the presence of visible algal growth; the tube with growth from the least inoculum is examined with the compound microscope and scored as to:

a. Prokaryotic, or bluegreen algae (Cyanophyceae)
  i. unicellular
  ii. filamentous with or without heterocysts
  iii. filamentous with or without branching

b. Eukaryotic algae
  i. greens (Chlorophyceae)
  ii. diatoms (Bacillariophyceae)
  iii. euglenoids (Euglenophyceae)

Within each category distinct types are scored by differences in size, form, and pigmentation. The minimal number present in the inoculum is determined for each algal type by multiplying the number observed in a given tube by the dilution factor of the tube (see Allen and Stanier 1968).

2. Qualitative analysis. Table 7-2 summarizes some qualitative aspects of such an enrichment with several different source materials. The total number of different prokaryotic (bluegreen) and eukaryotic algae occurring

TABLE 7-2. *Qualitative aspects of incubation temperature and nitrogen source on number of bluegreen and eukaryotic algae after enrichment*

	Conditions of incubation					
Temperature ▸	25 °C		35 °C		35 °C	
N-source ▸	NO_3^-		NO_3^-		N_2 (in air)	
Source material	Bluegreen	Eukaryotic	Bluegreen	Eukaryotic	Bluegreen	Eukaryotic
Shallow pond – January	5	6	9	1	1	0
– May	6	21	7	4	7	0
Freshwater lake	0	2	1	0	1	0
Soil – tropical, dried	2	9	7	2	7	0
– greenhouse	1	5	2	0	1	0
– mossy, moist	0	7	0	0	0	0
– rice paddy	2	4	3	0	3	0
Cumulative totals	16	54	29	7	20	0

The figures represent number of distinct morphological types observed per ml or per g in the sample, as inferred from microscope analyses of enrichment cultures inoculated with serial dilutions of the source material (see text; Allen and Stanier 1968).

TABLE 7-3. *The influence of nitrogen source on development of bluegreen algae enriched and grown at 35 °C*

		Type observed		
			Filamentous	
Source material	N-source	Unicellular	With heterocysts	Without heterocysts
Shallow pond – January	NO_3^-	10	1000, 100, 100, 10	100, 10, 10, 1
	N_2	0	1000	0
– May	NO_3^-	100, 100, 1	10000, 1000	1000, 1000, 100, 10
	N_2	0	1000, 1000	0
Rice paddy soil	NO_3^-	1000	1000	1000
	N_2	0	1000, 1000, 1000	0

Each figure represents minimal number of cells or filaments of a given type of bluegreen alga in source materials, as inferred from microscope examination of enrichment cultures inoculated with serial dilutions of the sample (see text; Allen and Stanier 1968).

TABLE 7-4. *Influence of incubation temperature and nitrogen source on recovery of bluegreen and eukaryotic algae*

	Conditions of incubation					
Temperature ▸	25 °C		35 °C		35 °C	
N-source ▸	NO_3^-		NO_3^-		N_2	
Source material	Bluegreen	Eukaryotic	Bluegreen	Eukaryotic	Bluegreen	Eukaryotic
Shallow pond – January	1 220	3 300	1 340	10	1 000	0
– May	2 210	7 460	2 310	30	2 030	0
Freshwater lake	0	2×10^4	1	0	1	0
Soil – tropical, dried	1 000	31 410	11 500	2 000	13 300	0
– greenhouse	1 000	4 100	2 000	0	100	0
– mossy, moist	0	10^5	0	0	0	0
– rice paddy	2 000	4 000	3 000	0	3 000	0

The figures represent minimal numbers of algae per ml or per g in the sample, as inferred from microscope analyses of enrichment cultures inoculated with serial dilutions of the source material (see text).

in parallel enrichment series are listed. In general, enrichment at 35 °C increased the variety of bluegreen algae present in any given sample with the eukaryotic algae greatly decreased. The absence of a source of combined nitrogen severely restricted the number of algae developing. The qualitative consequences of omitting a combined nitrogen source are shown in more detail in Table 7-3. Each time unicellular bluegreen algae or non-heterocyst-forming filaments were present in the enrichment cultures at 35 °C with a combined nitrogen-free source, they were absent in nitrogen-free media. In rice paddy soil, growth of heterocyst-forming filamentous algae was favored by the omission of NO_3. This specific enrichment was observed with many soil samples (Allen and Stanier 1968).

3. Quantitative aspects. Table 7-4 shows the quantitative results of the same series of enrichment experiments using water and soil samples for inocula as Table 7-2. In most samples no algae were detected initially by macroscopic examination. The data for the dilution series incubated at 25 °C show that all enrichments contained appreciable numbers of eukaryotic algae and that bluegreen algae were also detected in most, but in lower numbers. A major change in the enrichment cultures was produced by elevating the temperature of incubation to 35 °C. Eukaryotic algae were virtually eliminated (a small number of unicellular forms sometimes developed) and there was a great increase in the number of bluegreen algae recovered. The results obtained with the sample of tropical soil are worthy of note. After incubation at 25 °C, a much larger population of eukaryotic algae (31 400/g) were present when compared to the bluegreen algal population (1000/g). Incubation at 35 °C decreased the number of eukaryotic algae to 2000/g and revealed at least 11 500 bluegreen algal cells or filaments/g. This clearly shows that in samples containing large populations of eukaryotic algae, the recovery of bluegreen algae by enrichment at 25 °C is far lower than the true content of the sample. The development of many bluegreen algae is prevented, under non-selective growth conditions, by overgrowth of the large eukaryotic algal population also present in the sample.

Enrichment at 35 °C in medium free of combined nitrogen totally eliminated the few eukaryotic species which were able to develop at this temperature when NO_3 was present, as seen in Table 7-4. Nitrogen-fixing bluegreen algae were usually recovered. Table 7-4 also shows that garden soils such as the moist, mossy soil contain few bluegreen algae, although the content of eukaryotic algae was high.

B. *Isolation, purification*

Seven unicellular strains were easily purified using the techniques described (Allen 1966); six of these were essentially devoid of sheath material, while the seventh was heavily encapsulated. Over twenty more axenic strains have been obtained recently using this method (R. Y. Stanier, personal communication). Ten species of *Oscillatoria* and eight Nostocaceae were purified by use of liquid – 1% agar transfers (Allen 1966).

VI. POSSIBLE PURIFICATION PROBLEMS

A. *Slime-forming algae*

One filamentous strain producing abundant gelatinous sheath was readily enriched for, but could not be purified, by these techniques (Allen 1966). In such instances bacteria penetrate the sheath and cannot be removed by repeated transfer or by movement through the agar because the bacteria form an integral part of the sheath.

B. *Slow-moving filamentous forms*

The method described for filamentous forms could not be used to purify four slow-moving strains. The algae moved slowly through the soft agar and bacterial contamination was prevalent and outgrew the algae.

C. *Gas-vacuolated forms*

A temperature of 35 °C is too high to permit the growth of certain planktonic bluegreen algae which form gas vacuoles readily (G. Cohen-Bazire, personal communication).

VII. ALTERNATIVE TECHNIQUES

A. *Repeated liquid subculture*

This technique has been successfully used when a natural collection is particularly rich in a specific bluegreen alga (Gerloff, Fitzgerald and Skoog 1950).

B. *Fragmentation*

Homogenization of filamentous forms with a glass homogenizer for 5–10 min allows short filaments of 4–8 cells long, to be obtained. In some instances these are viable and form isolated, individual colonies when streaked on agar media (Allen 1966).

C. *Antibiotics*

See Chp. 3, IV.3.B.

D. *Ultraviolet irradiation*

This method has been widely used to obtain purified cultures of bluegreen algae (Gerloff *et al.* 1950; Bowyer and Skerman 1968; Koch 1964).

E. *Micromanipulation*

This technique has been used to obtain axenic cultures of soil-borne and endophytic bluegreen algae (Bowyer and Skerman 1968). Also see Chp. 8.

F. *Higher temperature incubation*

Thermophilic bluegreen algae (optimal growth range above 45 °C) can be readily isolated from thermal habitats by enrichment above 45 °C (Castenholz 1969, 1970). Most thermophilic cyanophytes grow poorly below 30–35 °C, whereas the temperature maximum appears to be near 73 °C (Castenholz 1969, 1970).

VIII. REFERENCES

Allen, M. M. 1966. Studies on the properties of some blue-green algae. *Ph.D. Thesis.* University of California, Berkeley.

Allen, M. M. 1968. Simple conditions for growth of unicellular blue-green algae on plates. *J. Phycol.* **4**, 1–4.

Allen, M. M. and Stanier, R. Y. 1968. Selective isolation of blue-green algae from water and soil. *J. Gen. Microbiol.* **51**, 203–9.

Bowyer, J. W. and Skerman, V. B. D. 1968. Production of axenic cultures of soil-borne and endophytic blue-green algae. *J. Gen. Microbiol.* **54**, 299–306.

Castenholz, R. W. 1969. Thermophilic blue-green algae and the thermal environment. *Bact. Rev.* **33**, 476–504.

Castenholz, R. W. 1970. Laboratory culture of thermophilic cyanophytes. *Schweiz. Z. Hydrol.* **32**, 538–51.

Fogg, G. E. 1956. The comparative physiology and biochemistry of the blue-green algae. *Bact. Rev.* **20**, 148–65.

Gerloff, G. C., Fitzgerald, G. P. and Skoog, F. 1950. The isolation, purification, and culture of blue-green algae. *Amer. J. Bot.* **37**, 216–18.

Hughes, E. O., Gorham, P. R. and Zehnder, A. 1958. Toxicity of a unialgal culture of *Microcystis aeruginosa*. *Can. J. Microbiol.* **4**, 225–36.

Koch, W. 1964. Cyanophyceenkulturen. Anreicherungs und Isolierverfahren. *Zentbl. Bakt. Parasitenk.* **1**. Suppl. 1, 415–31.
Kratz, W. A. and Myers, J. 1955. Nutrition and growth of several blue-green algae. *Amer. J. Bot.* **42**, 282–7.

8: Special methods–micromanipulators

JAHN THRONDSEN

Institute of Marine Biology, Section B,
University of Oslo, Blindern, Oslo 3, Norway

CONTENTS

I. INTRODUCTION

Micromanipulators in a wide sense comprise every type of equipment designed for manipulating objects of microscopic size, e.g., groups of cells, single cells, organelles. The term is usually restricted to equipment in which some mechanical parts are involved in operating the microtool in work on the object. Microsurgical methods are beyond the scope of the present handbook and the reader is referred to the works of Fonbrune (1949) and Seidel (1957) for an introduction to this type of work.

The equipment and technique used for isolating single algal cells range from simple needles and capillary tubes, manipulated by hand (see Chps. 3, 4), to complex micromanipulators with micropipettes of elaborate design, see e.g., Perfil'ev and Gabe (1969). In this paper the use of a simple micropipette is considered.

II. EQUIPMENT

A. *Inverted microscope* (e.g., Wild M 40 plankton microscope: Wild Heerbrug Instruments, Inc., 465 Smith St., Farmingdale, New York 11735; Wild of Canada, Ltd, 881 Lady Ellen Pl., Ottawa K1Z 5L4, Ontario) with long-working distance condenser, high-intensity lamp, low magnification objectives (e.g., 2.5–4×, 10×, 20–40×), and 10–15× oculars of wide field type. It may prove necessary to use short extension tubes on the objectives.

B. *Microtool holder*, consisting of plastic plate (Perspex) with hole for condenser, and oblique hole for microtool (shown in Fig. 8-1; see legend for details).

C. *Syringe* with screw adjustment (shown in Fig. 8-2, a left; see legend for details) and 400–500 mm of plastic tube for connecting it to the pipette.

D. *Pasteur-type pipette* (outer diameter 3 mm) with capillary end drawn in a flame to get a slim capillary end (Chp. 3, IV.A.1; see Fig. 8-3, and

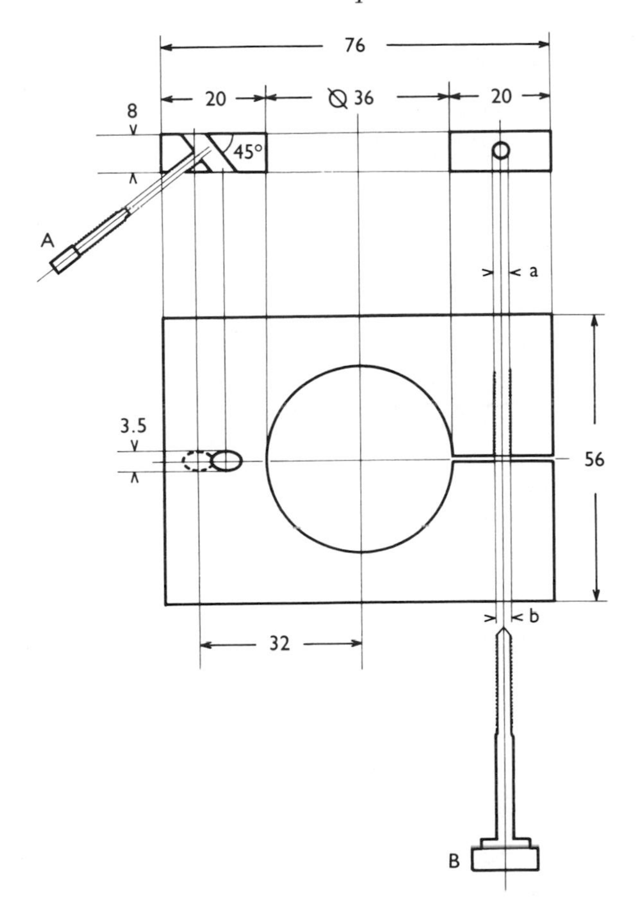

Fig. 8-1. Microtool holder for Wild M 40 long-working distance condenser, consisting of Perspex plate with holes for condenser and pipette. A and B screws. Diameter b must be large enough to permit free movement of the screw B. All measurement in mm.

legend for details). The oblique opening at the capillary end can be made by grinding on a fine wetstone.

E. *Cover glasses* (24 × 36 mm) with holder (Fig. 8-4, and legend) or standard microscope slides.

III. TEST ORGANISMS

Suitable cells are *Haematococcus* or *Chlorella* (Chlorophyceae) available through IUCC (see Introductory Chapter).

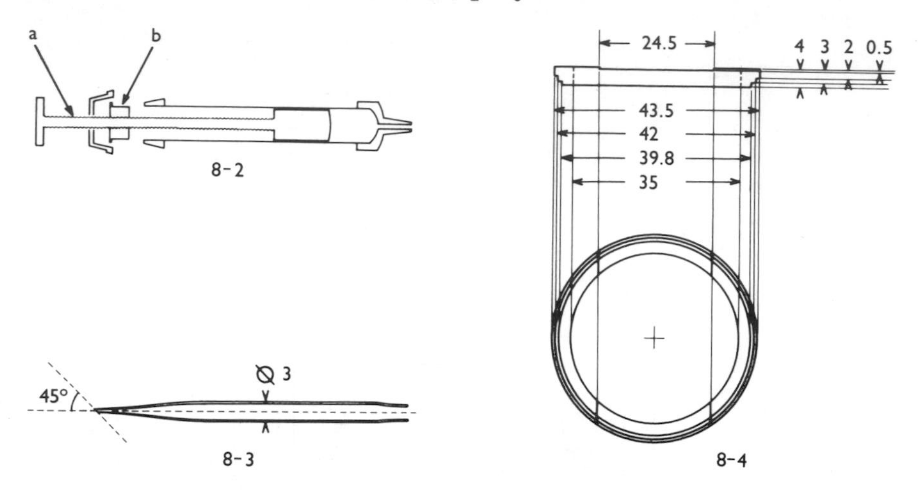

Fig. 8-2. Syringe modified for screw-controlled piston movement. a, piston rod with threads cut for fitting into the counterpart b which is fixed by assembling the syringe.

Fig. 8-3. Example of pipette for use with the microtool holder of Fig. 8-1. Diameter in mm.

Fig. 8-4. Holder for cover glasses (24 × 36 mm) consisting of Perspex ring with a ribbon-shaped depression to keep the cover glass in place. All measurements in mm.

IV. METHOD

Preferably use a room with a relative humidity high enough to prevent rapid evaporation of the culture medium. Temperature shocks should be avoided during the work.

A. Place the micropipette device on the mounted microscope below the condenser as shown in Fig. 8-5. Place a clean cover glass in the holder on the stage (or a standard slide if low magnifications only are to be used). Center the pipette by means of the centering screws of the condenser holder.

B. Apply separately some drops of culture medium and a drop of the material on the cover glass. Let the pipette fill with medium by capillary force.

C. Bring the material in focus and lower the pipette so it can be seen while looking through the microscope. Select the cell to be isolated, bring it close to the mouth of the pipette (by use of the mechanical stage) and suck it in (by turning the syringe screw).

D. Raise the pipette, move the stage so the pipette may be lowered in a drop of medium. Eject the cell in the medium.

E. Raise the pipette, empty it into a piece of (sterile) filter paper and refill it from a drop of medium.

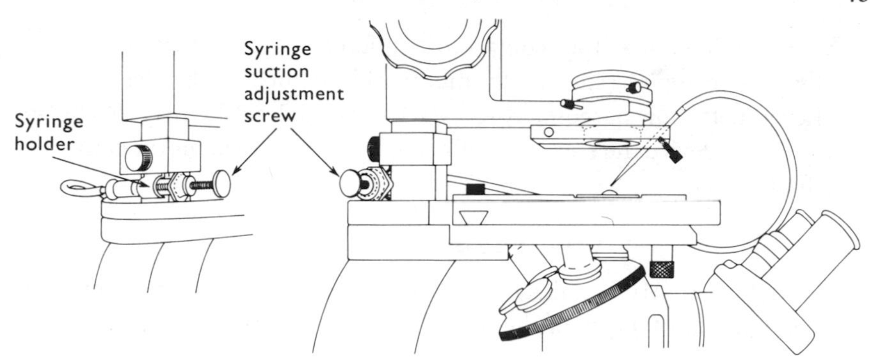

Fig. 8-5. Micropipette device mounted on Wild M 40 microscope.

F. Repeat the procedure as described in steps C to E until the cell in question is the only one present in the drop.
G. Then the cell is sucked into the pipette, the pipette holder is raised and if necessary the pipette is removed from the holder and the cell is ejected into the medium of a culture flask or tube.

V. POSSIBLE PROBLEMS

The precautions to be taken to prevent contamination of the organism and the medium during the isolation procedure will vary according to the laboratory conditions and the result desired. The pipettes may be dry-sterilized and kept in test tubes with cotton wool stoppers, the cover glasses or slides in petri dishes. Depending on the type of algal material used it may prove desirable to change the pipette before the final isolation step (G, preceding) is carried out.
Pipette centering. Straight pipettes will usually be nearly centered when mounted; if not, pipettes with a slight bend of the pointed end may prove to be more easily centered.
Sucking in the cell. For swimming cells a straight cut pipette end may be more convenient than the oblique type mentioned in II.D, preceding.

VI. ALTERNATIVE TECHNIQUES

A. Other pipette holders for inverted microscopes are described by Haller Nielsen (1950) and Throndsen (1969). Both are to be used without a condenser; the first has its own mechanical stage for centering the pipette, whereas the other replaces the condenser (of Wild M 40 and Zeiss plankton microscopes), the pipette being centered by means of the condenser holder's centering screws.

B. Micromanipulators for compound microscopes (not inverted) are usually more complex (e.g., Leitz micromanipulator), see Seidel (1957).
C. Micropipette for stereomicroscopes has been described by Nicholson (1957). In his device the suction is effected by an electrical motor controlled by a foot switch.

VII. ACKNOWLEDGMENT

I am indebted to Mr K. Nilsen (Institute of Physics, University of Oslo) for making the prototypes of the holders and the syringe modification.

VIII. REFERENCES

de Fonbrune, P. 1949. Technique de micromanipulation. *Monogr. Inst. Pasteur* **1949**, 1–203.

Haller Nielsen, P. 1950. An auxiliary apparatus for plankton studies by means of the sedimentation method. *J. Cons. Explor. Mer* **16**, 307–9.

Nicholson, H. F. 1957. Mechanical pipette for picking out diatoms. *J. Cons. Explor. Mer* **23**, 189–91.

Perfil'ev, B. V. and Gabe, D. R. 1969. *Capillary Methods of Investigating Micro-organisms.* Oliver and Boyd, Edinburgh. 627 pp.

Seidel, F. 1957. Bauweise und Verwendung von Mikromanipulatoren. In Freund, H., ed., *Handbuch der Mikroskopie in der Technik* **1** (1), 589–613. Frankfurt/Main.

Throndsen, J. 1969. A simple micropipette for use on the Wild M 40 and the Zeiss plankton microscopes. *J. Cons. Explor. Mer* **32**, 430–2.

9: Special methods – virus detection in Cyanophyceae

ROBERT S. SAFFERMAN
Environment Protection Agency,
National Environmental Research Center, Cincinnati, Ohio 45268

CONTENTS

I. INTRODUCTION

The circumstances which eventually led to the detection of the bluegreen algal (BGA) virus initially grew out of a need for a better understanding of the underlying factors responsible for algal decomposition (Safferman and Morris 1963). With the discovery of the BGA virus, a new concept in algal pathology evolved since there was no doubt that these viruses had the capacity to exercise a far-reaching effect on the development of algal communities. Our knowledge is still too limited to relate phycovirus activity directly with specific incidences of algal fluctuations. However, the magnitude and distribution of the LPP group of BGA viruses have led some to believe that these lytic agents are responsible for precluding susceptible algal species, such as *Plectonema boryanum*, from attaining objectionable proportions (Safferman and Morris 1967; Padan and Shilo 1969*a*).

This concept has been further strengthened from the search for viruses infecting the bloom-forming *Microcystis aeruginosa*. Rarely observed (Safferman *et al.* 1969*b*) and apparently not as virulent in nature as the LPP viruses, one can conceive of this viral type as exercising only a negligible degenerative effect on a susceptible algal population. At a time when the over-abundance of algae in our natural resources is of increasing concern, it is evident that laboratory cultivation of highly virulent strains of these rare viruses could eventually find application in regulating bluegreen algal development. Ecological considerations, however, afford only one view of the potential contribution of viruses to algal research. An even greater impact is anticipated with the elucidation of the cell's physiological reaction to viral infection, for from this will undoubtedly come fundamental advances analogous to those achieved with phage-bacteria systems.

It is essential in the study of BGA viruses to realize that these viruses more closely resemble bacteriophages than viruses infecting higher plants. Many of the procedures devised for their detection and enumeration have therefore had their origin in phage research.

II. TEST STRAINS

A. *Virus–host system*

The archetype of the BGA viruses, LPP-1, has been deposited with the American Type Culture Collection (ATCC 15581-P). There are presently 13 algal strains reported susceptible to LPP-1 and are available from the Indiana University Culture Collection. It is recommended that of these *Plectonema boryanum* (IUCC 594 or ATCC 18200) be adopted as the host test organism for the LPP group of viruses. Other susceptible algal strains are *Phormidium luridum* (IUCC 426); *Ph. faveolarum* (IUCC 427); *Phormidium* sp. (IUCC 485); *Lyngbya* spp. (IUCC 487, 488); *Plectonema notatum* (IUCC 482); *Pl. boryanum* (IUCC 581, 595, 596, 597, 790); *Pl. calothricoides* (IUCC 598).

The short trichome mutant of *Plectonema boryanum* strain 594 recently isolated (Padan and Shilo 1969*b*) is still susceptible to the virus. It is reported to have the same growth rate and yields in liquid medium as the wild type.

B. *Sampling location*

Sampling natural sources for BGA viruses may be helpful in standardization of procedural techniques. The LPP viruses are an excellent test group, readily detectable in samples collected from waste stabilization ponds in the United States (Safferman and Morris 1967). In any sampling scheme, a favorable climatic condition for the host organism and the examination of more than one pond are obvious considerations. If necessary, pond samples can be mailed without any significant loss in their LPP titer, for as seen in the following section, these viruses are relatively stable. Locations of waste stabilization ponds in the United States are listed in an inventory made on municipal waste treatment facilities by the U.S. Public Health Service (1963).

III. PROCEDURES

A. *Detection of LPP group: method 1*

The procedure to be described was first devised for establishing the existence of a BGA virus and shortly thereafter instituted in a screening program for the detection of other phycoviruses. Slightly modified, this method was later used to investigate the distribution of the LPP group (Safferman and Morris 1967). It is this revised method which is cited here. Although prepared for a specific viral group, it has also been successfully tested in surveying for other BGA viruses as evident by the isolation of the SM-1 virus (Safferman and Morris 1967; Safferman *et al.* 1969*b*).

Surveys carried out with this procedure have been limited to waste stabilization ponds but only because this is a unique environment. No special procedures were instituted in sampling these ponds. Composite samples consisted of surface waters collected in sterile polypropylene bottles from the influent and effluent areas of the ponds.

1. Positive presumptive test

a. Medium and growth conditions

i. Chu no. 10 medium (see Chp. 1, IV.A.5)

– 100 ml in 250 ml flask

– 50 ml in 125 ml flask (2).

ii. 160–180 ft-c (1700–1950 lux); continuous illumination (cool-white fluorescent lamps); at 20 °C.

b. Algal host

i. *Plectonema boryanum* IUCC 594; grown 1 week, 100 ml in 250 ml flask (suggest use 5 ml of 3 week old broth culture as inoculum).

ii. Stock culture – 10 ml (minimum).

c. Method

i. Collect sample in sterile, polypropylene bottle from waters believed to contain members of the LPP group.

ii. Add 50 ml to 1 week algal culture (b, preceding).

iii. Incubate 5–7 days (as in a, preceding).

iv. Centrifuge 40 ml for 5 min at 2000 *g*.

v. Transfer 10 ml of supernatant to glass-stoppered bottle; add 0.2 ml chloroform and shake 30 s (this destroys microbial contaminants).

vi. Let chloroform settle.

vii. Inoculate two 125 ml flasks containing 50 ml Chu no. 10 medium with 2 ml stock algal culture.

viii. To one flask of algae, add 1 ml of the aqueous phase of the chloroform-treated supernatant. The other flask is the control.

ix. Incubate both flasks 7 days (as in a, preceding).

The positive presumptive test for the presence of LPP viruses is based on reduction in turbidity of the algal culture. Thus flasks showing no decrease in algal density are discarded without further analysis. This scheme may prove undesirable in surveying for other BGA viral groups. In early SM-1 studies, with stocks of low virulence, virus development was not always apparent. Consequently it was necessary to assay such cultures on solid medium and examine for evidence of plaque development (see 2, following) before discarding. Cultures of *Plectonema boryanum* showing a decrease in algal density are assayed for LPP viruses.

2. *Assay for LPP virus* (done in duplicate)

a. Medium, equipment

i. Chu no. 10 agar (1.5% agar); 15 ml solidified in petri dish (see Chp. 1, II.B).

ii. Chu no. 10 agar (1%); melt, cool to 47 °C.

iii. Sterile tubes.

iv. Saline–magnesium solution

0.2 g $MgCl_2.6H_2O$
5.85 g NaCl
1 l distilled water.

v. Sterile, ultra-fine sintered glass filter; or 0.22 μm Millipore filter.

b. Algal host *Plectonema boryanum* stock growing in Chu no. 10 liquid.

c. Virus source culture of *Pl. boryanum* (prepared as in 1, preceding) showing decrease presumed due to LPP virus.

d. Method

i. Filter *Pl. boryanum* culture containing LPP virus through sterile ultra-fine filter.

ii. Dilute filtrate with tenfold dilutions of saline–magnesium solution; usually 10^{-6} or 10^{-7} most satisfactory.

iii. To sterile tube add 0.5 ml appropriately diluted filtrate, 2.0 ml *Pl. boryanum* stock, and 2.5 ml warm (47 °C) 1% Chu no. 10 agar (termed *inoculated agar* – 5.0 ml).

iv. Pour warm inoculated agar over solidified 1.5% Chu no. 10 agar in petri dish.

v. Incubate 3–4 days as described in 1, preceding.

vi. Examine for plaque development seen as clear circular 1–5 mm spots in the 'algal lawn' (Safferman and Morris 1967).

3. *Direct viral isolation.* Some waste stabilization ponds support substantial populations of the LPP viruses. When this is suspected, direct viral isolations are attempted simultaneously with the enrichment technique. For this assay, another 40 ml aliquot is removed from the sample bottle and centrifuged at 2000 *g* for 5 min. A 10 ml amount of supernatant is treated with chloroform as previously described (1.c), then immediately plaque assayed in duplicate. From the duplicate plates, viral titer is figured as the sum of the plaque counts multiplied by the dilution factor. Since the figures are only relative to the number of virus particles present, the results are given in plaque-forming units per ml. As a control measure, a saline–magnesium blank is processed with the test samples to detect possible introduction of extraneous phycoviruses into the system.

TABLE 9-1. *Population fluctuations of LPP viruses in waste stabilization ponds*

Location ▸ Month ▾	Smackover, Arkansas (PFU/ml)	Lancaster, California (PFU/ml)	Sunman, Indiana (PFU/ml)
January	23	+	0
February	8	+	0
March	16	1	0
April	42	+	+
May	52	7	+
June	151	+	+
July	36	11	1
August	19	+	+
September	4	5	1
October	25	+	1
November	22	+	+
December	35	+	+

PFU = Plaque forming units.
+ = presence of at least one virus in a 50 ml pond sample.

A number of waste stabilization ponds were studied for varying periods using the enrichment procedure alone or in conjunction with the direct assay method (Safferman and Morris 1967). In all but one of these ponds, LPP viruses could readily be detected by these techniques. Of those in which direct LPP counts were undertaken, all gave positive values at some time during the sampling period. Some typical data collected in these studies can be seen in Table 9-1. The highest natural LPP population densities so far established were those reported for a 30 acre waste stabilization pond in South Dakota from which two consecutive bimonthly samples gave 270 and 240 plaque-forming units per ml.

4. Purification. To obtain purified strains of LPP isolates, individual plaques are stabbed, and the virus, dispersed in the saline–magnesium solution (see 2.a, preceding), is replated at appropriate dilutions. Repeat this two or three times. The plaques thus isolated are serially transferred in liquid medium to obtain a high-titer stock.

The LPP-1 virus has been found relatively stable when stored under controlled conditions. In one such study, lysates kept at 4 °C for as long as three months showed no loss in activity and, in a number of instances, had higher titers after storage (Goldstein 1966). Losses were also negligible when kept at 20 °C for more than two weeks. Although data on similar comparisons with other LPP viruses are not available, all isolates have remained viable at 4 °C, including many which have been stored for 1–2 years.

B. *Detection of LPP group: method 2*

Inherent in the previously described procedure is uncertainty stemming from treatment of the lysates with chloroform. Although only a hypothetical consideration at present, a definite possibility exists that some BGA viruses may not have been detected because of chloroform sensitivity.

Padan and associates (1967, 1968, 1969*a*) have described an alternate, chloroform-free procedure for phycovirus detection.

1. Equipment

a. Whatman GF/C glass filter paper.

b. Saline–magnesium solution (see 2.a, preceding).

c. 80% (v/v) polyethylene glycol (MW 20000–40000) dissolved in saline–magnesium solution.

d. Ultra-fine sintered glass filter.

e. 100 ft-c (1100 lux) lights, continuous; at 26 °C.

2. *Algal host. Plectonema boryanum* (IUCC 594) – 10 day culture (in 100 ml Chu no. 10 medium).

3. Method

a. Clarify 3 l sample by passing through Whatman paper.

b. Concentrate to 10 ml by hydro-extraction: dialyze filtrate against 80% polyethylene glycol, then dialyze concentrate at 4 °C against saline–magnesium solution for 12 h.

c. Filter concentrate through sintered glass filter.

d. Plate with *Pl. boryanum* directly on solid medium (see A.2, preceding), or enrich with *Pl. boryanum* by adding the 10 ml concentrate to 10 day culture of *Pl. boryanum* and incubate 3 days (see 1.e, preceding).

e. If lysis present in enrichment culture, plate samples (as in A.2, preceding) and examine for plaque development.

Thus far, the most encompassing study with this procedure is a quantitative survey of LPP population of 5 fish ponds (Padan and Shilo 1969*a*). The same pronounced fluctuations seen in the stabilization pond data given in Table 9-1 were also observed. The magnitude of the LPP populations, however, gives an altogether different picture. Only one pond had titers greater than 0.3 plaque-forming units/ml and in only two samples, both giving LPP counts of less than 14 plaque-forming units/ml. In searching for other BGA viruses, this approach has so far proved unsuccessful, but at the same time, it has not been as extensively applied as the chloroform technique.

C. *Perspectives in detection methodology*

Factors such as media, incubation temperature and light intensity are to some extent arbitrary preferences which can be varied as long as the algal cell is maintained in an actively growing state. The condition of the host cell cannot be over-emphasized, particularly when screening for the rare phycoviral forms.

Only brief mention has been made of the SM-1 virus in development of detection procedures, as it is less satisfactory than the LPP-1 virus. Although not recommended as a standard, the greater demand placed on the skill of the investigator by SM-1 could lead to procedural refinements not considered from viral recovery studies with LPP-1. The SM-1 virus and its host, *Microcystis aeruginosa* NRC-1, are on deposit in the American Type Culture Collection under ATCC accession numbers 18800-P and 18800.

The two preceding sections provide the established methodology for detection of BGA viruses. Mention should be made of a report in which a virus for *Anabaena variabilis* was indicated by electron microscopy (Granhall and von Hofsten 1969). It is doubtful, however, that the electron microscope could be incorporated into a definitive procedure in phycoviral detection. On the other hand, techniques from other fields of virology may find application here as routine detection procedures. Several are discussed in the following section, although step by step procedures are not given.

IV. ALTERNATIVE PROCEDURES

The detection of BGA viruses by direct assay has been possible only within the LPP group of viruses. It is conceivable that other phycoviruses will be discovered by this procedure, but detection of these agents in such small samples undoubtedly negates any advantage gained from the simplicity. Already described in the prior sections are the enrichment and hydroextraction techniques (III.A, B). These procedures are well-adapted toward the search for the rare phycovirus; however, there is no rationale for limiting phycovirus detection to these two techniques or to considering them the most applicable of the methods presently available. Several alternate procedures have therefore been included. None of these procedures has thus far been instituted into a BGA virus detection program, but they do constitute established techniques for concentrating viruses from large volumes of water.

The relatively simple gauze pad technique allows examination of larger volumes of water and seems the most promising for detecting the rare BGA viral form, particularly if applied with enrichment procedures (see III.A.3; B, preceding). This technique is widely accepted in sampling natural and waste waters suspected of having low concentrations of animal viruses. A detailed description of the technique is given in numerous reports (Gravelle and Chin 1961; Berg 1966; Wiley *et al.* 1962; Melnick *et al.* 1954). The method depends on the capacity of the pads to concentrate viruses from the sampled environment and on the efficiency with which the virus can be eluted by expression of the liquid. The liquid is generally expressed from the gauze pad in a plastic bag with the pH of the absorbed water adjusted to 8.0. The expressed liquid is then clarified and the viruses that may be present are further concentrated usually by ultra-centrifugation. Other methods have been described by Berg (1966) and Lund and Hedstrom (1967). The method is restrictive with respect to measuring the viral density of a water and possibly to the groups of viruses which can be detected.

Apparently of greater efficiency is the membrane filter approach of Cliver (1967). Like the prior gauze pad technique, this membrane procedure is entirely dependent on the extent to which water-borne viruses can be absorbed and eluted from the system. Measured samples are passed through Millipore filters with a porosity of 0.45 μm and absorbed viruses are then recovered by eluting with serum (Cliver 1967; Wallis and Melnick 1967*a*), or beef extract (Berg *et al.* 1968; Rao and Labzoffsky 1969). Where samples first require clarification, membrane filtration has been preceded by prefiltration with microfiber glass pads (Cliver 1967). This approach, applied to the enteric viruses, affords a simple, rapid procedure for analyzing these viruses in sewage and natural water systems (Wallis and Melnick 1967*a*; Chang 1968; Rao and Labzoffsky 1969). Thus the membrane technique permits not only detection, as in the gauze pad procedure, but also the quantitation of water-borne viruses.

Ion exchange polymers constitute still another adsorption procedure successfully applied in enumerating low concentrations of enteric viruses from the aquatic environment. Early studies used 'Dowex' and 'Amberlite' resins (Kelly 1953; Gravelle and Chin 1961); since then other polymers of higher efficiency have been reported (Wallis *et al.* 1969).

The polymer two-phase system of polyethylene glycol and sodium dextran sulfate developed for the concentration and purification of viral suspensions (Philipson, Albertsson and Frick 1960; Philipson 1969) has been modified and applied to the purification of crude LPP-1 lysate (Luftig and Haselkorn 1968; Luftig 1967). By a stepwise addition of polyethylene glycol, the algal debris and then the virus are precipitated to obtain sufficient quantities of virus for structural studies. The processing, for example, of 40 l crude lysate with the specific infectivity of 8×10^9 PFU/ml yielded a concentrate having a final LPP-1 titer of 4×10^{12} PFU/ml. With further modifications the polymer system should be adaptable to BGA viral detection because as an approach to the monitoring of enteric viruses, the expectations are that 10 PFU/l or less will be detectable as concentration factors of at least 1000 are eventually anticipated (Shuval *et al.* 1967).

Reference has already been made to centrifugation in association with several of the previous methods. Although there are no reported studies directly incorporating centrifugation into BGA viral detection, systems adapted to continuous-flow have been extensively used in concentrating laboratory cultures of the LPP and SM phycovirus groups (Smith *et al.* 1966; Safferman *et al.* 1969*a*,*b*). The relatively recent achievement in zonal centrifugation has provided a high-resolution technique that is finding broad application in the separation of subcellular particles (Anderson 1966). Although presently available to few laboratories, the system's future in virus detection warrants its mention (Anderson *et al.* 1967).

There are several other prospective viral detection procedures that could be adapted to the BGA virus. These additional methods are cited in the bibliography (Rao *et al.* 1968; Gartner 1967; Wallis and Melnick 1967*b*; Bier *et al.* 1967).

V. TEMPERATE VIRUSES

At present there is no evidence that temperate viruses are harbored by any of the bluegreen algae. Nonetheless, procedures for their detection have been included here since the non-existence of these viruses seems as unlikely today as virus immunity of the bluegreen algae did a decade ago (Safferman and Morris 1963).

In deducing the probable nature of these viruses, it must be assumed that the temperate BGA virus will have biological and physiochemical properties that closely resemble those of the temperate bacteriophage (Lwoff 1953; Bertani 1958; Jacob and Wollman 1959; Hayes 1968). Thus the temperate phage should provide the methodology in developing this area of phycovirology just as the virulent phage has in past studies on the LPP-1 and SM-1 viruses.

In finding the temperate virus, detection is centered on the lysogenic population, which in the course of its development, always undergoes a low order of spontaneous lysis with a simultaneous release of viral particles. It is these particles, detected by their lytic action, which determine recognition of lysogeny. At the same time, one must bear in mind that the lysogenic strains are immune to infection by the same viruses that they release. Thus, detection of lysogeny and assay of the temperate virus are contingent on the availability of an indicator strain. Once attained, isolation of the temperate virus proceeds in the same manner as for virulent strains: particles are separated from parent cells by such previously described procedures as chloroform treatment or filtration and plated directly with the indicator strain on solid medium. The frequency of lysis can be increased in a number of lysogenic populations by inducing agents such as ultraviolet light, X-rays, mitomycin C or fluorodeoxyuridine. As a result, it is possible to achieve nearly total lysis of a population.

It is essential to show that lysogeny is maintained in the absence of any free virus forms. Pseudo-lysogeny has been described in association with some bacterial populations (Hayes 1968) and no doubt similar interactions will also emerge from studies on bluegreen algal populations.

VI. ACKNOWLEDGMENT

I wish to thank Mary-Ellen Morris for her kind help in the preparation of this manuscript.

VII. REFERENCES

Anderson, N. G., ed. 1966. *The Development of Zonal Centrifuges and Ancillary Systems for Tissue Fractionation and Analysis*. U.S. National Cancer Institute Monograph No. 21. 526 pp.

Anderson, N. G., Cline, G. B., Harris, W. W. and Green, J. G. 1967. Isolation of viral particles from large fluid volumes. In Berg, G., ed., *Transmission of Viruses by the Water Route*, pp. 75–88. Interscience Publishers, New York.

Berg, G. 1966. Virus transmission by the water vehicle. II. Virus removal by sewage treatment procedures. *Health Lab. Sci.* **3**, 90–100.

Berg, G., Dean, R. B. and Dahling, D. R. 1968. Removal of poliovirus 1 from secondary effluents by lime flocculation and rapid sand filtration. *J. Amer. Water Works Assn.* **60**, 193–8.

Bertani, G. 1958. Lysogeny. *Advances in Virus Res.* **5**, 154–93.

Bier, M., Bruckner, G. C., Cooper, F. C. and Roy, H. E. 1967. Concentration of bacteriophage by electrophoresis. In Berg, G., ed., *Transmission of Viruses by the Water Route*, pp. 57–73. Interscience Publishers, New York.

Chang, S. L. 1968. Waterborne viral infections and their prevention. *Bull. World Health Organ.* **38**, 401–14.

Cliver, D. O. 1967. Enterovirus detection by membrane chromatography. In Berg, G., ed., *Transmission of Viruses by the Water Route*, pp. 139–41. Interscience Publishers, New York.

Gartner, H. 1967. Retention and recovery of poliovirus on a soluble ultrafilter. In Berg, G., ed., *Transmission of Viruses by the Water Route*, pp. 121–7. Interscience Publishers, New York.

Goldstein, D. A. 1966. Some biological, chemical and physical properties of blue-green algal virus LPP-1. *Ph.D. Thesis.* University of Pittsburgh.

Granhall, U. and von Hofsten, A. 1969. The ultrastructure of a cyanophage attack on *Anabaena variabilis. Physiol. Plant.* **22**, 713–22.

Gravelle, C. R. and Chin, T. D. Y. 1961. Enterovirus isolation from sewage: a comparison of three methods. *J. Infect. Dis.* **109**, 205–9.

Hayes, W. 1968. *The Genetics of Bacteria and Their Viruses.* John Wiley and Sons, New York. 925 pp.

Jacob, F. and Wollman, E. L. 1959. Lysogeny. In Burnet, F. M. and Stanley, W. M., eds., *The Viruses*, vol. 2, pp. 319–51. Academic Press, New York.

Kelly, S. M. 1953. Detection and occurrence of coxsackie viruses in sewage. *Amer. J. Public Health* **43**, 1532–8.

Luftig, R. B. 1967. Studies on cyanophage LPP-1. *Ph.D. Thesis.* University of Chicago.

Luftig, R. B. and Haselkorn, R. 1968. Studies on the structure of blue-green algal virus LPP-1. *Virology* **34**, 664–74.

Lund, E. and Hedstrom, C.-E. 1967. Recovery of viruses from a sewage treatment plant. In Berg, G., ed., *Transmission of Viruses by the Water Route*, pp. 371–7. Interscience Publishers, New York.

Lwoff, A. 1953. Lysogeny. *Bacteriol. Rev.* **17**, 269–337.

Melnick, J. L., Emmons, J., Opton, E. M. and Coffey, J. H. 1954. Coxsackie viruses from sewage. *Amer. J. Hyg.* **59**, 185–95.

Padan, E. and Shilo, M. 1968. Spread of viruses attacking blue-green algae in freshwater ponds and their interaction with *Plectonema boryanum. Bamidgeh* **20**, 77–87.

Padan, E. and Shilo, M. 1969*a*. Distribution of cyanophages in natural habitats. *Verh. Internat. Verein. Limnol.* **17**, 747–51.

Padan, E. and Shilo, M. 1969*b*. Short-trichome mutant of *Plectonema boryanum. J. Bacteriol.* **97**, 975–6.

Padan, E., Shilo, M. and Kislev, N. 1967. Isolation of 'cyanophages' from freshwater ponds and their interaction with *Plectonema boryanum. Virology* **32**, 234–46.

Philipson, L. 1969. Aqueous polymer phase systems in virology. In Habel, K. and Salzman, N. P., eds., *Fundamental Techniques in Virology*, pp. 109–28. Academic Press, New York.

Philipson, L., Albertsson, P. Å. and Frick, G. 1960. The purification and concentration of viruses by aqueous polymer phase systems. *Virology* **11**, 553–71.

Rao, N. U. and Labzoffsky, N. A. 1969. A simple method for the detection of low concentration of viruses in large volumes of water by the membrane filter technique. *Can. J. Microbiol.* **15**, 399–403.

Rao, V. C., Sullivan, R., Read, R. B. and Clarke, N. A. 1968. A simple method for concentrating and detecting viruses in water. *J. Amer. Water Works Assn.* **60**, 1288–94.

Safferman, R. S. and Morris, M. E. 1963. Algal virus: Isolation. *Science* **140**, 679–80.

Safferman, R. S. and Morris, M. E. 1967. Observation on the occurrence, distribution and seasonal incidence of blue-green algal viruses. *Appl. Microbiol.* **15**, 1219–22.

Safferman, R. S., Morris, M. E., Sherman, L. A. and Haselkorn, R. 1969*a*. Serological and electron microscopic characterization of a new group of blue-green algal viruses (LPP-2). *Virology* **39**, 775–80.

Safferman, R. S., Schneider, I. R., Steere, R. L., Morris, M. E. and Diener, T. O. 1969*b*. Phycovirus SM-1: A virus infecting unicellular blue-green algae. *Virology* **37**, 386–95.

Shuval, H. I., Cymbalista, S., Fatal, B. and Goldblum, N. 1967. Concentration of enteric viruses in water by hydro-extraction and two-phase separation. In Berg, G., ed., *Transmission of Viruses by the Water Route*, pp. 45–55. Interscience Publishers, New York.

Smith, K. M., Brown, R. M. Jr., Goldstein, D. A. and Walne, P. L. 1966. Culture methods for the blue-green alga *Plectonema boryanum* and its virus with an electron microscopic study of virus infected cells. *Virology* **28**, 580–91.

U.S. Public Health Service. 1963. 1962 Inventory Municipal Waste Facilities. *U.S.P.H.S. Publication no. 1065*. 9 vols.

Wallis, C. and Melnick, J. L. 1967*a*. Concentration of enteroviruses on membrane filters. *J. Virology* **1**, 472–7.

Wallis, C. and Melnick, J. L. 1967*b*. Concentration of viruses on aluminium phosphate and aluminium hydroxide precipitates. In Berg, G., ed., *Transmission of Viruses by the Water Route*, pp. 129–38. Interscience Publishers, New York.

Wallis, C., Grinstein, S., Melnick, J. L. and Fields, J. E. 1969. Concentration of viruses from sewage and excreta on insoluble polyelectrolytes. *Appl. Microbiol.* **18**, 1007–14.

Wiley, J. S., Chin, T. D. Y., Gravelle, C. R. and Robinson, S. 1962. Enteroviruses in sewage during a poliomyelitis epidemic. *J. Water Pollution Control Fed.* **34**, 168–78.

NOTES ADDED IN PROOF

The presumed occurrence of lysogeny in bluegreen algae has now been confirmed. Cannon *et al.* (1971) reported lysogeny in a strain of *Plectonema boryanum* by BGA virus LPP-1D from spontaneous and mitomycin C induction after incubation in the presence of LPP-1D antiserum. A subsequent report of Padan *et al.* (1972) has concluded lysogeny of *Pl. boryanum* by an LPP-2 virus (strain SPI). They observed immunity to infection by viruses of the LPP-2 type but not to those of LPP-1. Ultraviolet, X-ray and mitomycin C induction were unsuccessful; however, induction occurred when the lysogenic strain of a thermosensitive SPI mutant was transferred from 26 °C to 37 °C.

Cannon, R. E., Shane, M. S. and Bush, V. N. 1971. Lysogeny of a blue-green alga, *Plectonema boryanum*. *Virology* **45**, 149–53.
Padan, E., Shilo, M. and Oppenheim, A. B. 1972. Lysogeny of the blue-green alga *Plectonema boryanum* by LPP-2-SPI cyanophage. *Virology* **47**, 525–6.

10: Special methods – dry soil samples

RICHARD C. STARR
Department of Botany,
Indiana University, Bloomington, Indiana 47401

CONTENTS

I. OBJECTIVE

The use of soil as a source of algal populations from which unialgal or axenic isolations can be made started as early as 1910 by Jacobsen (a student of Beijerinck who is generally credited with the first isolation of cultures). Jacobsen was interested in algae with heterotrophic tendencies and so combined with his soil cultures such organic materials as fibrin, casein and albumin. That the addition of certain organic compounds or the inorganic solution used to flood a soil sample will influence the composition of the resulting populations is a basic tenet of workers using soil samples as a source of algae.

Soil samples are collected and allowed to air dry without extra heat. They are stored in glass containers or plastic bags to protect them from changing levels of humidity that might cause premature germination of resistant stages or might result in death of the algae. Genera having a life history phase that is resistant to drying, such as zygospore or refractory asexual spore, are obvious candidates for isolation from soil samples. However, in many instances, algae appear in soil cultures which have no specialized resistant phase, but the vegetative cells are apparently capable of living in a partially dehydrated condition. Some soil samples remain a good source of algal cultures for several years after collection, but this will vary due to the nature of the soil, the storage conditions, and the algal species.

Soil samples can provide a source of a great variety of algae, if the source of the sample is taken into consideration. For example, random sampling from cultivated soils routinely gives many typical soil Chlorophyceae, as *Chlamydomonas*, *Chlorococcum*, *Bracteacoccus*, *Hormidium*, etc. More selective sampling of mud from the edge of small ponds containing species of the colonial *Volvocales* (Chlorophyceae) at the time of sampling, may yield many of these species when soil cultures are established from the dried sample. This latter method is used by phycologists interested in studying populations originating from widely divergent geographical locations, especially when the algae will not withstand shipment in the typical vegetative condition.

After obtaining the dried soil sample, there are almost as many methods and minor variations in obtaining populations as there are investigators. Certain of these methods given here serve as a starting point.

II. EQUIPMENT

Most of the materials required are standard for isolating algae. See Chp. 3, II for details as to preparation, etc. Specific materials required for isolating from dry soil samples include the following:

A. *Isolation*

1. Petri dishes, sterile, 20 × 100 mm.
2. Sterile nutrient, liquid, in flasks.
3. Sterile nutrient agar (1–1.5%) in petri dishes
   a. solidified,
   b. melted (45 °C) to be poured over soil.

B. *Culturing*

1. Spot plates, sterile, in petri dishes.
2. Soil water, in test tubes or bottles.
3. Sterile nutrient media – liquid or 1–1.5% agar.
4. Culture facilities as outlined in Chp. 11, II, III.

III. METHODS

The methods presented here have been successfully used to isolate the algae as indicated. Materials are listed in II, preceding, except when specialized items are required.

A. *Soil algae*

Harold C. Bold and his students have been engaged for many years in a study of soil flora especially those belonging to the families Chlorococcaceae and Chlorosarcinaceae in the green algae. In Bold's summary (1970) covering twenty years of work he recommends several primary treatments to achieve algal populations from soil samples: (1) dry soil is placed in petri dishes and moistened (not flooded) with distilled water or an inorganic nutrient solution; (2) moistened soil may be overlaid with sterile filter paper or nutrient agar; (3) soil suspension streaked on the surface of nutrient agar in a petri dish; and (4) soil introduced into a flask containing sterile nutrient solution. Any nutrient medium, such as Bold Basal medium (Chp. 1, IV.A.2) may be used to rewet the air-dried soil.

The samples are placed under conditions to ensure algae growth. This may include culture facilities as outlined in Chp. 11, II, III, or at room temperature where extra illumination (but not heat) is available. A window not receiving direct sunlight (north-facing in the Northern Hemisphere) is often quite satisfactory, as long as the temperature remains between 12 and 27 °C.

The samples are examined periodically for algal growth. This may begin as soon as 48 h after incubation and extend for several days or weeks. Once the populations of algae have appeared, they serve for isolation into unialgal cultures as outlined in Chp. 3, IV.A.

B. *Green flagellates* (*Chlorophyceae*)

The study of sexual isolation between populations of colonial green flagellates has been possible due to the use of soil samples as a means of transporting material from widely separated geographical areas to the laboratory for analysis of the contained population. Studies of *Gonium* (Stein 1958*b*), *Pandorina* (Coleman 1959), *Eudorina–Pleodorina* (Goldstein 1964), *Platydorina* (Harris and Starr 1969), *Astrephomene* (Pocock 1953; Stein 1959*a*; Brooks 1966) and *Volvulina* (Carefoot 1966) utilized soil samples in part to obtain material for study. Routinely the air-dried sample is collected from a habitat where the desired genus was known or believed to occur.

1. Place soil sample in petri dish. Use enough soil to cover partially the bottom of the dish, leaving a portion free. This enables observation of resulting populations with a stereomicroscope using refracted light.

2. Illuminate the sample with fluorescent light of *ca.* 350 ft-c (3700 lux) at 20–25 °C.

3. Within 48–72 h observe samples for presence of colonies in order to make single colony isolations prior to asexual multiplication. In this way it is hoped that multiple isolates represent different products of zygote germination rather than asexual multiplication from a single germination product.

Ordinarily the soil sample provides sufficient nutrients for several generations of autotrophic species. With *Astrephomene*, *Volvulina steinii*, certain heterotrophic species of *Gonium* (namely *G. octonarium* and *G. quadratum*), *Pyrobotrys* and some species of unicellular flagellates (*Diplostauron*, Lynn and Starr 1970) the addition of boiled dried pea (*Pisum sativum*) to the petri dish when rewetting the soil sample provides a heterotrophic medium in which these algae flourish. This is reminiscent of Jacobsen's (1910) use of fibrin and other proteinaceous materials in the soil cultures yielding *Chlorogonium* and other algae with heterotrophic tendencies.

Before establishment of axenic cultures (Chp. 3, IV.B) it is recommended that clonal populations be established in Pringsheim's soil water medium (Chp. 1, IV.B.5). In addition the mating types (where heterothallic) should be determined (refer to Coleman 1959; Goldstein 1964; Stein 1958*a*).

C. *Volvox* (*Chlorophyceae*)

Like many algae, this rather delicate organism is difficult to send through the mail, but the use of dried soil samples has provided Starr and his students (Starr 1968, 1969; McCracken and Starr 1970; Vande Berg and Starr 1971) with material from such widely divergent geographical areas as Japan, India, and South Africa. The large size of *Volvox* individuals, including the newly developed germling from the germinated zygote, permits isolation and purification directly from the material which appears in the soil culture. As with other autotrophic flagellates mentioned previously a small soil sample in a petri dish is flooded with distilled water. Within 48–72 h, germlings from germinating zygotes of *Volvox* may appear in the dish. The germlings are isolated into spot plates (Chp. 3, IV.A.1 and Fig. 3-1) containing sterile *Volvox* medium (Chp. 1, IV.A.7). The spot plates, placed in petri dishes, are surrounded by water and covered to prevent evaporation. The germlings are allowed to enlarge and just prior to the release of the young spherical colonies formed inside, the germling parental colony is removed and washed in a spot of sterile medium (Chp. 3, IV.A.1). Then drawing the colony into a Pasteur-type capillary pipette which is slightly smaller than the parent but large enough to accommodate the young colonies (Chp. 3, IV.A.1), the latter are released and transferred to another spot of sterile medium. By moving young colonies whose surfaces are free of bacteria through 4 or 5 spots of sterile media, bacteria in the surrounding medium are diluted out. The individual young colonies serve as inoculum for sterile tubes of medium and the eventual establishment of clonal, axenic populations by asexual reproduction.

Sexual reproduction in *Volvox* cultures is more likely to occur in the sterile *Volvox* medium than in soil water medium with bacteria, so it is recommended that this purification procedure be followed. It should also be noted that all species of *Volvox* grow in *Volvox* medium whereas soil water medium is not as effective in some instances.

D. *Sexual strains of Chlamydomonas, etc.* (*Chlorophyceae*)

Sexual populations of *Chlamydomonas*, *Haematococcus*, and other algae are often seen in the initial cultures started from soil samples. As a means of selecting sexual species from such cultures, Lewin (1951) recommends mixing some of the heterogeneous populations with acetone for 10 min. After evaporation of residual acetone the algal material is placed in a nutrient medium. The rationale is that during the initial growth strains becoming sexual produce zygotes that are to withstand the acetone treatment.

1. Materials

a. Acetone (equal volume to soil sample).
b. Soil sample.

2. Method

a. Place 2 g air-dry soil in 30 ml nutrient (mineral) medium in a flask.
b. Incubate 2 weeks at 23 °C.
c. Shake flask and decant 5 ml algal suspension.
d. Add 5 ml acetone; mix; let stand 10 min.
e. Filter acetone–algal mixture.
f. Dry residue in air 20 min.
g. Rinse residue into flasks containing nutrient medium (see a, preceding).
h. Incubate until growth appears (1–2 weeks).

For details of securing mating clones, refer to Lewin (1951) or use methods outlined for colonial Chlorophyceae (see B and C, preceding).

E. *Other methods*

The foregoing does not presume to include all the accounts of the use of air-dried soils for the production of algal populations for study and isolation in the laboratory. Other examples of the value of soil cultures are studies of sexual reproduction in desmids (Chlorophyceae) such as *Closterium* (Cook 1963; Lippert 1967), the great variety of Xanthophyceae algae in alpine soils (Vischer 1945) and the algae of desert soils (Cameron and Blank 1966). Finally, this account would be too incomplete were it not to include some reference to the work of Dr Mary A. Pocock whose account (Pocock 1962) of soil cultures from the De Klip area near Capetown, South Africa, and her work with *Volvulina* and *Astrephomene* (Pocock 1953), *Gonium* (Pocock 1955), and *Haematococcus* (Pocock 1960) has shown the excitement which awaits the investigator who turns to soil cultures as a source of varied and interesting algae.

IV. IMPORTATION OF SOILS

The introduction of harmful insects, nematodes, and microorganisms into most countries is an ever-present danger. Generally the importation of non-sterilized soil samples is prohibited except through special permission. Unless the investigator is willing to follow these precautions, it is suggested that he restrict his investigations to soils available in the home country.

A. *United States*

Application forms for permission to move soil may be requested from the following address:

Plant Quarantine Division
Agricultural Research Service
United States Department of Agriculture
Federal Center Building
Hyattsville, Maryland 20782

The required application forms after being filled out by the applicant will be referred to the state entomologist of the home state for his approval. As indicated in a letter of approval from the Plant Quarantine Division to the writer, the following safeguards were required:

1. All samples would be required to be shipped and stored in sturdy leak-proof containers (e.g., plastic bags or plastic vials in pliofilm) enclosed in cardboard cartons to prevent pest dissemination.
2. All samples, containers, and equipment would be required to be sterilized with steam or dry heat at the completion of the tests.
3. None of the material would be permitted to be transferred to another laboratory without prior approval from this office.

B. *Canada*

For information and necessary permits write:

Production and Marketing Branch
Plant Protection Division
Canada Department of Agriculture
Ottawa

The application is for a permit to import products other than plants.

Regulations under the Destructive Insect and Pest Act require that each importation shall be accompanied by a certificate of inspection issued, signed and dated by an authorized official of the country of origin. The certificate must indicate that the plants included in the shipment at or about the time of packing are apparently free from plants and diseases.

In some situations additional certification is required. In such instances the importer is provided with special instructions for transmittal to the shipper.

Material from outside continental United States must be certified that potato wart disease (*Synchytrium endobioticum* (Schilb.) Perc.) has not been recorded. In addition, based on official soil surveys and other precautionary practices, that golden nematode (*Heterodera rostochiensis* Woll.) is not known to occur.

An official mailing label is supplied. The shipment is sent directly to Plant Protection Division, Canada Department of Agriculture, where it is inspected before being forwarded to the importer.

V. ACKNOWLEDGMENT

Some of the work referred to in this paper as well as its preparation was completed under the auspices of PHS Grant FR-00141.

VI. REFERENCES

Bold, H. C. 1970. Some aspects of the taxonomy of soil algae. *Ann. New York Acad. Sci.* **175**, 601–16.

Brooks, A. E. 1966. The sexual cycle and intercrossing in the genus *Astrephomene. J. Protozool.* **13**, 367–75.

Cameron, R. E. and Blank, G. B. 1966. Desert algae: soil crusts and diaphanous substrata as algal habitats. *Tech. Report No. 32–971*, pp. 1–41. Jet Propulsion Laboratory, Calif. Inst. Tech., Pasadena.

Carefoot, J. R. 1966. Sexual reproduction and intercrossing in *Volvulina steinii. J. Phycol.* **2**, 150–6.

Coleman, A. W. 1959. Sexual isolation in *Pandorina morum. J. Protozool.* **6**, 249–64.

Cook, P. W. 1963. Variation in vegetative and sexual morphology among the small curved species of *Closterium. Phycologia* **3**, 1–18.

Goldstein, M. 1964. Speciation and mating behavior in *Eudorina. J. Protozool* **11**, 317–44.

Harris, D. O. and Starr, R. C. 1969. Life history and physiology of reproduction of *Platydorina caudata* Kofoid. *Arch. Protistenk.* **111**, 138–55.

Jacobsen, H. C. 1910. Kulturversuche mit einigen niederen Volvocaceen. *Z. Bot.* **2**, 145–88.

Lewin, R. A. 1951. Isolation of sexual strains of *Chlamydomonas. J. Gen. Microbiol.* **5**, 926–9.

Lippert, B. E. 1967. Sexual reproduction in *Closterium moniliferum* and *Closterium ehrenbergii. J. Phycol.* **3**, 182–98.

Lynn, R. I. and Starr, R. C. 1970. The biology of the acetate flagellate *Diplostauron elegans* Skuja. *Arch. Protistenk.* **112**, 283–302.

McCracken, M. D. and Starr, R. C. 1970. Induction and development of reproductive cells in the K-32 strains of *Volvox rousseletii. Arch. Protistenk.* **112**, 262–82.

Pocock, M. A. 1953. Two multicellular motile green algae, *Volvulina* Playfair and *Astrephomene*, a new genus. *Trans. Roy. Soc. South Africa* **34**, 103–27.

Pocock, M. A. 1955. Studies in North American Volvocales. 1. The genus *Gonium. Madroño* **13**, 49–64.

Pocock, M. A. 1960. *Haematococcus* in southern Africa. *Trans. Roy. Soc. South Africa* **36**, 5–55.

Pocock, M. A. 1962. Algae from De Klip soil cultures. *Arch. Mikrobiol.* **42**, 56–63.

Starr, R. C. 1968. Cellular differentiation in *Volvox. Proc. Nat. Acad. Sci.* **59**, 1082–8.

Starr, R. C. 1969. Structure, reproduction and differentiation in *Volvox carteri* f. *nagariensis* Iyengar, strains HK 9 and 10. *Arch. Protistenk.* **111**, 204–22.

Stein, J. R. 1958*a*. A morphological study of *Astrephomene gubernaculifera* and *Volvulina steinii. Amer. J. Bot.* **45**, 388–97.

Stein, J. R. 1958*b*. A morphologic and genetic study of *Gonium pectorale. Amer. J. Bot.* **45**, 664–72.

Vande Berg, W. J. and Starr, R. C. 1971. Structure, reproduction and differentiation in *Volvox gigas* and *Volvox powersii. Arch. Protistenk.* **113**, 195–219.

Vischer, W. 1945. Heterokonten aus alpinen Böden, speziell dem schweiz. Nationalpark. *Ergebnisse der wissensch. Untersuchung des Nat.-Parkes*, N. F. **1**, 479–512.

Section 11

General equipment and methods

11: Apparatus and maintenance

RICHARD C. STARR
Department of Botany,
Indiana University, Bloomington, Indiana 47401

CONTENTS

I. OBJECTIVE

In this section on apparatus and general maintenance of cultures, I will not presume to imply that there is a single best way of constructing facilities to achieve the necessary conditions for routine growth and maintenance of cultures. Rather I shall describe, for the most part, my experience in connection with the Culture Collection of Algae at Indiana University (IUCC) (United States), and with a research program that has involved the cultivation of algae for life history, physiology and genetics studies since 1951. As with all similar facilities that have been constructed or will be constructed in the future, the desire to build the best facility must be tempered with the often more deciding factors of available funds and multiple use of limited space. It is the very rare experience that the Culture Centre of Algae and Protozoa in Cambridge, England, has been able recently to design an entire building for the express purposes of research and maintenance of algae in culture.

II. CULTURE ROOMS – WALK-IN

A. *Temperature control*

The degree of temperature control required in a culture facility is determined by the requirements of the research. There are many commercial brands of culture chambers varying in size from 10 ft^3 to the large walk-in, fully-contained units. All claim to maintain the temperature at a very close setting. Where space is available, however, it may be converted to the purposes of culturing algae with less expense than the commercial self-contained units; with proper engineering, the level of control can be just as exacting.

The culture rooms used by the Collection at Indiana University (see Fig. 11-1 were part of the original construction of the building and so utilize compressors mounted on the roof outside the building proper. Heating strips situated in front of the cooling coils in each room are used to achieve temperatures showing little variance from the setting. These rooms are approximately 10 × 17 ft.

Temperatures at Indiana University are maintained at 20 °C in order

Fig. 11-1. One side of culture room showing on the left the lighted rack standing in center of the room. The walls are lined with unlighted shelving, on the right. Illumination received by the unlighted shelves is adequate for maintenance of cultures which have grown for *ca.* 10 days on lighted shelves, as seen in Fig. 11-2.

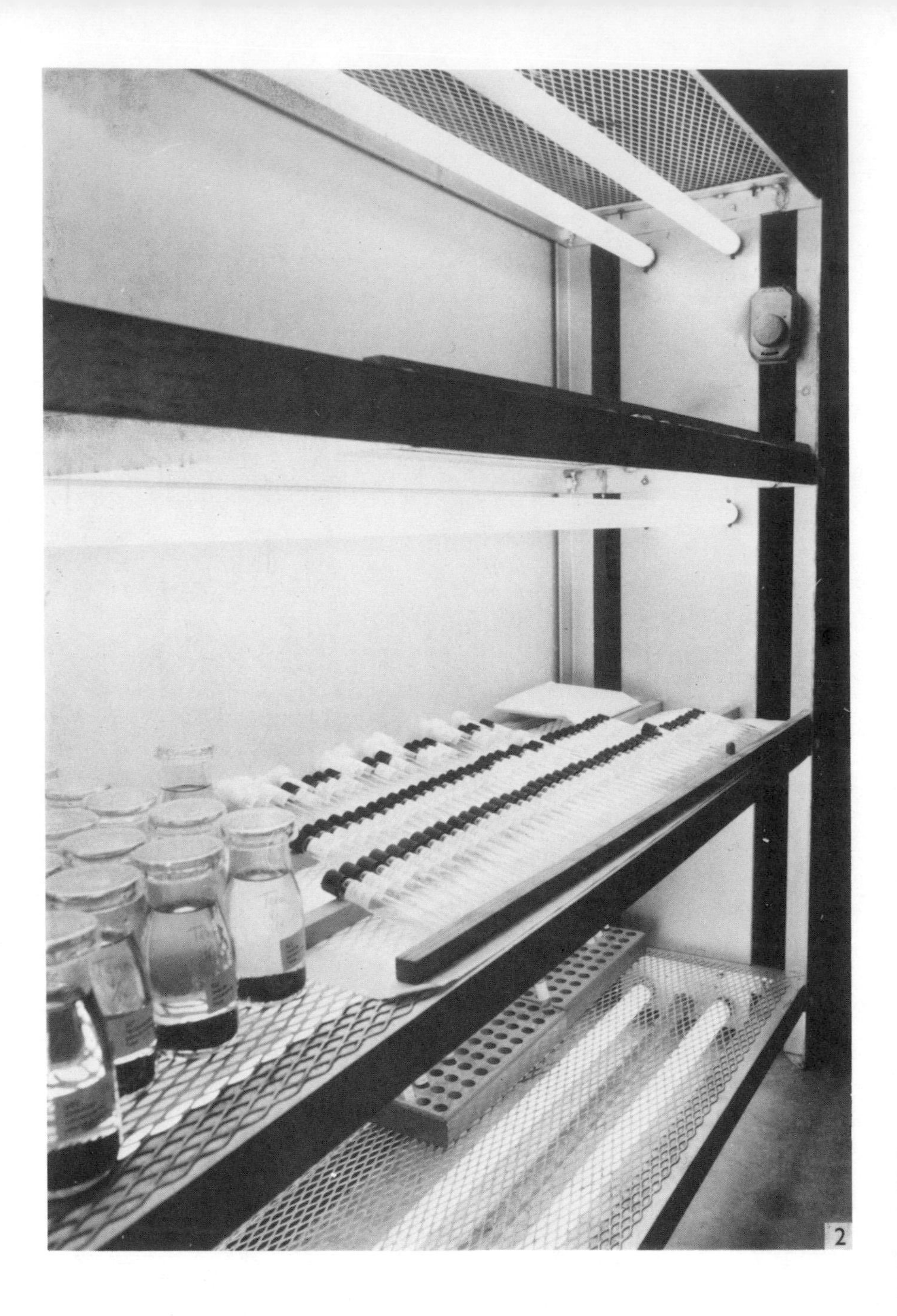

Fig. 11-2. Lighted shelves showing arrangement of lights above each shelf. Instrument mounted on right, below top set of fluorescent tubes is thermostatic switch. Bottles and tubes seen on the shelf are recent transfers, allowed to grow before being moved to unlighted shelves for storage.

that the rooms may be used for both storage and active growth of cultures. Where space is not a problem, 10 °C or 15 °C rooms would provide better for storage and for cultivation of certain marine algae.

It is not, however, necessary that investigations utilizing cultures be restricted to culture rooms under close control. Ordinary laboratories with reliable central air-conditioning that will cool the room to 20 °C, or even window units which will provide adequate cooling, will serve. Even with the more expensive type of culture facility, the use of lighted shelves in a temperature controlled laboratory will take the burden off the other facilities so that they may be used for other than routine work. The isolation of algae from natural populations collected in the field or grown from soil cultures often requires the examination of such populations for several days. It is such culture facilities as present in the open laboratory where field samples may be kept that prevent chance introduction of mites, etc., into the cultures.

B. *Shelving*

Two types of shelves can be useful for algal cultures. Some shelving should be provided with fluorescent lights in order to provide intensities of light (see C, following) high enough for routine cultivation of the algae. Other shelves for holding stock cultures can be arranged to receive illumination from lighted shelves or a light rack to provide intensities sufficient for maintenance but not so high as to cause harmful insolation of the algae. One common mistake in stock maintenance is over illumination. The 10 × 17 ft rooms accommodate an 8 ft long lighted unit easily. In addition it is possible to use unlighted shelving units, at right angles to the lighted unit, as shown in Fig. 11-1, to the right and in the background.

1. Unlighted. Any type of shelving can be used, but I prefer the metal shelving (e.g., Dexicon) whose surface is not harmed by the usual solvents (Figs. 11-1, 11-3). Such shelves can be wiped periodically with various compounds to clean and disinfect the surfaces and thus prevent or limit the problems with mite infestation and general mold contamination. Metal shelving also provides the strength necessary to support large numbers of cultures in bottles and flasks.

2. Lighted. The attachment of lights to shelves produces the added problem of heat and thus it is important this be considered in constructing the shelves. As will be indicated in the discussion on lighting, I prefer to use those fluorescent lights whose ballasts can be mounted away from the tubes (Fig. 11-5, right) to cut down excess heat on the shelves and prevent shading of the tube from the top, as occurs when the ballast is mounted in the tube holder.

Most of the lighted shelves at the Culture Collection of Algae are constructed to use 8 ft fluorescent tubes rather than 4 ft tubes, inasmuch as the same amount of ballasts and wiring are required for both tubes. Thus 8 ft tubes provide space at lower cost (Fig. 11-2). The shelves are made of 1 in angle iron welded at the corners, with a single brace through the center of the shelf. Each shelf is 18 in wide and the length is 8+ ft to accommodate the long tubes. The surface of each shelf is a sheet of 1 in expanded metal grid, spot-welded to the frame. This provides substantial support which will not stretch over years of usage and at the same time allow passage of reasonable percentage of the light from below. The use of expanded metal grid provides a good circulation, which is not possible with glass shelves, and prevents formation of hot pockets.

The fluorescent tubes are mounted in pairs 5 in below each shelf and *ca.* 1 ft apart (Fig. 11-2). With the standard tubes such proximity to the shelves results in an increase of approximately 2–3 °C on the shelf during the period the lights are on, assuming reasonable air circulation. High intensity, General Electric Power-Groove lamps (see Illumination, in C, following) are restricted to the top shelf as they will raise the temperature 8–9 °C.

The circuit coming to the lighted shelves should pass through a time clock and a thermostat switch. This latter is of special importance as a safeguard against overheating of cultures should the cooling of the room fail. By setting the thermostat switch to a non-critical level (*ca.* 10 °C above desired temperature) one can be assured that the lights will not continue to burn and add continued heat to the room if the cooling has failed and the pre-set temperature on the switch has been reached.

3. Lighted rack. The handling of large numbers of test tubes containing liquid can best be effected through the use of a rack so designed as to hold the tubes in rows for easy access (Fig. 11-1, left). This is especially useful for single cell isolation or growth of cultures where it is important to have the tube fully illuminated. The light rack at Indiana University is patterned after that observed many years ago in the Cambridge Collection of Algae established by Professor E. G. Pringsheim.

Fluorescent tubes are mounted in a single row, 6 in apart. The use of 4–8 ft tubes depends on the space available and the need of the investigator. On each side of the fluorescent tubes an open frame is mounted 0.5–1 ft from the tubes (Fig. 11-1). Across this open frame, iron rods ($\frac{1}{16}$ in diameter) are mounted 4 in apart. These rods are threaded on one end and held in place on that end by a wing nut; this allows for periodic tightening of the rod as the frame, if wood, may warp. The 4 in distance between the rods allows the test tubes to hang with some overlap, but not enough to shade the contents (Fig. 11-4). The test tubes are hung on wire hangers made by cutting stovepipe wire (a relatively soft iron wire available at most hardware stores) into 7 in lengths. A 5 in portion of these lengths is wound around a tube slightly smaller than the tube to be used. The remaining 2 in are bent to form a hook. The tube is placed in the coil and the hook placed over the iron rod (see Fig. 11-4). This coiled wire hook holds the tubes tightly, but is easy to slip on or off as desired.

As with the lighted shelves, a time clock and a thermostat switch should be installed in the electrical circuit leading to the light rack.

C. *Illumination*

For routine growth of algal cultures, cool-white fluorescent lights giving an intensity of approximately 350 ft-c (3700 lux) are used. For long term maintenance of cultures lower intensities (100 ft-c; 1100 lux) are used; whereas, some investigations use intensities near 600 ft-c (6500 lux).

1. Lamps. Intensities between 50 and 350 ft-c can be provided using fluorescent tubes of the type commonly employed in room lighting fixtures. In order to minimize the heat problem, it is advisable to employ single pin tubes of the type involving 2 tubes per ballast; the ballast may then be mounted some distance from the shelf or rack on which the algal cultures are to be grown. Two 8 ft, 40 W, cool-white fluorescent tubes will give light of an intensity of approximately 300 ft-c (3200 lux) on a surface 16 inches away. The exact intensity will vary, of course, depending on whether the reading is taken on the surface near the end of the tubes or at a midpoint. The efficiency of fluorescent tubes drops with time, so tubes must be replaced periodically. For intensities of less than 200 ft-c (2200 lux), cultures can be stored on shelves some distance from the light source, thus utilizing the same light source for the continuous turn-over of research cultures and the storage of stock strains.

For high intensity illumination, General Electric Power Groove fluorescent lamps are used. These are present in outdoor lighting at service stations and similar places. The tubes give twice the output of light as ordinary fluorescent lights, but there is also the concomitant increase in the heat output from the lamp surface. This latter point must be taken into consideration where such fixtures are to be installed (as mentioned previously, B.2).

2. Costs. The necessary fixtures, lamps, and ballasts (General Electric Stock) required for a single setup of two tubes (cool-white) are listed below:

No.		Estimated Cost (US)
1	G.E. BALLAST no. 7G1202, 120 volt, 2 lamp ballast for 1500 MA lamps	$30
2	G.E. F96PG17/CW, 1500 MA, 96 in fluorescent lamps	6
1	Radio Interference Suppressor. G.E. no. 6G3751	6
2	Sets each of G.E. ear mounted compressible lamp holders consisting of: 1 each no. ALF822-03 stationary end; 1 each no. ALF822-04 compressible end	8
		$50

3. Window illumination. For maintenance purposes, the light from a northern exposure (Northern Hemisphere) is excellent but it is rare to find building construction available where north windows do not have heating vents or radiators immediately below them. The Culture Centre of Algae and Protozoa, Storeys Way, Cambridge, England (formerly located in the Botany School of Cambridge University), moved into a new building especially designed for the Centre. Special culture rooms were designed to use natural light although during the winter months supplementary lighting must be provided. The Collection at the University of Göttingen, West Germany, also utilizes the low light intensities from north windows in maintaining cultures.

III. CULTURE CHAMBERS – REACH-IN

(J. Stein, personal communication)

Culture chambers may be the commercial products or modified units made by the laboratory personnel. The units are variable depending upon the needs. It is possible to have less variation in temperature and lighting.

A. *Commercial units*

Several companies supply these and the list following is not complete. When inquiring about units, details as to size and usage should be given to the manufacturer. The specifications should include *all* the uses of the unit, including the variables to be controlled along with their ranges and tolerance (see AIBS 1971). Mention of companies does not imply endorsement of the product. Each unit must be judged on its own merit. Units cost from $1200 (US).

1. Controlled Environments, Inc., 601 Stutsman St., Pembina, North Dakota 58271.
   Canada: Controlled Environments, Ltd, 661 Madison St., Winnipeg 21, Manitoba.
2. Environmental Growth Chambers Division, Integrated Development and Manufacturing Co., Box 407, Chagrin Falls, Ohio 44022.
3. Hotpack Corporation, 5086A Cottman Ave., Philadelphia, Pennsylvania 19135.
   Canada: 385 Phillip St., Waterloo, Ontario.
4. Instrument Specialities Co. (ISCO), Box 5349, Lincoln, Nebraska 68505.
5. Lab-Line Instruments, Inc., 15th and Bloomington Ave., Melrose Park, Illinois 60160.
6. National Appliance Co., Box 23008, Portland, Oregon 97223.
7. Percival Manufacturing Corp., Box 249, Boone, Iowa 50036.
8. Sherer-Dual Jet Division, Kysor Industrial Corp., 910 W. Industrial Rd, Marshall, Michigan 49068.

B. *Non-commercial units*

These units are modified from materials available but not primarily designed as growth chambers. Refrigerator units may easily be converted and generally can maintain the temperature within a narrow range. A brief discussion is given in Klein and Klein (1970, Chp. 3, especially pp. 212–18). The lights may be mounted below the individual shelves as described previously (see II.B.2). It is also possible to mount the lights vertically on the sides of the unit or on the door. With vertical lights the amount of light reaching the cultures is not the same.

IV. TRANSFER UNITS

A. *Rooms*

The most versatile sterile rooms for transferring or working with axenic cultures have proved to be small (5 ft wide × 4 ft deep) rooms constructed in the corner or along the sides of an air-conditioned laboratory (Fig. 11-6). The rooms have walls with the upper portion made of glass in order to prevent the feeling of claustrophobia when working in the room for relatively long periods of time. Louvers in the topmost part of the walls provide some escape of the hot air in the room, but do not jeopardize the sterility of the room. A gas jet is provided in the room, and a bunsen burner with pilot light provides a means by which the flame needed for sterilizing can be turned off and on when desired. The pilot flame adds little heat to the room while providing a large enough flame for the pulling of fine glass pipettes used in isolating single cells ('Touch-o-Matic', Bellco 1981, is an example). Electrical outlets are provided so that microscope work needed in handling and purifying some algae can be carried on in the room. The size of the room provides enough space for a laboratory table (5 × 2 ft). For extended periods of work it is suggested that the table be approximately 32 in high so that a comfortable laboratory chair can be used rather than the typical high chair needed for high tables.

A fixture containing a germicidal UV lamp (see Chp. 12, II.F.2) is mounted on the wall above the level of the eyes and arranged so that its rays cannot be seen at any time the person is either standing or seated in the transfer room. To ensure sterility the irradiance must exceed 1000 mJ/cm^2 (Klein and Klein 1970). See Chp. 12, II.F.2, for further details.

The greatest asset towards preventing contamination in handling cultures lies in having such a room in which the surfaces may be kept wiped clean with a damp cloth and thus prevent the dirt and dust found in unprotected areas.

B. *Transfer chambers* (J. Stein, personal communication)

Cost of space may negate construction of sterile transfer rooms. However, transfer units that sit on laboratory benches are satisfactory (see Chp. 4, II.C). The main purpose is to create an area with little dust and movement of air. These units may be made in the laboratory (see Klein and Klein 1970, pp. 330–2); or commercial units (often known as transfer or microbiology hoods) are available from companies as: Labconco, 8810 Prospect St., Kansas City, Missouri 64432; G-F Supply Division, Standard Safety Equipment Co., 431 North Quentin Rd, Palatine, Illinois 60067.

Units that filter air through fine filters are also available. These force air by use of fans (e.g., a bathroom fan) through fine filters (e.g., Volvo Automobile Filter Type PC1). The air which is free of most microorganisms, flows continually over the work surface. These units which may be constructed in the laboratory or are available commercially from such companies as: Conn Environment R & D Corp., Subsidiary of Intertech Enterprises Ltd, New Britain, Connecticut; Germfree Laboratories, Inc, Equipment Manufacturing Div, 2600 SW 28th Lane, Miami, Florida 33133.

A pilot bunsen burner (as described previously, IV.A) is used to sterilize needles, loops, etc. It is necessary to add a protective shield around the flame due to the velocity of air moving through the unit.

A germicidal lamp (see IV.A, previously) may be used to sterilize the surface. Also, wiping the surface with 70% ethanol helps remove sources of contamination (but see Chp. 12, II.A).

V. REFERENCES

AIBS. 1971. Controlled environment enclosure guidelines. *BioScience* **21**, 913–14.

Bickford, E. D. and Dunn, S. 1972. *Lighting for Plant Growth*. Kent State Univ. Press, Kent, Ohio. 222 pp.

Klein, R. M. and Klein, D. T. 1970. *Research Methods in Plant Science*. Natural History Press, Garden City, New York. 756 pp.

Fig. 11-3. Detail of sturdy metal shelving used for storage of stocks.

Fig. 11-4. Detail of lighted rack. Note wire hooks holding tubes on metal rods, to provide maximum use of space with no shading of tubes by row above.

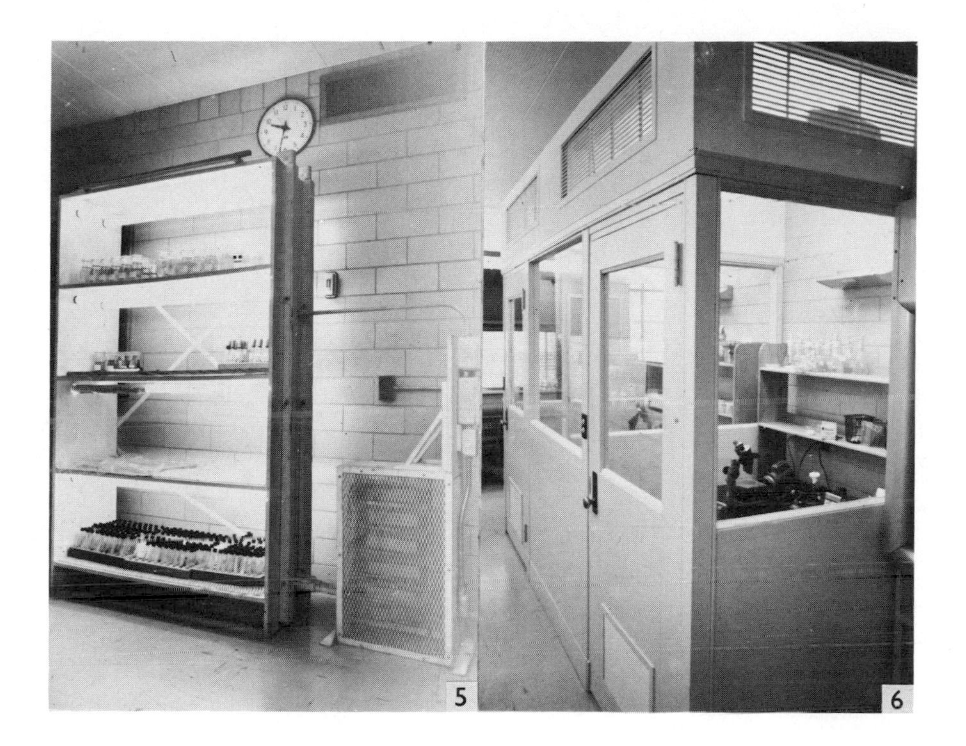

Fig. 11-5. Unit of lighted shelves utilizing 4 ft fluorescent tubes, showing the ballasts mounted on the right in a cage made of expanded metal grid.

Fig. 11-6. Two transfer rooms showing large windows and louvers at top.

12: Sterilization

R. D. HAMILTON

Freshwater Institute, Fisheries Research Board of Canada,
501 University Crescent, Winnipeg, Manitoba R3T 2N6, Canada

CONTENTS

I. INTRODUCTION

At the outset one must decide if strict sterility and the use of aseptic techniques are really essential to one's methods. In many instances it may be that simple chemical cleanliness would suffice. This may be true particularly where sterilization may actually produce toxic substances. However, if the procedure requires sterility, it should be noted that chemical cleanliness forms the basis for most sterilization techniques; the autoclave is not a panacea for all problems.

While the terms *sterilization* and *disinfection* are often assumed to be synonyms, there is a very fundamental difference between them. *Sterilization* is normally reserved for processes which ensure the total inactivation of all microbial life – bacteriocidal (note that these processes do not ensure the termination of all enzymatic activity). However, *disinfection* is taken to mean the reduction of bacterial numbers to some arbitrary 'safe' or 'acceptable' level – bacteriostatic. In this chapter an effort will be made to provide a varied selection of techniques for both sterilization and disinfection.

II. METHODS

A. *Non-volatile disinfectants*

1. Applicability. The use of these compounds is usually confined to the reduction of the microbial load on surfaces and instruments. These agents should be employed in the phycological laboratory only in emergencies or where other methods of reducing the microbial load would be impractical, such as on bench tops and other surfaces.

2. Chemicals, methods. While there is a wide range of non-volatile disinfectants (McCulloch 1945; Reddish 1957; Sykes 1965) available to the researcher, the following compounds are the ones which are most commonly used. All of these agents are solutions which may be applied directly to surfaces or into which instruments may be immersed. The time of application varies widely but is summarized with each disinfectant.

a. Lysol: 3.5% v/v. This is rapidly effective to vegetative cells and quite ineffective to spores.

b. Chlorine (hypochlorite): this would probably be the compound of choice. It is usually used as a solution containing 400 ppm available chlorine (Sykes 1965) or as a solution of hexachlorophene, an organic chlorine compound that may be banned by the Canadian or United States governments. As hypochlorite it is effective on spores in 15–70 min (Sykes 1969). To be effective, organic chlorine compounds such as hexachlorophene require a build-up on surfaces over a period of time.

c. Formaldehyde: concentrations which have been reported range from 0.1 to 10% v/v formalin. However, a 5% solution is considered to be effective at normal temperatures (Sykes 1969). Formaldehyde is reported to require from minutes to 24 h to be effective. It should be noted that very small quantities of organic matter severely restrict the action of this chemical.

d. Mercurial compounds: 0.001–0.005% w/v $HgCl_2$ or sodium ethyl mercurythiosalycilate used at concentrations of 0.01–0.1% w/v. Mercurial compounds are largely bacteriostatic rather than bacteriocidal. They are occasionally used to maintain bacteriostasis in solutions which are to be highly diluted in subsequent use.

e. Alcohol, usually used as 70% isopropanol v/v, or 70% ethanol v/v. Acidified alcohol is also used (70% v/v, pH 2). Alcohols in spite of their wide use in the medical field are not good disinfectants and appear to be used more for effect than efficiency. Acidified alcohol will, however, kill spores within 4 h (Sykes 1969), but it is not in wide use.

3. Problems. As a class, these agents are described as non-volatile. However, it should be noted that at room temperature the use of almost all can be detected by the human nose. The term non-volatile is therefore to be accepted with reservation and those culturing sensitive algal species are well advised to keep stocks divorced from areas in which these compounds are being used. Specific problems are as follows.

a. Lysol in its concentrated form can cause severe chemical burns to the skin. Lysol leaves oily residues on surfaces and instruments. Subsequent autoclaving will result in the chemical contamination of all material in the load as well as the autoclave itself, jeopardizing future loads. Lysol is a common constituent of industrial detergent preparations, and the volatile phenols may affect cultures. It is suggested that the preparation used on laboratory floors be investigated.

b. Hypochlorite preparations may corrode metal instruments. Do not use inorganic chlorine preparations in the presence of ultraviolet lamps, as free chlorine gas may be released and is an extremely serious health hazard.

c. Formaldehyde is quite volatile at room temperature. It causes irritation to the eyes and mucous membranes and can cause damage to the olfactory nerves. Polymerization of formaldehyde will cause a white film to appear on surfaces.

d. Mercurial compounds are difficult to remove from glassware. Human tolerances of mercury are still being debated but the currently accepted level in foods is as low as 0.5 ppm. This may be a standard of convenience for mercuric chloride is considered to be 'infinitely toxic' to fish (less than 0.01 ppm, Bligh 1970).

e. Acidified alcohol may corrode metal instruments.

B. *Volatile disinfectants*

1. Applicability. By definition these compounds are classed as disinfectants yet two of the most commonly used substances could be classified as sterilizing agents for all practical purposes. These are ethylene oxide and β-propiolactone. The former compound is now widely used for the sterilization of disposable plastic material.

2. Chemicals, methods

a. Ethylene oxide: this is available either as liquified gas in the pure form or as a mixture of *ca.* 12% ethylene oxide with *ca.* 88% dichlorodifluoromethane ('Cryoxcide': product of Ben Venue Laboratories, Bedford, Ohio; 'Pennoxide': product of Pennsylvania Engineering Co. Philadelphia, Pennsylvania).

The mixtures are designed to be used with specialized gas-tight enclosures (a desiccator or polyethylene bag with walls 4 mm thick is quite satisfactory). When using the pure form, a brine ice bath, glass ampules, and an ampule-sealing torch are needed. If commercial handling equipment is available follow the manufacturer's instruction. If using the pure gas, chill the cylinder of gas, delivery line and ampules in the ice bath. Deliver the requisite amount of liquid into each ampule and *quickly* heat seal. The ampules may be stored at room temperature but must be chilled to below 5 °C before use.

Wrap the materials as for autoclaving (see C.2.c, following) and place in any gas-tight container. If working with commercial gas mixtures, it is helpful to evacuate the chamber before filling with gas. If working with pure gas, the chilled ampule is opened and introduced into the chamber which is then quickly sealed. Under these latter conditions precautions should be taken to ensure that the volume of gas so released is not enough to cause the chamber to leak. If plastic bags are used, they may be evacuated and the ampule broken inside the sealed bag through mechanical pressure. The final concentration of ethylene oxide should be *ca.* 750 mg/l of chamber volume (Bruch 1961).

The chamber should remain undisturbed for at least 8 h at 37 °C, or 24 h at room temperature. If necessary the relative humidity should be adjusted to 25–50 % (Bruch 1961).

After sterilization is complete the container may be opened in a fume hood or it may be re-evacuated through the use of a water aspirator. Ethylene oxide is quite soluble in water and this procedure effectively disposes of any free gas which might remain.

b. β-propiolactone: this is usually sold as a 96–99 % pure solution. Its use generally is restricted to aqueous solutions to which it is added directly.

This liquid is added to the solutions to be sterilized to a final concentration of 0.2–0.5 % v/v. Himmelfarb, Read and Litsky (1961) used a 0.2 % v/v solution to sterilize 20 % sugar solutions by incubating treated solutions at 37 °C for 2 h period. The mixture was shaken every 10–15 min. Solutions so sterilized were diluted 1:40 in culture media on the following day.

c. Chloroform: analytical reagent grade chloroform is usually used. Chloroform is definitely a disinfectant for while it kills some organisms very quickly its activity with regard to other organisms is less consistent (Meynell and Meynell 1970). It has been used at a final concentration of 1.5 % v/v with concentrated culture media. The media is then diluted for subsequent use. Chloroform vapors have also been used to sterilize solid surfaces.

3. Problems

a. Ethylene oxide is very reactive chemically so its effect upon media is to some extent unknown. It will not penetrate some materials (Sykes 1969). It is explosive and toxic. It hydrolizes in some materials and may persist for long periods of time (Bruch 1961).

b. β-propiolactone is a very powerful lachrymator (tear-producer). It hydrolyzes in solution to give β-hydroxy-propionic acid (Hoffman and Warshowsky 1958).

c. Prolonged exposure to chloroform can cause tissue damage, and the effects are cumulative. Concentrations causing damage can be detected by odor.

c. *Moist heat*

1. Applicability. This is the most widely used method of producing sterility. It is applicable to all media and materials which will withstand a temperature of 100–120 °C.

2. Equipment, methods

a. Boiling water bath: fill the bath with distilled water and raise to 100 °C. Place media or instruments in the bath, allow them to come to temperature, then begin timing and allow the treatment to proceed 5–10 min. This will ensure the death of all vegetative cells but will not affect spores. This method should be used only in emergencies. Its efficiency may be increased by the use of a 2 % w/v solution of sodium carbonate instead of distilled water, but this solution may corrode instruments.

b. Commercial steamer: this may be used in two ways, either to produce a single exposure for 90 min, or to provide intermittent exposures of 30 min duration on three successive days (steaming or 'tyndallization', of some workers). The latter process kills the vegetative cells produced by spores which germinate in the intervals between treatments. In order to ensure that media attains temperature rapidly it is strongly recommended that media volumes should not exceed 500 ml. If tyndallization is used, the containers must be cooled to room temperature for at least 18 h between successive steam treatments. An autoclave heated to 100 °C (no build-up of pressure) may be used as a steamer.

c. Autoclave or pressure cooker: sterilization may be accomplished by autoclaving at 121 °C in pure saturated steam at 15 lb/in^2 above atmospheric pressure. It is important to ensure that all air is removed from the chamber before timing is commenced. If this is not done sterilization will not be attained (Meynell and Meynell 1970; Sykes 1969). The removal of air is accomplished automatically in commercial autoclaves. However, if a pressure cooker is being used, do not close the escape valve until a steady stream of pure steam is observed. Once the valve is closed the pressure is allowed to reach 15 lb/in^2 at which point timing is begun.

In order to ensure sterility the steam must penetrate the materials, as 121 °C by itself will not sterilize. Therefore wrap the instruments in materials which permit easy access of steam. Unwaxed Kraft paper is considered the material of choice. Aluminium foil is in common use but it should be noted that it is quite impermeable to steam except by reason of poor wrapping procedures. Its use is, therefore, not recommended.

The time required for the attainment of sterility depends upon the individual volumes of media to be treated. An autoclave may be purged of air and its gauges may indicate that it has reached operating temperature and pressure without the interior of the load itself being more than warm to the touch. A volume of 100 ml requires 10 min to reach sterilizing temperature, while 2 l requires 20 min and 5 l requires 35 min (Meynell and Meynell 1970).

3. Problems

a, b. Boiling water bath and commercial steamer are not effective at high elevations (above 6000 ft).

c. Autoclave or pressure cookers may cause precipitation in some media, notably those prepared with natural or artificial seawaters. This precipitate will contain almost all the phosphate originally present in the media. The precipitates may be reduced by the addition of sterile phosphate, trace metal and silicate solutions after autoclaving. In addition, some reduction is observed if the media is autoclaved in small lots.

Solutions containing disinfectants, particularly formalin, should not be autoclaved for fear of contaminating not only the load but also the autoclave itself. 'Water conditioners' added to steam supplies may contaminate media (Kordan 1965). Certain plastics may become toxic upon autoclaving or may release toxic products into associated media. Indeed all plastic, cotton or cheesecloth closures should be tested for toxicity after autoclaving (see Chp. 14 on toxic materials).

D. *Dry heat*

1. Applicability. Sterilization by dry heat requires the use of temperatures which are higher than those employed during the moist heat methods. Therefore, its use is usually restricted to heat stable materials such as glassware.

Hot-air sterilization has the added advantage that if a high enough temperature is employed, there is a complete destruction of most organic matter. This is of benefit to those employing bioassay procedures or attempting to culture forms which are particularly sensitive to trace organic contaminants.

2. Equipment, methods

a. Bunsen burner, propane torch, alcohol lamp or torch: heating to incandescence in a flame is the choice method for sterilization of inoculating loops or needles. It may also be used for forceps and other metal instruments which will not be damaged by heat. If an alcohol lamp is being used care must be taken to ensure that adequate heating is obtained as the alcohol flame is cooler than that obtained from a Bunsen burner or propane torch.

Dipping in alcohol, followed by flaming, is frequently employed for instruments which would be damaged by red heat. This method does not necessarily ensure sterility.

b. Hot-air oven: this must have adequate heating-up time, and air circulation and be thermostatically controlled within 1 °C. Load a cold oven with the dry materials to be processed. Do not pack the materials in tightly, but leave space between each item of load. Pipettes and tubes may be plugged with non-absorbent cotton wool. Sterilization is obtained within 2 h of the load (not the oven) reaching 160 °C. For baking of glassware for vitamin bioassays temperatures of up to 400 °C are recommended. Again, 2 h at these high temperatures appears to be adequate (Carlucci and Silbernagel 1966). Glasswool must be substituted for cotton wool if glassware is to be baked.

3. Problems

a. Flaming causes boiling in the initial stages resulting in spattering of material carried on the inoculating loop or needle. This spattering can cause aerosol formations which contain living cells.

b. Syringes which have metal parts may be damaged by dry-heat sterilization as well as any other instruments constructed of materials which have differing coefficients of expansion. Electric or gas-fired hot-air ovens should provide identical temperatures throughout the whole oven. In the gravity convection ovens which are commonly used this even distribution of heat seldom occurs. It is important to understand that the loading of materials markedly affects heat distribution and that ovens and pipette cannisters must not be overloaded. Equipment should not be closely packed but spaced upon the shelves. Sykes (1969) quotes some experimental data which shows the relation of load size and arrangement to rate of heating.

E. *Filtration*

1. Application. Filtration is used for the production of sterile, as well as particle-free solutions which might be affected by heat treatment. It is also used to sterilize air supplies for cultures and for the production of particle-free air for clean rooms.

2. Equipment, methods

a. Sintered glass disks of ultra-fine porosity: these are used with suitable holder, vacuum flask and a vacuum pump with gauges. A convenient integrated all-glass filter unit is marketed as the Morton filter apparatus (Corning 33990).

Plug the side arm with non-absorbent cotton wool or glass wool. Separate the ground glass joints slightly by means of a fine wire or wick of filter paper. Wrap unit in Kraft paper and sterilize in a hot-air oven (D.2.b, preceding). If units are to be baked at high temperatures they may be wrapped in aluminium foil and the use of paper or cotton products must be avoided.

Filtration is begun under $\frac{2}{3}$ atmosphere. Should less than $\frac{1}{3}$ atmosphere be required to produce flow, the filter is considered clogged and should be replaced. High vacuum systems can disrupt cells. After filtration is completed the receiving flask should be allowed to reach equilibrium with the atmosphere by permitting a gentle flow of air through the cotton wool filter.

Cleaning of glass filters may be accomplished by inverting the unit over a rubber stopper which is connected to a vacuum source plus a large volume trap. Distilled water may then be drawn backwards through the filter. Should this procedure prove inadequate, soaking overnight in a dilute solution of nitric acid usually clears the filter. After using nitric acid rinse with distilled water as described previously. Use of chromic acid cleaning solution should be avoided, as it is impossible to render treated units unquestionably non-toxic. Chromic acid may be acceptable under situations in which the individual investigator has demonstrated a lack of toxicity to his own satisfaction.

b. Membrane filters of 0.5 μm or less: these should be of a size suitable to the amount and type of liquid being filtered. In addition, suitable filter holders, vacuum flasks or other receivers and a vacuum/pressure pump equipped with gauges are needed. (Pressure vessels are useful for some procedures.) Membrane filter units are available from: plastic: Nalge Co., Nalgene Labware Division, 75 Panorama Creek Dr., Rochester, New York 14625; glass or metal: Colab Laboratories, Inc., 3 Science Rd, Glenwood, Illinois 60425; Gelman Instrument Co., 600 S. Wagner Rd, Ann Arbor, Michigan 48106: Millipore Corp., Ashby Rd, Bedford, Massachusetts 01730: Sartorius-Membranfilter GmbH., P.O. Box 142, 34 Göttingen, West Germany.

Plastic membrane filter holders may be gas sterilized. Glass or metal membrane filter holders are assembled and autoclaved with the membrane, and pre-filter if necessary, in place. The membrane filters must be handled with blunt, flat forceps. The units should be completely dry and wrapped in Kraft paper or its equivalent. Attached hoses should be uncoiled and all stopcocks open. If it is necessary to re-tighten the sterile units, this should be deferred until after the filter has been wetted with the solution being filtered.

Large volumes of media should be filtered through large diameter membrane filters under pressure rather than under vacuum. While vacuum can be used, the use of pressure simplifies the collection of large volumes and eliminates the possibility of contamination during the equilibrium of a vacuum collector with the atmosphere. Very small volumes of solution can also be pressure filtered by means of membrane filtration units which can be attached to a syringe (produced by companies listed previously). Intermediate volumes and most laboratory filtrations are carried out using all glass vacuum operated membrane filtration units. In no instance should more than 5–10 lb pressure be imposed upon a membrane filter.

It is strongly recommended that investigators contemplating the use of membrane filters read the excellent manuals published by the Millipore Filter Corporation (Millipore 1964).

Process air for cultures may be conveniently sterilized by the in-line use of small diameter membrane filter units such as those which can be attached to a syringe (preceding). The air input to these filters must be absolutely dry or immediate loss of efficiency will occur.

For an outline of the procedures and problems involved in the treatment of air see the review by Elsworth (1969).

3. *Problems*

a. Sintered glass filters are notoriously hard to clean. They are best discarded when mild acid treatment will not clean them; yet, they are very expensive. They can only be used for very small quantities of solution.

b. Membrane filters have been known to leach compounds into the filtrate (Parker 1967 and references). This may be overcome in part by boiling the filters in several changes of distilled water. Filters so treated must be autoclaved and stored under water or they will become brittle. They must be aseptically assembled into the holder after sterilization. This may be extremely difficult, if not impossible, with the large diameter filters required to handle large volumes of media.

Many disposable (plastic) integrated membrane filter systems are currently being marketed. They have much to recommend them in matters of convenience but the investigator should be aware that impurities have been known to leach from such units (Simpson 1966; also see Chp. 14).

F. *Radiation*

1. Applicability. While X- and γ-radiation find application in industrial situations for the sterilization of a wide variety of products, there are few practical applications in the average laboratory. In addition, the safety hazard association with these treatments renders them extremely dangerous in unskilled hands. They are therefore not recommended. Such ionizing radiation is employed in the treatment of disposable plasticware and some pharmaceuticals. The investigator should be aware that the accepted sterilizing dose is based upon the assumption that the initial concentration of bacteria is quite low. In industrial environments this may not be always true, and it has always been our personal practice to subject samples of any batch of disposable equipment to a suitable test for sterility.

Ultraviolet radiation finds wide application in the laboratory, particularly in the disinfection of solid surfaces. Its action is uncertain, especially when penetration is required (see Sykes 1969 for a discussion). While the use of ultraviolet radiation is usually restricted to the disinfection of table top surfaces in culture transfer rooms, its combined chemical and disinfecting action has some unique possibilities in the preparation of culture media (see Hamilton and Carlucci 1966).

2. Equipment, methods. The effective lethal wavelength is in the range of 240–280 nm. The ratings of the lamp required will depend upon the application. Usually a 20–40 W lamp is deemed appropriate for transfer hood or transfer room applications. The procedure suggested by Hamilton and Carlucci (1966) requires a 1200 W lamp equipped with a cooling fan and quartz tubes.

a. Transfer hoods and rooms: ensure that all surfaces are directly exposed to the radiation for UV has a very poor penetrating power. The presence of a flask in a transfer hood is enough to prevent adequate disinfection for the area in the shadow of the flask will be completely unaffected. If dealing with a whole transfer room, place the lamps beneath the tables as well as above them. The duration of exposure should be determined for each individual situation.

b. Culture media: the natural water used to make the media is dispensed into quartz test tubes and irradiated for 2–4 h. All subsequent manipulations are done under aseptic conditions, including the addition of the nutrients required.

3. Problems. Ultraviolet radiation can cause severe eye damage, therefore protective glasses should be worn. With the very high power UV sources used by Hamilton and Carlucci (1966) skin tanning can occur at remarkable distances from the lamp so adequate shielding must be provided to ensure the safety of laboratory personnel. Certain people appear to intensely dislike the ozone which is produced during UV treatments.

Quartz tubes are easily marked by finger prints and are easily scratched. Scratches particularly can drastically affect the efficiency of UV penetration. Efficient exposure times can be twice as long for old tubes as for new ones.

III. REFERENCES

Bligh, E. G. 1970. Mercury and the contamination of freshwater fish. *Fish. Res. Bd Canada.* Manuscript Rept. Ser. 1088.

Bruch, G. W. 1961. Gaseous sterilization. *Ann. Rev. Microbiol.* **15**, 245–62.

Carlucci, A. F. and Silbernagel, S. R. 1966. Bioassay of seawater. I. A ^{14}C uptake method for the determination of concentration of vitamin B_{12} in seawater. *Can. J. Microbiol.* **12**, 175–83.

Elsworth, R. 1969. The treatment of process air for deep cultures. In Norris, J. R. and Ribbons, D. W., eds., *Methods in Microbiology* **1**, 123–36. Academic Press, New York.

Hamilton, R. D. and Carlucci, A. F. 1966. Use of the ultra-violet irradiated seawater in the preparation of culture media. *Nature* **211**, 483–4.

Himmelfarb, P., Read, R. B., Jr and Litsky, W. 1961. An evaluation of β-propiolactone for the sterilization of fermentation media. *Appl. Microbiol.* **9**, 534–7.

Hoffman, R. K. and Warshowsky, B. 1958. β-propiolactone vapor as a disinfectant. *Appl. Microbiol.* **6**, 358–62.

Kordan, H. A. 1965. Fluorescent contaminants from plastic and rubber laboratory equipment. *Science* **149**, 1382–3.

McCulloch, E. C. 1945. *Disinfection and Sterilization*, 2nd ed. Henry Kimpton, London. 472 pp.

Meynell, G. G. and Meynell, E. 1970. *Theory and Practice in Experimental Bacteriology*, 2nd ed. Cambridge Univ. Press, London. 346 pp.

Millipore Filter Corp. 1964. *Sterilizing, Filtration and Sterility Testing*. Application Data Manuals – ADM 20.

Parker, B. C. 1967. Influence of method for removal of seston on the dissolved organic matter. *J. Phycol.* **3**, 166–73.

Reddish, G. F. 1957. *Antiseptics, Disinfectants, Fungicides and Chemical and Physical Sterilization*, 2nd ed. Henry Kimpton, London. 975 pp.

Simpson, L. 1966. Toxic impurities in Nalgene filter units. *Science* **153**, 548.

Sykes, G. 1965. *Disinfection and Sterilization*, 2nd ed. E. and F. N. Spon Ltd, London. 486 pp.

Sykes, G. 1969. Methods and equipment for sterilization of laboratory apparatus and media. In Norris, J. R. and Ribbons, D. W., eds., *Methods in Microbiology* **1**, 77–121. Academic Press, New York.

13: Preservation by freezing and freeze-drying

OSMUND HOLM-HANSEN

Institute of Marine Resources,
University of California, San Diego, La Jolla, California 92037

CONTENTS

I. INTRODUCTION

The primary purpose of freezing or lyophilizing (freeze-drying) cells as described here is to maintain algal cultures in a viable state for long periods of time. Advantages of maintaining cultures in the frozen or dry state include: (1) a great reduction in time, equipment, and space required to maintain algal stock collections as compared to that required to maintain them by routine transferring; (2) the genetic stability of the preserved strains over long time periods; and (3) in the instance of dried cultures, there is no refrigeration required during storage. The ability of dried cultures to survive at room temperature for many years greatly facilitates storing large numbers as well as ease of distribution to other laboratories.

During freezing and freeze-drying of organisms there are many physical and chemical factors which may cause cell damage. Studies on diverse microbial cells have demonstrated that no one experimental procedure is satisfactory for all. The proper conditions for freezing or drying any particular algal species must often be determined empirically. The following introductory comments on freezing and freeze-drying will indicate the main factors which must be considered.

When algal cells are subjected to freezing, low-temperature storage, and subsequent thawing, they are exposed to a sequence of chemical and physical stresses, any one of which may result in loss of cell viability. The major dangers which may cause cellular damage are as follows: (1) decreasing temperatures down to and often a few degrees below the suspension freezing point may alter overall metabolic patterns due to differential effects of temperature on rates of various enzyme reactions; (2) formation of ice crystals subjects cells to physical stresses, particularly if the rate of freezing is sufficiently rapid to result in intracellular ice crystals; (3) growth of intercellular ice crystals causes all free water to diffuse out of cells and change to the crystalline state with extreme dehydration of the cells resulting; (4) with dehydration, there are increasing salt concentrations in the cell that can be lethal by denaturation of most proteins, solubilization of lipoproteins, and alteration of cell membrane permeability; (5) very low temperatures (below −100 °C) are required for storage to reduce metabolic reactions to negligible levels as cells stored at higher temperatures (about

–30 °C) have in time a steady and measureable decline in viability; and (6) thawing a frozen sample and returning it to room temperature exposes the cells to potentially lethal effects of changing electrolyte concentrations and temperature effects on enzymatic rates.

In the process of lyophilization, the sample is first frozen and then all the ice is removed by sublimation from the crystalline state, without passing through the liquid state. The vapor pressure of ice is directly proportional to the absolute temperature; the gaseous water molecules which are in equilibrium with the ice crystals at the temperatures commonly employed for lyophilization (–20 to –40 °C) can easily be removed from the system by vacuum pumping, chemical desiccants, or by cold traps. When all the ice has been removed from the sample, the cells are devoid of water except for the small amount of 'bound' water which is firmly held by hydrophilic ions and molecules. As lyophilized cultures are hygroscopic and also show some loss in viability when exposed to molecular oxygen, they are generally sealed in glass ampules while under high vacuum. Such samples show little or no metabolic activity, and are capable of maintaining viability for many years during storage at room temperature.

II. EQUIPMENT

Work on the freezing and freeze-drying of algal cultures uses much of the laboratory glassware and chemicals found in most laboratories. One must have the apparatus and chemicals for growing algae, for concentrating algal samples by filtration or centrifugation, and for quantitative estimates of cell viability. Other items which are generally required are listed following.

A. *Freezing*

1. Cryoprotective agents such as glycerol, dimethyl sulfoxide, or dried milk solids.
2. For short-term freezing of cells it is convenient to use any deep-freeze with a temperature of –20 °C or lower; glycol/dry ice mixtures for temperatures down to –78 °C; or liquid nitrogen at –196 °C. For long-term preservation of cells in the frozen state, a deep-freeze at –25 °C may be satisfactory for some cultures, but for most work special cryogenic refrigerators with temperatures below –40 °C must be employed.

B. *Freeze-drying*

Equipment can be secured through any of the following dealers: Bellco Glass Inc., 340 Edrudo Rd, Vineland, New Jersey 08360; Dental Vacuum

Inc., Cherry Hill Industrial Center, Cherry Hill, New Jersey 08034; Edwards High Vacuum. Inc., 3279 Grand Island Blvd, Grand Island, New York 14072 (430 South Service Rd W., Oakville, Ontario, Canada); Vacudyne Corp., 371 E. Joe Orr Rd, Chicago Heights. Illinois 60411; VirTis Co., Inc., Route 208, Gardiner, New York 12525.

1. A vacuum system containing:

a. a vacuum pump capable of maintaining at least 100 μm Hg and preferably close to 1 μm Hg;

b. a vacuum gauge such as the McLeod gauge;

c. a cold-trap of sufficient size that the ports will not be obstructed by the amount of water to be removed from the samples (a glycol/dry ice mixture is convenient to use in the trap);

d. a vacuum chamber for holding the tubes during the drying process.

2. Small disposable Pyrex glass tubes or ampules which can be sealed under high vacuum.

3. A vacuum manifold for holding any convenient number of the glass tubes during the sealing process.

4. A gas–air or gas–oxygen cross-fire torch for sealing the tubes while under vacuum.

5. A spark-coil for testing sealed lyophil tubes for vacuum.

III. TEST ORGANISMS

There has not been sufficient work with freezing of algal cells for any particular strain of alga to be employed as a 'test organism'. It is suggested, however, that two strains which might be useful in this regard are: (1) the Cyanophyceae *Nostoc muscorum* – Wisconsin strain 1013; and (2) the Chlorophyceae *Chlorella pyrenoidosa* – Wisconsin strain 2005. Quantitative determination of viability may be made with the latter organism, whereas only semi-quantitative data may be obtained from the filamentous *N. muscorum*. (Available from Botany Department, University of Wisconsin, Madison, Wisconsin 53706. Other clones available from IUCC and Cambridge should be tested.)

IV. PROCEDURE

A. *Freezing*

1. Freshwater species

a. Suspending medium – dense cultures of algae may be frozen directly in the nutrient solution in which they were growing; or, the cells may be concentrated by filtration or centrifugation and resuspended in distilled water with or without the addition of organic adjuvants. Cryoprotective agents that are very useful in decreasing cell mortality during freezing and thawing include glycerol (5–10% v/v), dimethyl sulfoxide (10% v/v), and dry milk solids (10% w/v). Aliquots of the algal suspension should be pipetted into convenient sized test tubes or flasks, which are then stoppered with cotton plugs or screw caps. If axenic stock cultures are being frozen for preservation, all glassware must be sterile and aseptic techniques used throughout.

b. Freezing – the rate of cooling of the algal suspension is extremely important with regard to the number of cells surviving the freezing process. Some cultures show maximum survival with slow cooling rates (*ca.* 1 °C/min) while other cells survive best with very rapid freezing (over 1000 °C/min). It is suggested that relatively slow freezing rates be employed first, and if cultures do not show good survival, then faster cooling rates used. The simplest method of freezing is to place the sample tubes into a deep freeze at a temperature between −20 and −30 °C. Faster cooling rates may be obtained by immersion of sample tubes in glycol/dry ice mixtures (down to −79 °C) or in liquid nitrogen (−196 °C). When freezing cell suspensions, one must be alert to the possibility of getting supercooled suspensions down as low as −15 °C.

c. Storage temperature – degradative enzymatic reactions continue to occur in all frozen cells, resulting in a continual loss of viability with time. The rates of most chemical reactions decrease exponentially with decreasing temperatures, so that great differences in metabolic activity can be expected in cells stored at −20 °C as compared to −100 °C. Nearly all available data indicate that the lower the storage temperature, the greater the viability of the cells. For very long periods of storage, it is desirable to use temperatures below the eutectic point (lowest solidifying point of any component) of the mixture; in cells treated with glycerol or dimethyl sulfoxide, the temperature should be low enough to cause complete solidification (about −90 °C). There should be no significant chemical changes occurring at liquid nitrogen temperature, with most evidence indicating there is little or no detectable decline of viability of cells during many years storage at −196 °C. For relatively short-term storage (less than one year) of cultures, the use of a deep-freeze at *ca.* −25 °C might be satisfactory for most. Storage of cells for more than one or two years requires cryogenic refrigeration equipment to maintain the temperature below −100 °C. If cells are to be stored for long periods of time at very low temperatures (e.g., −196 °C), samples should first be frozen at a relatively slow cooling rate, then transferred to the storage container after complete freezing.

d. Rehydration – the rate of warming during the thawing process is also very important in regard to cellular damage incurred during this period of changing temperatures. For most algal cultures it has been found satisfactory to merely place the frozen sample in a water bath at room temperature.

e. Testing for viability – although many investigators report viability on the basis of plasmolytic observations, the uptake of vital stains, or by microscopic examination, it is recommended that viability be determined by actual reproduction of the cells. If quantitative data are desired regarding the ability of cells to survive freezing and storage at low temperatures, the number of cells in the initial suspension must be determined by electronic particle counting techniques (see Chp. 22 by Parsons), or by direct microscopic counting in a haemacytometer or other counting chamber (see Chp. 19 by Guillard). For most algae the number of viable cells in the thawed sample can be obtained by serial dilutions of the sample and the use of agar pour plates or streaking techniques with subsequent counting of algal colonies (Holm-Hansen 1963*a*). A few species of Cyanophyceae may be tested for viability by using agar plate techniques (Allen 1968). Most Cyanophyceae cannot be quantitatively plated out in this fashion, however. A semi-quantitative estimate of viability can be obtained by obervation of time length for visible growth in the nutrient solution that has been inoculated from the thawed suspension.

2. Marine species. The methodology described above for freshwater cultures should also be applicable to marine species. The relatively high salt content of seawater, however, may result in greater cell destruction through osmotic effects during the freezing process. In order to minimize such effects, it is suggested that either dense suspensions be frozen and/or the cells be resuspended in diluted seawater (*ca.* 80% seawater). It might also be beneficial to use increased concentrations of the cryoprotective agents mentioned before.

B. *Freeze-drying*

1. Freezing. Before attempting to freeze-dry a culture, it is advisable to first find the freezing conditions which result in high cell viability. To do this, algal suspensions are prepared as described preceding, either with or without addition of cryoprotective agents, before being pipetted into the lyophil tubes or ampules. The use of dried milk solids has proven to have a very beneficial effect on cell viability; glycerol cannot be employed as it cannot be removed from the cellular material by vacuum drying, but can be removed only by special low-temperature substitution procedures. The

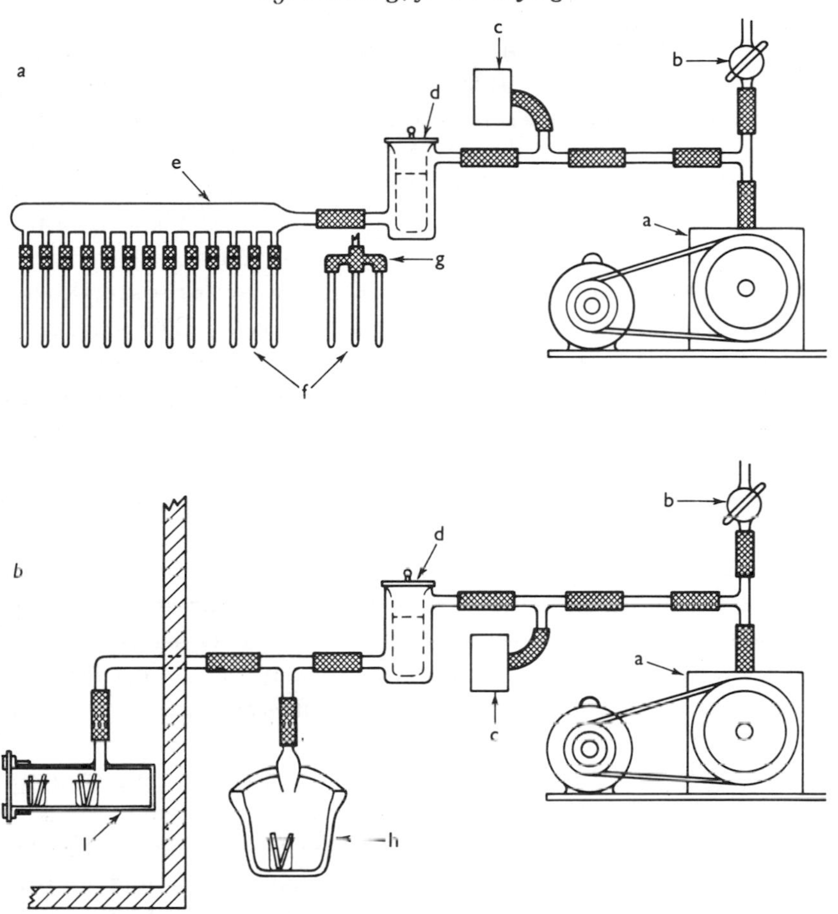

Fig. 13-1. Arrangement of vacuum apparatus for the freeze-drying of algal samples: *a*. lyophil tubes attached to a manifold; *b*. lyophil tubes placed in vacuum chambers. (a) vacuum pump; (b) vacuum release; (c) vacuum gauge; (d) cold trap; (e) manifold for 36 tubes; (f) lyophil tubes; (g) end view of manifold; (h) desiccator with tubes; (i) plexiglas vacuum chamber inside deep freeze.

size of the lyophil tube is not critical, but a tube with a diameter of 6–10 mm is convenient to use and easy to seal under vacuum. The tubes should be at least 100 mm long, with the algal suspension occupying only the bottom 10 or 20 mm. Tubes should be stoppered with a fairly loose cotton plug as a tight cotton plug slows down the rate of drying significantly. All samples are then frozen at the rate found to be optimum for that culture.

2. *Drying*. During the drying period it is convenient to have the tubes on a manifold (Fig. 13-1*a*) or in a vacuum chamber (Fig. 13-1*b*). Samples must

not be allowed to thaw during this drying period. If the tubes are on a manifold, they must be kept frozen by refrigeration or immersion of the ends of the tubes in a cooled glycol/dry ice bath. When samples are frozen and then quickly transferred to a vacuum chamber or desiccator (h in Fig. 13-1*b*) at room temperature, the vacuum pump must be started immediately upon placing the samples in the chamber. Thawing of these samples is prevented only by a sufficient rate of sublimation of ice from the samples. It is important to maintain a high vacuum over the samples until all ice has been removed and the cellular material is dry. This stage can be estimated by visual examination, but it can also be determined by (a) observing the degree of vacuum in the system; or (b) following the temperature of one of the samples which has a small thermistor frozen into it.

3. *Sealing of tubes*. As lyophilized samples must be protected against molecular oxygen and atmospheric moisture, samples are generally sealed while under high vacuum. Though there are some commercial units which do this operation automatically with glass vials and rubber and metal tops, the easiest way of doing this in the laboratory is to seal the Pyrex tubes or ampules with a cross-fire torch. If the samples have been dried directly on a manifold (Fig. 13-1*a*) the tubes are sealed one at a time by equal heating on all sides with the cross-fire torch. When the tube is melting and tending to seal itself, the lower end of the tube is grasped (it is recommended that you use a glove in case any glass should drip down) and slowly twisted and pulled downward. The result should be that both the lyophil tube and the glass stub remaining on the manifold are sealed and have no air holes. If air does leak into the manifold the faulty glass plug must be replaced with a leak-proof plug. In the event of any such leak, do not seal any more tubes until the system has been completely evacuated again. For tubes dried in a vacuum chamber, they must be removed, placed on the manifold, and evacuated for at least 10 min before sealing as above. The sealed sample tubes are then tested with a spark-coil (high frequency generator) to detect any tubes which may have air leaks. A tube with a vacuum will show a fluorescent-type discharge throughout the interior of the tube, while a tube with an air leak will not show any visible light.

4. *Storage*. Any residual metabolic activity in lyophilized algal cultures is so low that samples may be stored at room temperature. Storage at lower temperatures most likely will result in greater viability with time, but the decline of viability with time in dried samples is so slow that refrigeration is not essential.

5. Rehydration. To rehydrate a lyophilized sample, the Pyrex tube is scored with a file, snapped open, and the contents emptied into sterile nutrient medium at room temperature. In working with axenic cultures, the tube should be first wiped with alcohol. Any residue left in the tube may be rinsed out with nutrient solution by use of a capillary Pasteur-type pipette. The inoculated flasks are then placed in the proper light and temperature conditions as required by cultures.

V. SAMPLE DATA

The Cyanophyceae *Nostoc muscorum* survives under nearly all freezing conditions, though slow freezing results in more viable cells than rapid freezing. *Chlorella pyrenoidosa* (Chlorophyceae) also is a fairly hardy organism regarding ability to survive freezing, showing viabilities of 41%, 33%, 26% and 5% after being frozen at −25, −30, −70, and −196 °C, respectively.

Nostoc muscorum shows remarkable abilities to survive freeze-drying, having survived in over 1000 samples which have been tested over a five year storage period. There has been no detectable decline of viability of these samples with time. *Chlorella pyrenoidosa*, as with nearly all the eukaryotic algae tested, does not survive lyophilization as well as most bluegreen algae. Usually less than 1% of the cells survive the freeze-drying process; during five years of storage at room temperature the number of viable cells declined to about 1% of that occurring immediately after freeze-drying.

Additional data for *N. muscorum* and *C. pyrenoidosa*, as well as for a variety of other algae, are reported in the papers by Holm-Hansen (1963*a*, 1963*b*, 1964, 1967), Burns (1964), Leibo and Jones (1963), Daily and McGuire (1954), Watanabe (1959), and Hwang and Horneland (1965).

VI. POSSIBLE PROBLEMS

As stated previously, the chemical and physical stresses on a cell during the freezing process are very complex and the optimum conditions for any particular culture must be ascertained by trial and error. Cellular damage caused by temperature effects on rates of chemical reactions should be minimized by ultra-rapid freezing; such rapid freezing, however, is apparently quite damaging in intracellular ice crystal formation. The freezing conditions causing least cellular death thus reflects an interaction of physiological stresses, all of which may respond quite differently to decreasing temperatures.

In the freeze-drying of cultures, one area of particular interest and importance involves the length of the drying period between the time when all ice crystals have been removed and when the tubes are sealed. During this period there is evidence that varying amounts of the residual 'bound' water may be removed from the cells, particularly if the samples are allowed to warm to room temperature after ice crystal removal. The effects on cell viability of such prolonged drying under elevated temperatures and high vacuum are not fully known; some data indicate cell damage, whereas other data indicate beneficial effects on viability and temperature sensitivity by this secondary drying period.

For more detailed information on all aspects of freezing and freeze-drying, refer to the reference works of Meryman (1960, 1966), Harris (1954), Mazur (1970), Smith (1961), and Troshin (1967).

VII. ACKNOWLEDGMENT

This work was supported in full by the United States Atomic Energy Commission, Contract no. AT(11-1)GEN 10, P.A. 20.

VIII. REFERENCES

Allen, M. M. 1968. Simple conditions for growth of unicellular blue-green algae on plates. *J. Phycol.* **4**, 1–4.

Burns, M. E. 1964. Cryobiology as viewed by the microbiologist. *Cryobiology* **1**, 18–39.

Daily, W. A. and McGuire, J. M. 1954. Preservation of some algal cultures by lyophilization. *Butler Univ. Botan. Studies* **11**, 139–43.

Harris, R. J. C. 1954. *Biological Applications of Freezing and Drying*. Academic Press, New York. 415 pp.

Holm-Hansen, O. 1963*a*. Viability of blue-green and green algae after freezing. *Physiol. Plant.* **16**, 530–40.

Holm-Hansen, O. 1963*b*. Effect of varying residual moisture content on the viability of lyophilized algae. *Nature* **198**, 1014–15.

Holm-Hansen, O. 1964. Viability of lyophilized algae. *Can. J. Bot.* **42**, 127–37.

Holm-Hansen, O. 1967. Factors affecting the viability of lyophilized algae. *Cryobiology* **4**, 17–23.

Hwang, S. W. and Horneland, W. 1965. Survival of algal cultures after freezing by controlled and uncontrolled cooling. *Cryobiology* **1**, 305–11.

Leibo, S. P. and Jones, R. F. 1963. Effects of subzero temperatures on the unicellular red alga *Porphyridium cruentum*. *J. Cell. Comp. Physiol.* **62**, 295–302.

Mazur, P. 1970. Cryobiology: the freezing of biological systems. *Science* **168**, 939–49.
Meryman, H. T. 1960. Freezing and drying of biological materials. *Ann. New York Acad. Sci.* **85**, 501–734.
Meryman, H. T. 1966. *Cryobiology*. Academic Press, New York. 775 pp.
Smith, A. U. 1961. *Biological Effects of Freezing and Supercooling*. Williams and Wilkins Co., Baltimore. 462 pp.
Troshin, A. S. 1967. *The Cell and Environmental Temperature*. Pergamon Press, New York. 462 pp.
Watanabe, A. 1959. Some devices for preserving blue-green algae in viable state. *J. Gen. Appl. Microbiol.* **5**, 153–7.

Reprinted in U.S.A. from
'Handbook of Phycological Methods - Culture Methods
and Growth Measurements' edited by Janet R. Stein

14: Toxic and inhibitory materials associated with culturing

WILLIAM F. BLANKLEY*
Scripps Institution of Oceanography, University of California, La Jolla, California 92037

CONTENTS

* Present address: Duke University Marine Laboratory, Beaufort, North Carolina 28516.

I. OBJECTIVES

A. *Introduction*

Biologists are increasingly aware of the importance of determining possible deleterious effects of materials used in or around living systems. Such effects have been extensively investigated in several areas, especially industrial safety (Lefaux 1968) and medical applications (Autian 1964). There are several comments on noxious effects of materials in the literature on algal culture, but very few studies in which materials have been systematically tested for their effects on algal growth (Bernhard, Zattera and Filesi 1966; Dyer and Richardson 1962). A standard method (ASTM 1964) is available for testing industrial waste water for inhibitory effects on diatom (Bacillariophyceae) cultures, but is not readily applicable to other materials. Although some generalizations on the noxious effects of particular materials have been made, variations in methodology and organisms make interpretation and comparison of results difficult. This chapter presents a standard method of testing for the presence of toxic and inhibitory effects of materials on algal cultures. The method is designed for application to liquid cultures of algae; it should be applicable as well to other microorganisms and to invertebrates.

A list of materials which should be safe, inhibitory, or toxic to algal cultures is given in section VI, following, and Table 14–1; these are intended to serve only as guides. Although it might be reasonable to avoid the use of a material listed as toxic or inhibitory, each investigator should test the materials in his own situation. Sources of contamination are not always obvious; some situations in which noxious compounds might be accidently introduced into cultures are discussed in section VI, following.

This chapter is concerned only with materials which are not meant to have direct physiological function or to act as a nutrient. Trace metals, medium precipitates, biochemical inhibitors, and other compounds added to cultures are not considered. The emphasis is on detecting deleterious effects; for a discussion of mechanisms of action of noxious substances consult standard works as Albert (1968).

TABLE 14-1. *Effects of materials on algal cultures*

Material	Safe*	Inhibitory*	Toxic*
Acrylic (Lucite, Perspex, Plexiglas)	*abdde*	—	—
Aluminium alloy	*eeeeeee*	—	—
Cadmium	*e*	—	—
Charcoal, activated	—	*bg*	—
Copper alloy	*ee*	*e*	*eee*
Cotton	*b*	—	—
Epoxy resin	*ee*	—	—
Iron	—	*e*	—
Magnesium	*e*	—	—
Membrane filter (Millipore, Membranfilter)	*ab*	—	*a*
Nickel	*e*	—	—
Nylon	*be*	*a*	*ab*
Paraffin	*d*	—	—
Plywood	—	*d*	—
Polycarbonate	*f*	—	—
Polyethylene, black	*e*	*a*	—
Polyethylene, white, clear	*ab*	—	—
Polypropylene	*abe*	*aa*	*e*
Polystyrene	*b*	—	—
Polytetrafluoroethylene (Teflon, etc.)	*abe*	—	—
Polyurethane foam	—	*e*	—
Polyvinyl chloride	*aa*	*aab*	*aaaabee*
Polyvinyl chloride (Tygon – clear)	*abcee*	*d*	—
Polyvinyl chloride (Tygon – black)	—	—	*b*
Rubber – white, black, green, Buna N, neoprene, gum, latex, etc.	*e*	*abd*	*aaaaaaaaa bcee*
Silicone (stoppers, tubing, stopcock grease)	*abeee*	—	—
Silicone (cement, sealant)	*e*	*b*	*e*
Solder, silver	*ee*	—	—
Solder, soft	*e*	—	—
Stainless steel	*eeeeee*	*a*	—
Tin	*e*	—	—
Titanium	*e*	—	—
Zinc	—	*e*	—

* Each letter represents the result of one test of a specific formulation or product, as reported by the indicated author (see reference list). Where more than one species was tested by an author, the most adverse result is tabulated.

References, annotated with species used

a. Bernhard, M., Zattera, A. and Filesi, P. 1966. Suitability of various substances for use in the culture of marine organisms. *Pubbl. Sta. Zool. Napoli* **35**, 89–104. (*Leptocylindrus danicus, Chaetoceros danicus, Prorocentrum micans, Coccolithus huxleyi;* marine.)

TABLE 14-1. (*cont.*)

b. Blankley, W. F. Unpublished observations. (*Cricosphaera carterae*, *Coccolithus huxleyi*; marine.)
c. Davis, E. A., Dedrick, J., French, C. S., Milner, H. W., Myers, J., Smith, J. H. C. and Spoehr, H. A. 1953. Laboratory experiments on *Chlorella* culture at the Carnegie Institution of Washington, Department of Plant Biology. In Burlew, J. S., ed., *Algal Culture from Laboratory to Pilot Plant*. pp. 105–53. Carnegie Inst. Washington. Publ. 600. (*Chlorella*; freshwater.)
d. Doty, M. S. and Oguri, M. 1959. The carbon-fourteen technique for determining primary plankton productivity. *Pubbl. Sta. Zool. Napoli* **31** (Suppl.), 70–94 (marine phytoplankton.)
e. Dyer, D. L. and Richardson, D. E. 1962. Materials of construction in algal culture. *Appl. Microbiol.* **10**, 129–32. (*Synechococcus lividus*, *Chlorella pyrenoidosa*; freshwater.)
f. Lewin, J. 1966. Physiological studies of the boron requirement of the diatom, *Cylindrotheca fusiformis* Reimann and Lewin. *J. Exp. Bot.* **17**, 473–9. (*Cylindrotheca fusiformis*; marine.)
g. Ryther, J. H. and Guillard, R. R. L. 1962. Studies of marine planktonic diatoms. II. Use of *Cyclotella nana* Hustedt for assays of vitamin B_{12} in seawater. *Can. J. Microbiol.* **8**, 437–45. (*Thalassiosira pseudonana*; marine.)

B. *Definitions*

1. Inhibitory. Causing reduction in rate or yield of any physiological or growth parameter.

2. Material. Any substance (solid, liquid, or gas; compound, solution, or mixture) which can be tested for inhibitory or toxic effects on cultures.

3. Median inhibitory limit (ILm). The concentration (variously measured) of material at which the magnitude of the parameter being measured is reduced to 50% of that of the control.

4. Noxious. Deleterious, harmful, inhibitory, or toxic.

5. Toxic. Causing death, lethal.

II. EQUIPMENT

Since the amounts and exact types of equipment depend on the particular material, these are usually not specified.

A. *Basic equipment*

1. Culture vessels. 15 × 125 mm test tubes, to be used directly in a Spectronic 20 Spectrophotometer (Bausch and Lomb Inc., Analytical Systems Division. 820 Linden Ave., Rochester, New York 14625; 1790 Birchmont Rd, Scarborough, Ontario).

2. *Tube closures*. Small glass beakers, vials or 'Cencaps' (Central Scientific Company, 2600 S. Kostner Ave., Chicago, Illinois 60623; 2200 S. Sheridan Way, Clarkson, Ontario); Identiplugs (Gaymar Industries, Buffalo, New York) are non-inhibitory to all Haptophyceae tested.

3. *Growth measurements*. Bausch and Lomb Spectronic 20 Spectrophotometer (see I, preceding).

4. *Stirring rods, forceps*. Glass, clear polyethylene or teflon.

5. *Autoclave*. See Chp. 12, II.C.2.C.

6. *Growth*. See Chp. 11, II.A. and III; temperature control should be within ±1 °C.

B. *Auxiliary equipment*

Additional requirements, depending on modifications of method used, may include 50 ml flasks for powders, separatory funnels or centrifuge tubes for immiscible fluids, pipettes, and a pH meter.

III. TEST ORGANISMS

No standard assay organism is suggested. Because of the species variability, all testing should be done with the species to be cultured in the presence of the material. If the material will be used with several species, and all cannot be tested individually, the most delicate should be used. These would be the ones that grow most poorly, require the largest inoculum to initiate a culture, tolerate the narrowest ranges of salinity, pH, etc.

An alga–diatom (Bacillariophyceae) or flagellate (Chlorophyceae, Chrysophyceae, Haptophyceae, Dinophyceae, etc.) – that normally grows reasonably well but is not a laboratory 'weed' should be chosen. Planktonic forms from clean-water environments generally give more dramatic results than benthic species or those typical of euthrophic environments.

IV. METHOD

The method is designed to estimate the inhibition or toxicity caused by substances leached from a standard amount of material allowed to interact with the medium prior to inoculation. This is relatively simple and not time consuming, but can yield sufficient information to determine whether the material is safe to use. If detailed information is needed on the noxious effect of a material as a function of amount of material, the procedure for finding the median inhibitory limit should be followed (section VII, following).

The method has been divided into sections, most of which are identical for all classes of materials. Those sections which differ according to the class of material are given in detail for solids with definite shapes. Modifications necessary for handling other classes of materials follow.

A. *Materials*

1. Classes. These categories are based on the amounts to be tested and the procedures for handling the materials.

a. Solids with definite shapes – this includes all non-soluble materials which are sufficiently structured to maintain shapes. Almost all metal, glass, plastic, paper, and rubber objects are in this category. Such things as steel wool, glass wool, and non-rigid porous plastic or rubber foams are not, because their surface areas are either not practical to measure or not fixed in relation to their volumes. (Note that paper, cloth, etc., although porous, have relatively fixed nominal area : volume ratios.)

b. Solids without definite shapes – this class includes all non-soluble, generally porous, materials whose true surface areas are difficult to measure with accuracy and whose nominal surface areas bear no fixed relationships to their volumes (glass wool, foam rubber, etc.). These materials, however, have enough structure to permit manipulation by stirring rods, forceps, etc.

c. Immiscible fluids – this category includes all immiscible liquids (oils) and insoluble powders (charcoal, sand, etc.). Some solids which melt during autoclaving (silicone gum, waxes, etc.) are also included in this category. The fluid properties of these materials necessitate handling procedures different from those used for the other classes.

d. Miscible liquids and soluble solids – materials in this category dissolve completely in the medium rather than simply release a soluble component. For convenience of treatment, soluble substances are grouped into three subcategories:

i. Nutritive or physiological agents; such as carbon sources, trace elements, vitamins, etc. These substances are outside the scope of this chapter.

ii. Non-nutritive, non-metabolic agents; such as agar, pH buffers, etc. Substances in this category also are generally outside the scope of this chapter.

iii. Contaminants; such as detergents, cleaning solutions, etc., which may find their way into media through incomplete rinsing of glassware; cooling solutions and other liquids or solids which may be used in sufficient proximity to the cultures that the chance of accidental introduction into the medium is high; noxious vapors, etc.

2. Amount. The amount of material tested depends on two considerations:

a. The degree of contact between the material and the medium. Materials actually immersed in the medium can be expected to release larger quantities of substances into the medium. Such materials should be tested at a greater material:medium ratio than those which are not in contact with the culture medium.

b. The physical nature of the material. The amount of noxious substance released from a material depends primarily on the surface area available for reaction with the medium or for diffusion of substances into the medium. The surface area is of primary importance as in most non-porous solids only the outer few microns form or release substances into the medium. Almost all of a porous solid and all of a liquid may be so available. The following amounts per 5 ml medium should be used for each class of material:

i. Solids with definite shapes – 100 mm^2 surface area.
ii. Solids without definite shapes – 0.1 g.
iii. Immiscible fluids – 0.1 g.
iv. Soluble substances – dilution series.

B. *Preparation*

1. Selection and cleaning. Select clean, unused material for testing to minimize the possibility of noxious compounds having been absorbed or leached during previous usage. When materials are cleaned prior to testing, the procedure used should be only as thorough as that routinely employed. Care must be taken that noxious compounds are neither removed from the bulk of the material nor added by adsorption.

a. Solids with definite shapes – many solid materials come from the manufacturers with surface contaminants; remnants of the manufacturing process, powders or oils added for protection during transit and storage, or paints or inks in the form of imprinted labels. etc. In addition, surfaces become contaminated during storage and handling in the laboratory. All surface contaminants should be removed as completely as possible before the material is actually used or tested. Non-porous materials should be rinsed thoroughly with water. More vigorous cleaning with non-inhibitory detergents or with organic solvents may be used if necessary, provided that the solvent does not attack the surface of the material. The cleaning agents must be thoroughly rinsed off. Porous materials, such as paper, which are routinely used without prior cleaning, should not be cleaned prior to testing; selection of clean, unused material is therefore most important.

b. Solids without definite shapes – the selection of clean, unused material for testing is important since solid materials in this category are apt to acquire contaminants by adsorption or absorption. As with other materials, how to clean should depend on the procedure employed prior to routine use. When cleaning is attempted, the use of solvents or detergents should be avoided if possible because of the difficulty of thorough rinsing.

c. Immiscible fluids – since immiscible fluids are normally used only once, no problems should be encountered in finding new material to test. Generally cleaning prior to testing is neither necessary nor desirable.

d. Soluble substances – the same considerations apply as for immiscible fluids.

2. Measuring

a. Solids with definite shapes – prepare six pieces having a nominal surface area of 100 mm^2 each. Use a clean cutting tool to avoid contamination by oil, rust, or chips from previous work. Later operations will be simplified if the material can be cut into shapes (e.g., rings from tubing) which are easy to pick up with a stirring rod.

b. Solids without definite shapes – prepare two 0.5 g portions of material; single pieces weighing 0.5 g each are easier to handle than several smaller pieces.

c. Immiscible fluids – weigh out two portions of 0.5 g each.

d. Soluble substances – see section D.3.C, following, for setting up dilution series.

C. *Medium*

1. Type. The medium used for testing material should, ideally, have the same composition as that employed routinely in growing the alga. This is important because noxious effects of a material may be influenced by constituents of the medium. For surveys of a large number of materials, however, it is generally best to use a simple medium.

2. *Amount*. Prepare at least the following amounts.

a. Solids with definite shapes – 45 ml.
b. Solids without definite shapes and immiscible fluids – 75 ml.
c. Soluble substances – 30 ml per concentration and control.

D. *Leaching*

Two leaching procedures may be employed together or only one used. In both, the material and the medium are sterilized by autoclaving (see Chp. 12, II.C.2.c) rather than dry heat or chemical sterilization, which involve excessive temperature or toxic chemicals. The gentler treatment consists of separate autoclaving of test material and medium, followed by 24 h soaking together. The more drastic treatment requires autoclaving of material and medium together prior to the 24 h soaking.

1. Solids with definite shapes

a. Label twelve tubes, three each:
'A' – medium with material added after autoclaving
'B' – medium with material added before autoclaving
'C' – control medium
'M' – material for later addition to tubes A.
(For making repetitive measurements on the cultures, it is helpful to further label the three tubes of each set 1, 2 and 3.)

b. Dispense 5 ml portion of medium into each of tubes A, B, and C.

c. Cap tubes A and C; autoclave. (In order to avoid the possible absorption of volatile substances from materials in tubes B and M into medium in tubes A and C, autoclave separately from B and M. The autoclave should be drained and thoroughly cleaned prior to use, especially if it is of the type which generates its own steam supply.)

d. Place one piece of material into each of tubes B and M.

e. Cap tubes B and M; autoclave.

f. Cool all tubes to incubation temperature.

g. Transfer aseptically the test material from M to A. (This can be done by inverting M over A letting the material slide or fall in. Do not transfer any droplets of liquid that have condensed in M during autoclaving, as these might contain large amounts of leachates. Alternatively, the materials can be removed using clean, sterile, inert stirring rod or forceps.)

h. Place A, B, and C in incubator for 24 h.

i. Remove the material from A and B, allowing the medium to drain from the material.

j. Measure the optical density (see Chp. 21, VII). Check for colored substances leached from the material.

2. *Immiscible fluids, solids without definite shapes.* Immiscible fluids are difficult to separate completely from the medium, and solids such as cotton are difficult to remove completely from individual tubes because of loose fibers. Furthermore, removal of such materials from the medium may involve an appreciable loss of medium. The replicate samples of such materials are therefore autoclaved together in a single flask and separated into individual tubes after leaching. Two extra portions each of medium and of material are used to allow for some loss of medium during the separation step. The basic method is essentially the same as for solids with definite shapes.

a. Label four 50 ml flasks, one each A, B, C, and M (see 1.a. preceding), Label nine tubes, three each A, B, and C, and from each set, one each 1, 2, and 3.

b. Dispense 25 ml portions of medium into A, B, and C.

c. Cover A and C with suitable caps (50 ml beakers); cover all (empty) tubes; autoclave.

d. Place a 0.5 g portion of material each in B and M.

e. Cover flasks B and M; autoclave separately from A and C preceding.

f. Cool flasks to incubation temperature.

g. Aseptically transfer the material from M to A (see section g, preceding).

Some viscous or sticky fluids including powders dampened by autoclaving, may require transfer of the medium (A) to the material M, but this should be avoided if possible.

h. Place A, B, and C in incubator for 24 h.

i. Aseptically pipette 5 ml portions (free of test material) of media from A, B, and C into the correspondingly labeled sterile tubes. It may be necessary to remove loose particles or insoluble powder. This should be done aseptically. While pipetting do not disturb the pellet. Immiscible fluids may present the greatest obstacles to separation. If the test fluid is immiscible and more dense than the medium, the medium can be pipetted off. If the test fluid is less dense than the medium use a sterile separatory funnel (free of stopcock grease) to separate the layers. Two additional sterile flasks are needed for each fraction.

j. Measure the optical density as in 1.j, preceding.

3. Miscible fluids, soluble solids. This method differs from the preceding as known concentrations of a known compound or mixture are being tested. There are no materials to be removed from the medium prior to inoculation and therefore no leaching period is needed.

a. Label nine tubes, three each A, B, and C. Then label one of each set 1, 2, and 3. Label one tube M (the letters are as previously used, 1.a, preceding).

b. Dispense medium for dilution series (final volume per tube: 5 ml) into A. Put 5 ml into C; cap each tube; autoclave. (The amount of medium in A is determined by number of steps in dilution series.)

c. Prepare a dilution series in B using the same concentrations of material as those to be used in dilution series A. Most soluble substances entering the medium result from accidental contamination, thus prediction of concentration is not possible. However, an estimate of maximum probable concentration can be made. The highest concentration used in the dilution series should be one step higher than the maximum concentration expected. If this cannot be predicted, use a series ranging from 1 mg/ml medium down.

d. Place sufficient material plus some excess in M for preparation of dilution series A. For solids which absorb water and dissolve during autoclaving, use the exact amount necessary for the most concentrated member of the series.

e. Cap tubes B and M; autoclave. Ideally, each concentration should be autoclaved separately to avoid possible transfer by distillation of the material or its reaction products from tubes containing higher concentrations to lower concentrations. Autoclave in succession starting with the most dilute concentration.

f. Cool tubes to incubation temperature.

g. Aseptically prepare dilution series in A.

h. Place tubes in incubator for 24 h.

i. Measure the optical density.

E. *Inoculation*

1. Inoculum. The size and the physiological condition of the inoculum influences the magnitude of deleterious effects of a material; therefore careful standardization of the inoculum is desirable. If a standard inoculum is already in use for other experimentation, use this for testing. In standardizing the inoculum, consider the following factors:

a. Cells growing exponentially exhibit the shortest lag phase in the new culture. The lag phase in control cultures should be kept as short as possible in order to show any effects of the material in prolonging the lag phase.

b. Apparent sensitivity to noxious substances may be inversely related to inoculum size. This may be due to adsorption onto or absorption by the cells of a noxious substance (e.g., copper ion) or to the inactivation of a noxious substance by a product of the cells (e.g., peroxide inactivation by catalase). The effect of these processes on the observed inhibitions of a material depends on the number of cells relative to the amount of noxious substance. The inoculum should be kept small without prolonging lag phase in the control. Too large an inoculum may well overcome and obscure inhibitory effects.

c. The length of the initial lag phase with some algae increases with decreasing inoculum size and the inoculum used should be large enough to avoid this.

d. If growth curves are desired the inoculum should be sufficiently large that a measurable parameter for growth is obtained in early growth stages.

2. Single-cell inoculum. A single-cell inoculum is especially sensitive to small amounts of noxious substances. In experimental work with single cells the inoculum used in testing materials should be as small as possible to maximize sensitivity.

3. Procedure

a. Inoculate all tubes immediately after the 24 h leaching or equilibration period.

b. Measure the optical density if growth curves are to be constructed (see G, H, following).

F. *Incubation*

The incubation conditions for testing materials should duplicate conditions to be used for experiments. Maximal growth rate is essential as slight inhibitory effects produced by a material can be masked if the overall growth rate is depressed by environmental limitations.

1. Temperature. Temperature should be near optimum for growth. Temperatures near either extreme may change the sensitivity of the alga.

2. Light intensity. Light intensity should be such that cultures are not inhibited by insufficient light. Inhibitory effects also result from excess light, but normally this is not a problem except for some Cyanophyceae.

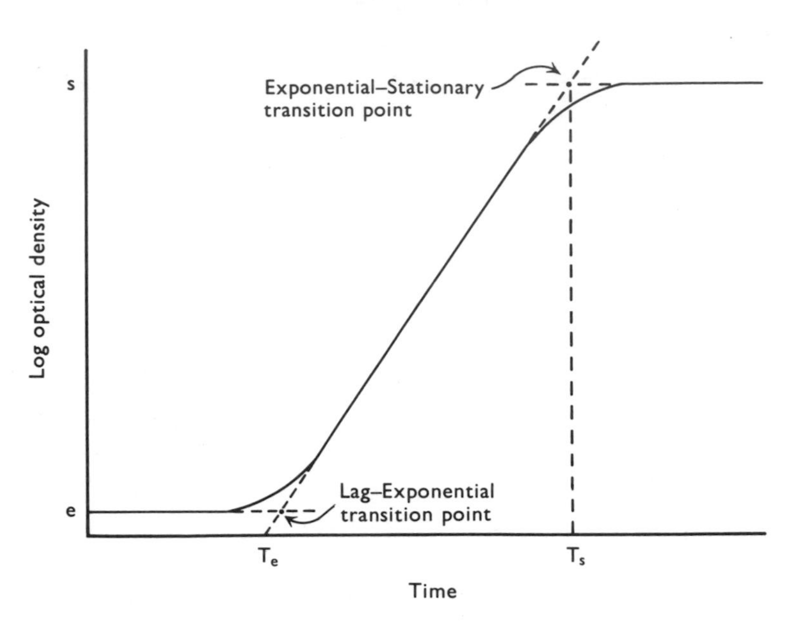

Fig. 14-1. Typical growth curve of algal cells in batch culture.

3. Aeration. The use of non-aerated test cultures is recommended as aeration or mixing tends to remove volatile compounds whose effects would then be missed. If aeration or mixing is used routinely in other experiments, aerate the medium for 24 h prior to inoculation.

G. *Measurement*

1. Method. Optical density measurements are recommended. See Chp. 21, VII, for details of this method.

2. Frequency. In order to distinguish inhibitory effects on the lag phase, growth rate, and final yield, sufficient points must be obtained (at least one, and preferably two or more per doubling time) to construct a growth curve, plotting log optical density as a function of time (see Fig. 14-1). Construction of a growth curve gives details essential for quantitative determination of the effect of a material on a particular growth aspect.

A composite relative inhibition can be obtained from one measurement. This measurement, taken at the exponential-stationary transition point (T_s), marks the onset of the stationary phase in control cultures. To do this, a growth curve must be constructed for control cultures.

H. *Calculations*

Calculations are made the same whether the test material is added before or after autoclaving; only the former are shown here. The calculations are independent of the method used for measuring growth. The appropriate blank values are subtracted from the data before beginning any calculations.

1. Inhibition expressed as specific effect on growth curve. (See Chp. 19 for details.)

a. Construct growth curves (see Fig. 14-1) for material added before (B) and after (A) autoclaving for each concentration and for the control (C).

b. Determine the length of the lag phase and the final yield; calculate the growth rate during exponential growth.

c. Calculate the percent inhibitions observed as:

lag phase:

$$\left(1 - \frac{\text{lag phase C}}{\text{lag phase B}}\right) \times 100, \tag{1}$$

final yield:

$$\left(1 - \frac{\text{final yield B}}{\text{final yield C}}\right) \times 100, \tag{2}$$

growth rate:

$$\left(1 - \frac{\text{growth rate B}}{\text{growth rate C}}\right) \times 100. \tag{3}$$

2. Inhibition expressed as a composite relative inhibition

a. Calculate the inhibition observed using the growth (yield) at the exponential stationary transition point (T_s) of the control (see Fig. 14-1):

$$\left(1 - \frac{\text{growth (yield) at } T_s \text{ in B}}{\text{growth (yield) at } T_s \text{ in C}}\right) \times 100. \tag{4}$$

I. *Interpretation*

A detailed discussion of physiological mechanisms susceptible to noxious effects of materials is beyond the scope of this chapter. However, presented here are some causes of noxious effects and factors which might influence their expression in an algal culture, as well as a brief guide to interpretation of the results in terms of toxicity or inhibition.

1. Possible causes of inhibitory or toxic action

a. Direct release of one or more inhibitory or toxic compounds or ions.

i. Dissolution of the material itself.

ii. Release of additives, monomers, etc.

iii. Release of contaminants present on the surface or absorbed within the material.

b. Interaction with the medium of a compound or ion released from the material.

i. Formation of noxious soluble compound, ion, or insoluble precipitate.

ii. Solubilization, chelation, etc., of potentially noxious, but previously unavailable, ion or compound.

iii. Elimination, or reduction in concentration, of a required compound or ion through competition, chelation, or precipitation.

iv. Change in pH or eH.

c. Adsorption of a required compound or ion on the surface of the material.

2. Determination of the exact cause of a noxious effect. This is frequently a complicated undertaking. One effect which is easily determined, however, is a change in pH or eH, which can be determined from an extra set of tubes carried through the entire leaching procedure (section IV.E, preceding) and tested for pH or eH changes after the initial 24 h incubation period.

Adsorption of nutrients onto material as a cause of inhibition can also be tested. The material is leached in a medium lacking essential nutrients, vitamins, etc., which are added after the material has been removed from the medium. Growth in this medium is compared with that in controls in which the nutrients were present during leaching. This procedure might also eliminate noxious effects due to interactions of substances from the material with certain nutrients (see 1.b, preceding). An alternative procedure, which counteracts adsorption without eliminating interactions, is to leach as usual, and then add extra portions of nutrients, vitamins, etc. after removal of the materials.

3. Factors affecting inhibition, toxicity

a. Medium components – organic compounds or soil extract, as well as the ionic content and pH of the medium, can affect both the action of noxious compounds on the organisms and the extent of leaching (see 1.b, preceding). The effect of a material in marine media may be quite different from that in freshwater media.

b. Presence of other organisms – xenic cultures are often somewhat less sensitive than are axenic cultures, possibly because the other organisms help to use or otherwise lessen the effects of the noxious compounds.

4. Differentiating between inhibition and toxicity. Lack of growth in test media usually indicates toxicity, but may alternatively indicate complete inhibition (i.e., cells would resume growth if transferred to fresh control medium). The results of these two possibilities on growth are identical for most practical purposes except when inhibition could be eliminated or reduced by driving off a volatile inhibitor by aeration.

A reduced growth rate or a reduced yield usually means that some physiological process is being inhibited. A reduced yield is often due to limitation by some essential nutrient, its supply available to the algae being reduced by chelation, precipitation, or adsorption.

An apparent lengthening of the lag phase may result from either of two causes:

i. An inhibitor whose effect is eventually overcome by the inoculum or which is gradually destroyed or dissipated from the medium.

ii. A toxic agent which kills some but not all of the inoculum.

These two possibilities can be distinguished, for example, by vital staining to determine the proportion of viable cells in the test media after inoculation.

5. Application of test results. If a material is found to be even moderately inhibitory, its use should be avoided if possible. If no innocuous alternative exists, it might be possible to detoxify the material (see section VI.C, following). The severity of the tests in relation to the proposed use of the material should be considered at this point.

V. SAMPLE DATA

For the standard method, Fig. 14-2 illustrates the expression of inhibition and toxicity which may occur. Note that in Fig. 14-2, parts *a*, *b*, *c*, as drawn, show combination relative inhibitions of 50% at time T_s. At time b, however, inhibitions of 0, 55, and 100% (apparent toxicity) are indicated respectively, whereas at time c the corresponding figures are 50, 20, and 0%. This illustrates the need for careful control of the sampling time and culturing procedures. Combinations of parts *a*, *b*, and *c* are also possible.

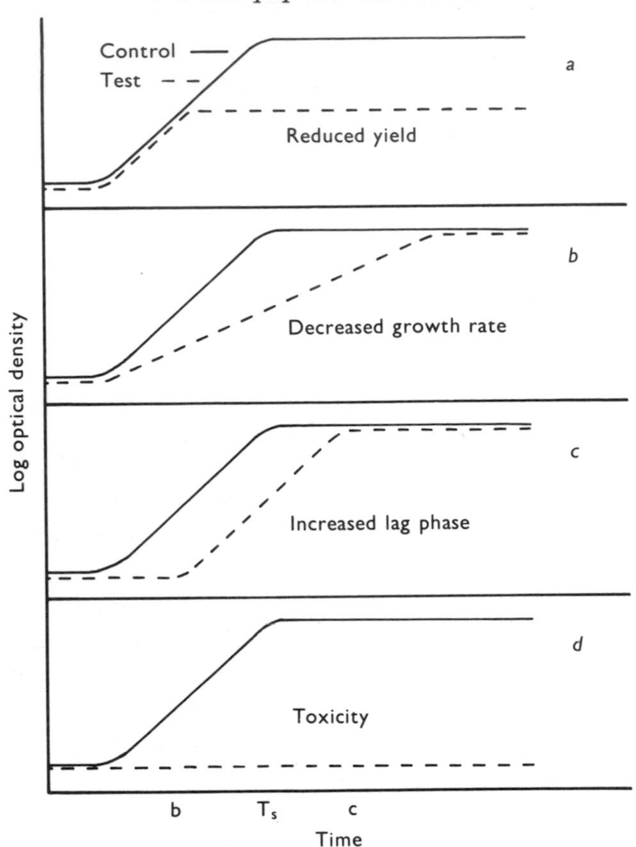

Fig. 14-2. Expression of inhibitory (*a*, *b*, *c*) or toxic (*d*) effects of a material by growth of algal cultures. Any combinations of the effects shown in *a*, *b*, and *c* are also possible.

VI. POSSIBLE PROBLEM AREAS

A. *Materials*

Table 14-1 is derived from literature reports and my unpublished work on marine coccolithophorids (Haptophyceae). The results were obtained from a variety of testing procedures with various species, and are only a general guide, primarily for avoiding certain materials rather than for assuming the others to be safe.

Some likely effects of metals can be adduced from the lists and from the review of Brookes (1969). The ionic content and pH of the medium will, of course, have a very great effect on the extent of corrosion and leaching of metals.

The situation with plastic and rubber is somewhat more complicated. Most commercial products made of natural and some artificial rubbers have noxious properties toward algal cultures, silicones being the main exception. The effects of plastics depend on their individual formulations, methods of manufacture, coloring agents, etc. In general, unpigmented, unplasticized plastics are most apt to be safe, and the degree of risk increases with the number and amount of additions to the pure polymer.

B. *Problems*

Some common problem areas, whereby noxious substances may be introduced into media, are discussed in the following to alert investigators to such situations and to suggest ways of avoiding them or altering the conditions so as to minimize undesirable effects.

1. Metal apparatus. Almost any metal in contact with a culture medium will tend to corrode, although some (certain stainless steels, monel, etc.) are relatively inert. The danger is greatest in marine media, and with plated metals on which surface corrosion may expose a more reactive, toxic metal to the medium.

2. Glassware. Although glass generally has no intrinsic noxious properties, glassware is a common source of contamination by virtue of its past history. It should be thoroughly cleaned and rinsed after washing. Glassware (and even more definitely, plasticware) used for preservation or histological work should be kept separate from that used for culturing, because of the difficulty of eliminating residues of various noxious substances used in such work.

3. Vapors. Volatile solvents, paints, etc. should not be used near cultures or areas of medium preparation. A 'No Smoking' rule should also be in effect in these areas, especially if they are in closed rooms.

4. Aeration. The source of air is of special importance. Aerating with laboratory air entails the risk of pumping volatilized solvents, tobacco smoke, etc., directly into the cultures. A good procedure with any air source is to prefilter the air through a column of activated charcoal dispersed in cotton or glass fibre, or through a commercial cartridge filter, before the air is sterilized and humidified. Such prefiltration also helps to remove foreign particles, pump oil and condensation water.

5. Sterilization. (See Chp. 12 for details of methods.)

a. Filtration – sterilization of liquids by membrane filters is often the method of choice, as this does not cause chemical changes as may be caused by autoclaving. However, filtration may introduce other problems. Filtration apparatus with elaborate parts or fritted glass supports may be difficult to clean and rinse efficiently. Noxious substances may also leach from certain metals or plastics in the assembly, or from the filter itself (see Simpson 1966; Cahn 1967). When certain medium components are present in small concentrations, significant quantities may be adsorbed onto the apparatus, padding, or filters, leading to reduced yields.

b. Autoclaving – the major problem encountered with autoclaving is that vapors evolved from a material during autoclaving may be absorbed and retained in media or on apparatus in the same autoclave. Therefore, it is important to clean the autoclave frequently. (Some sources of steam may be contaminated with anti-corrosion chemicals which could be absorbed by media, Kordan 1965.)

Some materials shown to be noxious under the test conditions may be used safely if autoclaved separately and kept from physical contact with the medium during use. This may be realized when differences in degree of inhibition obtained from addition of material before and after autoclaving indicates contact with the medium is important. For example, I have safely used bakelite screw caps (with inert liners) when autoclaved separately from media or other apparatus and dried before use. If such caps are sterilized at the same time as the medium, the latter is toxic or inhibitory although not actually in contact with the caps.

c. Dry sterilization – dry sterilization of some materials may initiate or accelerate processes which would increase subsequent leaching of noxious substances.

6. *Drying*. It is advantageous to dry apparatus prior to storage in order to reduce possible absorption of vapors and possible growth of microorganisms on moist surfaces. Culture apparatus should be dried separately from other materials and from analytical evaporations, etc.

7. *Adsorbates*. Noxious substances may accumulate on plugs and caps, in manifolds, and on other parts of apparatus which are not cleaned thoroughly after each use. Such substances may come from previous media, may be metabolic products of cultures, or may originate from external sources such as residues from cleaning procedures or vapors from various sources.

8. *Solution storage*. Glass vessels, although capable of release of trace amounts of heavy metals and other ions into solutions stored within them (Thiers 1957), generally are safe to use. Similarily unpigmented plastics usually are safe if thoroughly cleaned.

9. *Cleaning methods.* Cleaning procedures should be such that all cleaning agents can be completely rinsed off. Most detergents and other cleansers give trouble only if there are areas within the apparatus which resist adequate rinsing. Some detergents (and possibly other cleaning agents) may be absorbed by certain plastics and subsequently leach out slowly through many rinses.

Use of chromic acid cleaning solutions should be avoided because glass both adsorbs and absorbs the toxic chromic ion (see Chps. 1, II.C; 2, II.C. 2; Thiers 1957). Mixtures of hydrochloric and sulfuric acids or nitric and sulfuric acids are equally effective in cleaning glassware and their use avoids the chromic ion problem.

C. *Reducing noxious effects by leaching*

Some potentially inhibitory or toxic material can be made harmless by repeated leaching of noxious substances until testing shows the material safe for use. Repeated autoclaving of the material in successive washes of water, nutrient medium, or solutions of sodium chloride, dilute acids, or chelators is most likely to be effective. Obviously no single procedure will work for all materials, and sometimes all attempts may fail. Many materials, especially porous or powdery solids and certain immiscible liquids can be successfully treated, e.g., black dacron cloth (W. F. Blankley, unpublished) and activated charcoal (Ryther and Guillard 1962).

VII. ALTERNATIVE TECHNIQUE (MEDIAN INHIBITORY LIMIT)

The median inhibitory limit (ILm, the concentration of material causing 50% inhibition) is a more precise estimate of the noxious effect of a material than is that obtained by the standard method of section IV, preceding. More accurate comparisons of the deleterious effects of materials can be made, based on the amount of material required to produce a particular effect. Decisions regarding the use of a desirable but somewhat inhibitory material may then be made considering the likely effects of the concentration to be used.

A. *Equipment, medium, preparation*

The amounts of equipment and medium must be adjusted in relation to the number of concentrations tested; otherwise refer to sections II and IV.A, preceding. For a discussion of selection and cleaning of material, see IV.B.1, preceding.

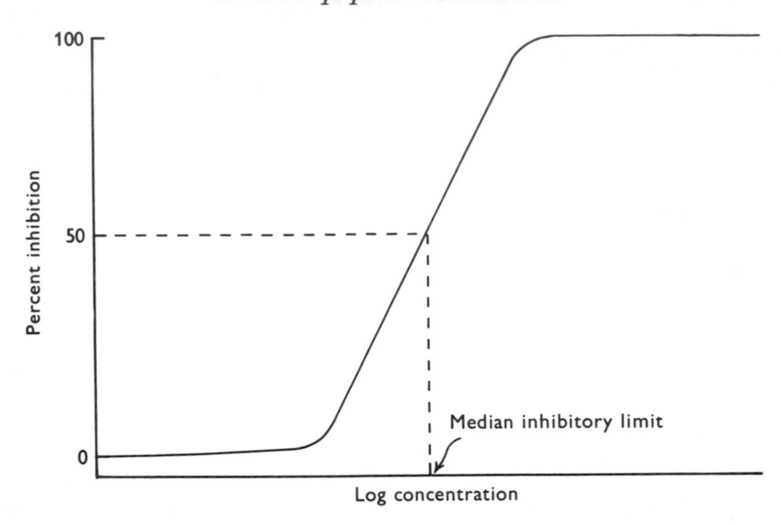

Fig. 14-3. Determination of the median inhibitory limit (ILm).

Concentrations of material should be chosen so that inhibitions obtained range from 0–100%, with at least two points between. In many cases a second test series may be needed in order to obtain the necessary points. For the first test series, surface areas of 1000, 100, 10, and 1 mm^2, or weights of 1000, 100, 10, and 1 mg/5 ml medium, should suffice.

B. *Leaching*

Use the appropriate procedure indicated in section IV.D, preceding. If the smaller amounts of insoluble materials are difficult to handle, leach all replicates, together with extras, using the procedure for solids without definite shapes (section IV.D.2, preceding).

C. *Inoculation, incubation, measurement*

These procedures follow sections IV.E–G, preceding.

D. *Calculations*

1. Calculate inhibitions as in section IV.H.
2. Plot percent inhibition as a function of log concentration of the material (see Fig. 14-3).
3. Find the ILm by interpolation at 50% inhibition.

E. *Interpretation*

Classify the material as 'safe' if there is no inhibition even in the presence of the highest concentration tested (1000 mm^2/5 ml medium or 1 mg/5 ml medium). For materials with an ILm within the range tested, the experimenter must decide if the material is safe to use.

VIII. ACKNOWLEDGMENT

I thank Professor Ralph A. Lewin for reading the manuscript and for many helpful suggestions.

IX. REFERENCES

Albert, A. 1968. *Selective Toxicity and Related Topics*, 4th ed. Metheun, London. 531 pp.

ASTM. 1964. Tentative method of test for evaluating inhibitory toxicity of industrial waste waters: D2037-64 T. pp. 517–25. In *1964 Book of Standards*. Part 23: Industrial Water; Atmospheric Analysis. American Society for Testing and Materials, Philadelphia, Pennsylvania.

Autian, J. 1964. Toxicity, untoward reactions, and related considerations in the medical uses of plastics. *J. Pharm. Sci.* **53**, 1289–301.

Bernhard, M., Zattera, A. and Filesi, P. 1966. Suitability of various substances for use in the culture of marine organisms. *Pubbl. Sta. Zool. Napoli* **35**, 89–104.

Brookes, R. 1969. Properties of materials suitable for the cultivation and handling of micro-organisms. In Norris, J. R. and Ribbons, D. W., eds., *Methods in Microbiology* **1**, 21–75. Academic Press, New York.

Cahn, R. D. 1967. Detergents in membrane filters. *Science* **155**, 195–6.

Dyer, D. L. and Richardson, D. E. 1962. Materials of construction in algal culture. *Appl. Microbiol.* **10**, 129–32.

Kordan, H. A. 1965. Fluorescent contaminants from plastic and rubber laboratory equipment. *Science* **149**, 1382–3.

Lefaux, R. 1968. *Practical Toxicology of Plastics*. Iliffe Books Ltd, London. 580 pp.

Ryther, J. H. and Guillard, R. R. L. 1962. Studies of marine planktonic diatoms. II. Use of *Cyclotella nana* Hustedt for assays of vitamin B_{12} in seawater. *Can. J. Microbiol.* **8**, 437–45.

Simpson, L. 1966. Toxic impurities in Nalgene filter units. *Science* **153**, 548.

Thiers, R. E. 1957. Contamination in trace element analysis and its control. In Glick, D., ed., *Methods of Biochemical Analysis*. **5**, 273–335. Interscience Publishers, New York.

Section III

Special culture methods

15: Continuous culture – a method for the production of unicellular algal foods

RAVENNA UKELES

National Marine Fisheries Service Experimental Biology Investigations, Milford Laboratory, Milford, Connecticut 06460

CONTENTS

I. INTRODUCTION

A. *Purpose*

Algae that are dietary delicacies in certain cultures are not acceptable foods in other parts of the world, consequently, utilization of algae as a new food source (Spoehr 1953; Tamiya 1960) is severely limited. Unicellular algae, however, can be functional as a nutritional supplement in rearing more conventional foods. One such possibility is feeding shellfish on marine or brackish water algae in hatcheries (Davis 1969). A dependable method of culturing sufficient quantities of phytoplankton foods for different stages of development is essential for commercial rearing of shellfish. Our objective was to develop a suitable method for production of unicellular marine algae as foods in rearing bivalves, both for research investigations and pilot plant hatcheries. The requirements of such an apparatus are the following: (1) inexpensive; (2) easily available standard items of laboratory equipment; (3) simple to assemble and operate; (4) accommodation of several different species in simultaneous culture; (5) adaptable to rapid expansion or decrease in culture capacity; (6) only chemically inert materials in contact with media and cultures; (7) axenic or near axenic cultures; (8) populations in a steady-state; and (9) cultures continually available.

Continuous or semi-continuous culture was adopted as it seemed most suitable for fulfilling the requirements. Continuous culture is characterized by a continuous supply of medium to the culture allowing optimum coordination of growth rate, nutrient flow, and outflow of cultured microorganisms. In semi-continuous culture, a continuous system is achieved in effect by withdrawing a standardized portion of the culture at periodic intervals and replacing this with medium to retain the original volume. Essentially, a culture is held at some chosen point in its growth curve by the regulated addition of fresh medium (Fogg 1965).

B. *Scope, adaptability, limitations*

There are numerous accounts of continuous culture devices for growing microbial populations under steady-state, but each has some deficiency in scope and adaptability for a particular application. The fundamental theory for continuous culture with bacteria has been reported (Monod 1950; Málek and Fencl 1966) and numerous assemblies described (for review see James 1961). Apparatus for continuous culture of bacteria generally are not satisfactory because of the long generation times and special requirements for illumination, aeration and agitation of algae.

Continuous culture for photosynthetic species was first designed for *Chlorella* (Chlorophyceae) (Myers and Clark 1944) and other devices were then developed to provide small volumes of organisms for biochemical and growth studies (Bach 1960; Pipes and Koutsoyannis 1962; Maddux and Jones 1964; Eppley and Dyer 1965; Droop 1966; Howell, Tsuchiya and Fredrickson 1967; Carpenter 1968; Cook 1968; Hare and Schmidt 1968; Taub and Dollar 1968; Fuhs 1969). Continuous culture also has been used as a method of gas exchange for air revitalization in submarines and space flights (Hannan and Patouillet 1963; Benoit 1964; Matthern and Koch 1964; Ammann and Fraser-Smith 1968). For these applications continuous culture provides a greater output, better quality control, and a more consistent cell population than batch culture. For food production, however, these devices do not afford large enough culture capacity, are often composed of parts toxic to species feeding on the algae, and require special handling and selected personnel for complex construction and maintenance that is impractical.

The method to be described has numerous applications but is probably best suited for the production of unicellular algal foods. Although reasonably good performance can be expected, it may not be useful in studies where highly accurate quantitative data on growth rates are needed. Maintenance of complete sterility is dependent on careful technique and is not assured for an extended length of time, although cultures will yield dense algal populations for an average length of 6 months and as long as 24.

II. EQUIPMENT

A. *Glassware*

1. Aseptic filling bells, 20 and 70 mm (Bellco 2330, Bellco Glass, Inc., 340 Edrudo Rd, Vineland, New Jersey 08360).
2. Bottles, aspirator with outlet for tubing, Pyrex, 4 l, 3.5 and 5 gal (US) (Corning 1220).
3. Centrifuge tubes, modified Hopkins 15 ml (Bellco, special order).
4. Drying tubes, Pyrex, straight single bulb, 100 and 200 mm (Corning 7775).

5. Erlenmeyer flasks, Pyrex, 125 ml.
6. Erlenmeyer flasks, Pyrex, 500 ml, wide mouth.
7. Fernbach flasks, Pyrex, 2800 ml.
8. Glass tubing, flint, 5 mm O.D.
9. Media storage and dispensing jar, 9 or 13 l (Bellco 2300).
10. Serum bottles, Pyrex, 4 or 9 l (Corning 1585).
11. Shell vials, plain tops 21 × 70 mm (60930-L, Kimble Products Division, Owens-Illinois, Inc., P. O. Box 1035, Toledo, Ohio 43601).
12. Solution bottles (carboys), Pyrex, 5 gal (Corning 1595).

B. *Hardware*

1. Air compressor and reducing valves (Bell and Gossett, 8234 Austin Ave., Chicago, Illinois 60620).
2. Autoclave, large capacity.
3. Centrifuge, equipped with head and appropriate tube shields for 15 ml tubes.
4. Cork boring tool.
5. Flowmeter, Flowmaster (National Appliance Co., P. O. Box 23008, Portland, Oregon 97223).
6. Fluorescent lamps, cool-white (see IV.D, following, for details).
7. Peristaltic pump (CRC 'Vibrostaltic', Chemical Rubber Co., 18901 Cranwood Pkwy, Cleveland, Ohio 44128: Durrum Multichannel, Model 12A, Durrum Instrument Corp., 3950 Fabian Way, Palo Alto, California 94303).
8. Controlled temperature room, or water bath (see IV.C for details of construction).
9. Vacuum pump.
10. Voltage regulator with variable transformer.
11. Water demineralizer (Bantam, Barnstead Model BD-1, Barnstead Co., 225 Rivermoor St., Boston, Massachusetts 02132).

C. *Miscellaneous*

1. Air filter (Deltech Engineering, Inc., 1226 Dargue Rd, New Castle, Delaware 19720).
2. Brass tubing, 1 in, fittings and needle valves, 0.25 in.
3. Clamps; screw compressor, Hoffman with open sides; miniature worm-drive clamps; hemostat (Chemical Rubber Co.).
4. Cotton wool and gauze.
5. CO_2; gas cylinder and regulating valves.

6. Filters; polypropylene core and filament, 1 μm and 15 μm (Sethco Manufacturing Corp., 1 Bennington Ave., Freeport, New York 11520).
7. Pleated membrane cartridge filters, 0.45 μm (Gelman Instrument Co., 600 S. Wagner Rd, Ann Arbor, Michigan 48106).
8. Flexible Briskeat insulated heating tapes; 0.5 × 24 in (BIH 2–$\frac{1}{2}$, Briscoe Manufacturing Co., P. O. Box 628, Columbus, Ohio 43216).
9. Interval timer (Singer Industrial Timer Corp., US Highway 287, Parsippany, New Jersey 07054).
10. Stone air breakers (Turtox 205A6322, Turtox/Cambosco, 8200 S. Hoyne Ave., Chicago, Illinois 60620; or local aquarium supply).
11. Stopcocks, polypropylene, 4 mm bore (6460 Nalge, Nalgene Labware Division, 75 Panorama Creek Dr., Rochester, New York 14625).
12. Sterilization bag (Laboratory Research Co., P. O. Box 36509, Los Angeles, California 90036).
13. Stoppers, polypropylene, 'Bacti-Caps', 18 mm (Bel-Art Products, Industrial Rd, Pequannock, New Jersey 07440), or rubber serum bottle stoppers.
14. Rubber stoppers, no. 1, 1-hole; and no. 4, 2-hole.
15. Stoppers, 'Steril-Kap', 60447 (BioQuest Division, Becton, Dickinson and Co., P. O. Box 243, Cockeysville, Maryland 21030; 2464 S. Sheridan Way, Clarkson, Ontario).
16. Tubing, amber latex rubber, $\frac{1}{4}$ in bore × $\frac{1}{16}$ in wall; gum rubber suitable for vacuum, $\frac{3}{16}$ × $\frac{1}{4}$ in; polyvinyl bubble tubing, with integral connectors at 36 in intervals, $\frac{9}{32}$ in I.D.
17. Wrapping paper and masking tape.

III. TEST ORGANISMS

The method described has been used for continuous culture of the following species. All are available in the Biological Laboratory National Marine Fisheries Service, Milford, Connecticut, in addition to the indicated source (Indiana University = IUCC; Marine Biological Association, United Kingdom = MBA; see Introductory Chapter).

A. *Chlorophyceae*

1. *Chlorella autotrophica* Shihira and Krauss 1965. IUCC 580 (as *Chlorella* sp.).
2. *Dunaliella euchlora* Lerche 1937 (strain may be identical to *D. tertiolecta* Butcher 1959; McLachlan 1960) MBA 83; IUCC 999.

B. *Chrysophyceae-Haptophyceae*

1. *Isochrysis galbana* Parke 1949 MBA 'I'; IUCC 987.
2. *Monochrysis lutheri* Droop 1953 Scottish Marine Biological Station Culture Collection, Millport, Isle of Cumbrae, Scotland No. 60; IUCC 1293.
3. *Pavlova gyrans* Butcher 1952 MBA 93; IUCC 992.
4. *Dicrateria inornata* Parke 1949 MBA 'B'; IUCC 988.

C. *Bacillariophyceae*

Phaeodactylum tricornutum Bohlin 1897 (*Nitzschia closterium* f. *minutissima*) IUCC 646.

IV. LAYOUT, SERVICES

A. *Layout*

The apparatus is located in a temperature controlled cold room, or cultures may be incubated in a cold water bath (see D, following). Three 18 in wide, white formica shelves, spaced 30 in apart, hold the apparatus.

The bottles containing sterile media are placed on the top shelf (30 in above middle shelf); the culture collecting bottles on the floor shelf (30 in beneath middle shelf) and the culture carboys on the middle shelf; (see Fig. 15-1 for semi-continuous culture; Fig. 15-2 for continuous culture). In the continuous culture (apparatus no. 2) the medium reservoir bottle, culture carboy, and culture harvest vessel are placed on the middle shelf. The medium reservoir bottle is left *in situ* and refilled with sterile medium from bottles on the top shelf as the reservoir is emptied to avoid transporting heavy bottles and repeatedly breaking connections to the culture carboys.

A 1 in wire mold with electric outlets, as well as a brass line with 0.25 in needle valves for aeration control, is mounted on a frame behind the middle shelf to provide services for each carboy.

B. *Aeration*

Air and CO_2 are delivered to a Flowmaster air/CO_2 mixing chamber and the ratio of CO_2/air adjusted by the flow gauges to about 2.0% CO_2. The air source is an oil-less air compressor with reducing valves to prevent dangerously high or low pressures (brought to about 12–15 lb/in^2) and filtered to remove particles and moisture. The source of CO_2 is from a pressurized tank (50 lb) reduced to 12–15 lb/in^2. The mixed gas is supplied to the cultures via gas lines and needle valves, adjusted to deliver a rate of bubbling in the culture of about 500 ml/min. Air saturation with water is not necessary since the volume decrease in the culture due to evaporation is

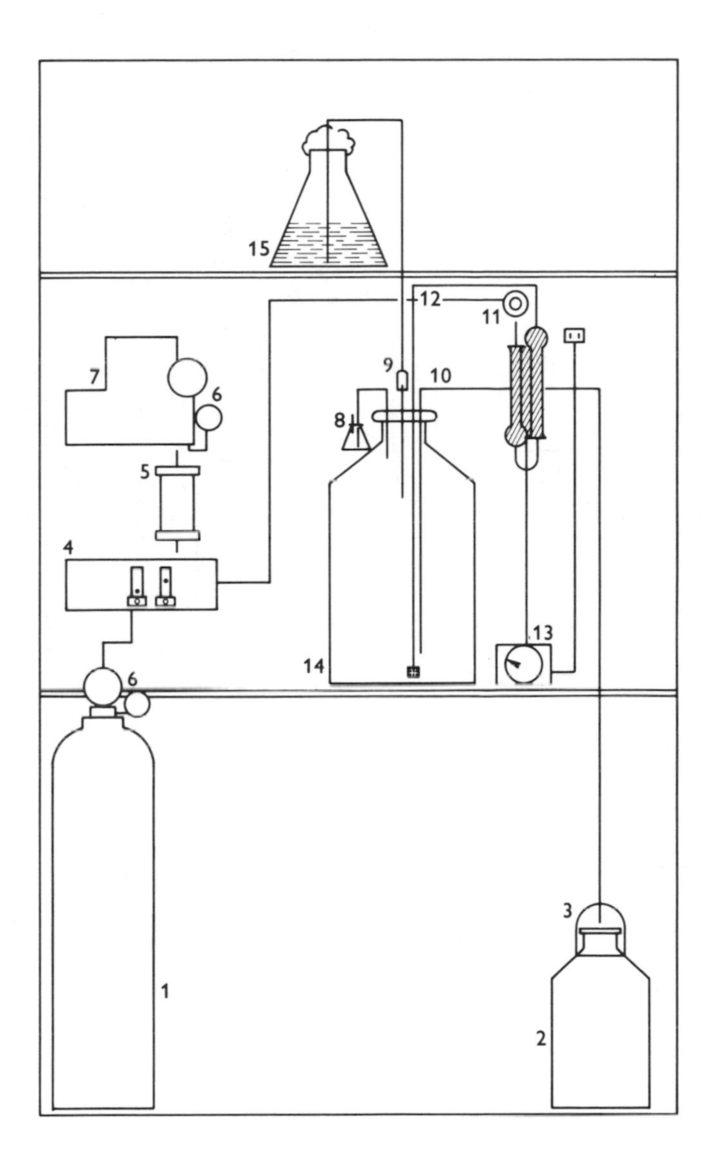

Fig. 15-1. Schematic diagram for semi-continuous culture (apparatus no. 1). (1) CO_2 cylinder; (2) culture collecting bottle (4 or 9 l serum bottle); (3) 70 mm aseptic filling bell; (4) gas flowmeter (Flowmaster); (5) air filter (Deltech); (6) pressure regulator; (7) air compressor; (8) effluent gas outlet; (9) 20 mm aseptic filling bell on inoculating port; (10) harvest siphon; (11) needle valve for gas pressure regulation; (12) influent gas tube with cotton filters and stone air breakers; (13) voltage regulator; (14) growth unit, or carboy (5 gal Pyrex solution bottle); (15) Fernbach flask with inoculum.

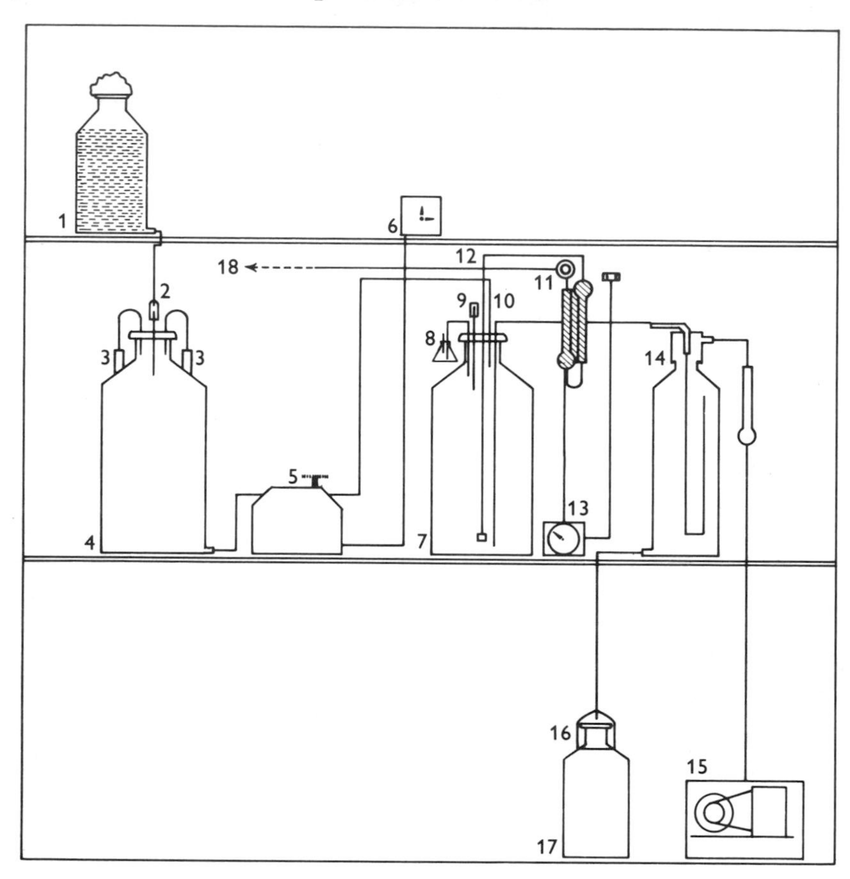

Fig. 15-2. Schematic diagram for continuous culture (apparatus no. 2). (1) medium 3.5 gal aspirator bottle; (2) 20 mm aseptic filling bell on media intake port; (3) effluent cotton air filters; (4) medium reservoir (5 gal aspirator bottle); (5) peristaltic pump (Durrum); (6) interval timer; (7) growth unit, or carboy (5 gal Pyrex solution bottle); (8) effluent gas outlet; (9) inoculating port (100 mm Pyrex drying tube); (10) media filling tube; (11) needle valve for gas pressure regulation; (12) influent gas tube with stone air breaker and cotton filters; (13) voltage regulator; (14) culture harvest vessel (Bellco, media storage and dispensing jar); (15) vacuum pump; (16) 70 mm aseptic filling bell; (17) culture collecting bottle (4 or 9 l serum bottle); (18) to aeration and CO_2 system, as in Fig. 15-1 (items 1, 4, 5, 6, 7).

restored with the addition of medium; however, an air saturation device as described by Conger (1969) may be used. Mixed gas is filtered through drying tubes packed with cotton wool before entering each culture carboy. The cotton filter is kept dry by gentle heating with insulated Briskeat heating tapes around the drying tubes; a variable transformer adjusts the degree of heating. Filters that become damp from water vapor of the culture carboys provide a troublesome source of contamination.

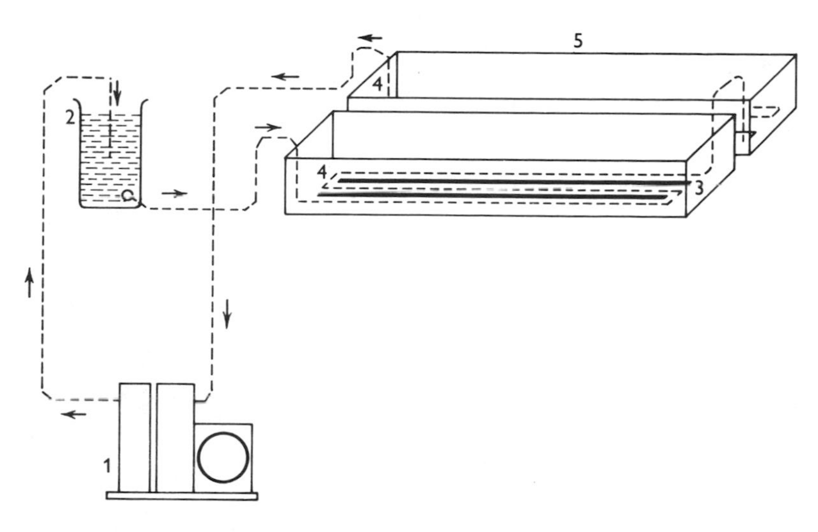

Fig. 15-3. Water bath for holding culture carboys (see text IV.C, for details). (1) force-flow liquid water cooler; (2) constant overflow jar for maintaining centrifugal pump; (3) carboy (growth unit) supports; (4) 0.5 in copper tubing for cooling; (5) pine box filled with *ca.* 4 in tap water.

C. *Temperature*

In a temperature controlled room the thermostat should be adjusted to 19–20 °C where the differential temperature control is ±2 °C. A cold water bath (48 × 6 × 6.5 in) for cultures may be constructed of 1 in pine wood boxes with plywood bottoms painted white inside and outside; four of these accommodate sixteen 5 gal carboys or twenty 9 l serum bottles immersed *ca.* 4 inches in water. The temperature is regulated to 16–19 °C by a ¼ h.p. Remcor force-flow liquid cooler unit that circulates refrigerated water through copper tubing in the bottom of each box (Fig. 15-3).

D. *Illumination*

In a temperature controlled room with a large culture facility, about 500 ft-c (5400 lux) illumination is provided by a series of 8 ft 80 W cool-white fluorescent tubes 12 in from the carboys and mounted behind glass. The ballasts are removed to a duct supplied with cool air to dissipate the heat. In a smaller unit, where temperature control is by a water bath, light is obtained from four 40 W cool-white fluorescent tubes mounted behind glass 12–20 in from growth chambers and three 40 W tubes suspended above to give illumination of *ca.* 380 ft-c (4100 lux). Adequate safety devices for automatic shut off of illumination should be built into the electrical system in the event of temperature control failure.

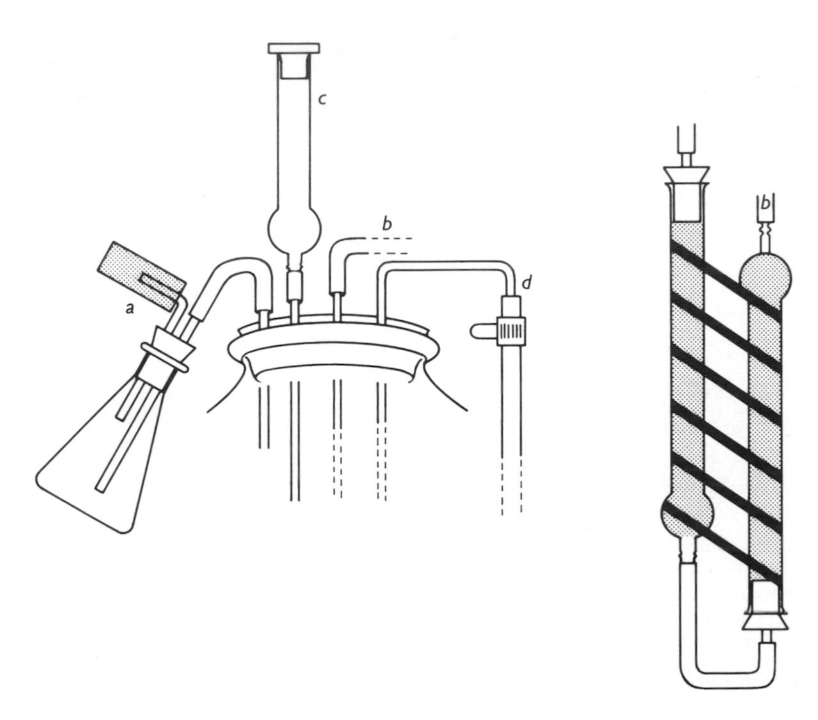

Fig. 15-4. Left: Steril-Kap on culture carboy. *a.* effluent gas escape through 125 ml Erlenmeyer flask; *b.* influent gas tube and 200 mm cotton-plugged drying tubes wound with flexible heating tape (see right for details); *c.* inoculating port with rubber serum bottle stopper; *d.* harvest siphon with polyvinyl bubble tubing attached with worm-drive clamp.

V. PROCEDURES

A. *Preparation of apparatus*

1. Growth chamber. The culture vessel consists of a 5 gal Pyrex solution bottle, with a Steril-Kap stopper. Four holes (five for apparatus no. 2) are bored to accommodate 5 mm glass tubing for the following: (*a*) effluent gas; (*b*) influent gas; (*c*) inoculating port; (*d*) harvest tube (Fig. 15-4). In apparatus no. 2 an additional opening is made for the medium intake line.

a. Effluent gas is allowed to escape through a short glass rod inserted in the carboy cap. This is attached via latex tubing and a glass tube through a rubber stopper to a 125 ml Erlenmeyer flask where the moisture collects at the bottom. Gas escapes from this through a short glass tube in the two-hole rubber stopper, the protruding end of which is inserted in a shell vial plugged with cotton wool (Fig. 15-4*a*).

b. Small, aquarium type stone air breakers are prepared by removing plastic or stainless steel rings that are often attached. The opening must be further enlarged so that it will accept the aerator tube made to adhere to the air stone with the aid of a ring of thin-walled rubber tubing. The aerator consists of an 18 in glass tube that holds the air breaker at the bottom. At the end protruding from the cap a long (*ca.* 36 in) tubing of amber latex rubber connects two 200 mm drying tubes that have been previously attached in series and packed with cotton wool (Fig. 15-4*b*).

c. The inoculating port consists of a 100 mm drying tube capped with a polypropylene Bacti-Cap or rubber serum stopper. The drying tube is attached with latex tubing to a short piece of glass tubing in the culture cap. In apparatus no. 1 this also serves as a media filling port (Fig. 15-4*c*).

d. The harvest siphon reaches to the bottom of the carboy leaving *ca.* 6 in outside of the cap. For apparatus no. 1 the protruding end is bent into a wide U shape. To this is affixed a 36 in length of bubble tubing with a worm-drive clamp. The tubing is cut *ca.* 24 in and a polypropylene stopcock inserted followed by another piece of tubing (8 in) that is attached to a 70 mm aseptic filling bell. For apparatus no. 2 the harvest tube protruding from the cap is bent into an L shape, attached with vacuum tubing in which a polypropylene stopcock has been inserted midway to the culture harvest vessel.

e. For apparatus no. 2, only, a short length of glass tubing is inserted into the fifth opening in the carboy cap. On the outside is attached *ca.* 3 ft of amber latex rubber tubing with a polypropylene stopcock inserted midway.

2. Medium reservoir. The medium reservoir is used only in apparatus no. 2. It consists of a 5 gal Pyrex solution bottle with a 3-hole Steril-Kap and an aspirator outlet for tubing. One hole in the stopper receives the inoculating port for media addition, whereas the other two receive the air vents that consist of 200 mm drying tubes packed with cotton wool. A long length of amber latex tubing terminating in a short glass tube is attached to the aspirator opening at the bottom of the carboy.

3. Culture harvest, collecting vessels. A 9 or 13 l aseptic media storage jar is used as a harvest vessel for culture overflow in apparatus no. 2 (Figs. 15-2, 15-5). A U shaped glass tubing is attached to the medium inlet aperture in the interior of the jar. On the exterior, a short piece of vacuum tubing is connected to the harvest siphon of the culture carboy. When in use, vacuum aperture is plugged with cotton, attached to a drying tube

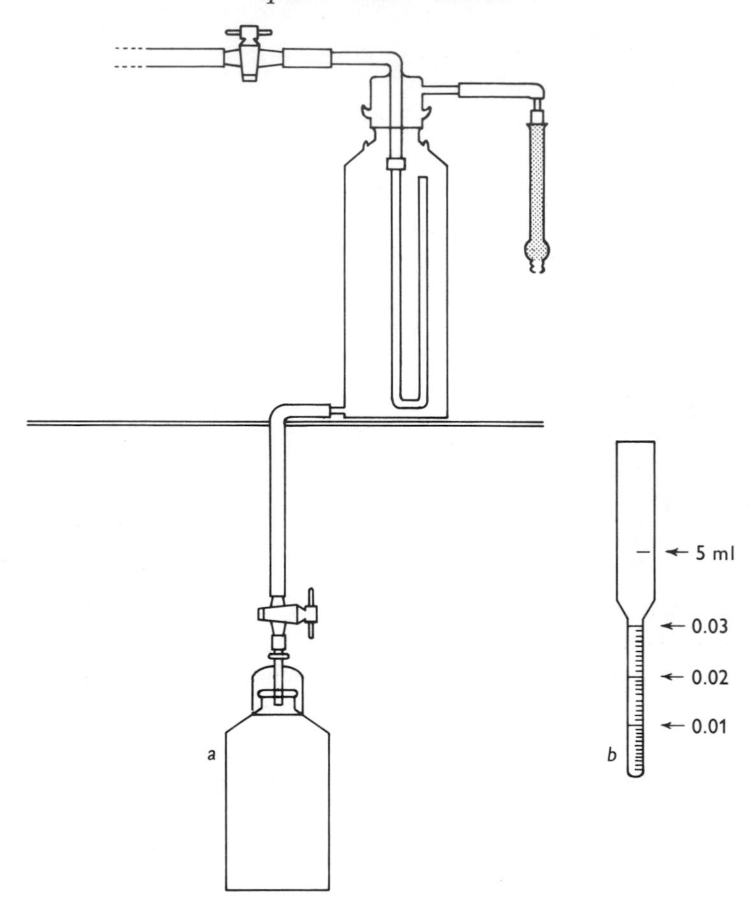

Fig. 15-5. *a*. Details of culture harvesting and collecting vessels for continuous culture (apparatus no. 2); *b*. modified Hopkins centrifuge tube for measuring packed cell volume.

packed with cotton wool, and to the vacuum pump. The tubulation outlet at the base of the jar is fastened to a length of bubble tubing into which is inserted a stopcock and then 70 mm aseptic filling bell. The filling bell rests on a 4 or 9 l culture collecting bottle (Fig. 15-2, item no. 17). In apparatus no. 1 it is the aseptic filling bell at the end of the harvest siphon that rests on the culture collecting bottle (Fig. 15-1, item no. 2).

B. *Preparation of media*

1. Medium. Seawater used for the growth medium is clarified by passing through a 15 μm and 1 μm filter consisting of orlon wound on a polypropylene or polyvinyl core and then through a Gelman cartridge pleated membrane filter (0.45 μm).

The medium consists of:

KH_2PO_4	20 mg
$NaNO_3$	300
NaFe Sequestrene (13% Fe, Geigy Chemical Corp., Saw Mill River Rd, Ardsley, New York 10502)	5
Cyanocobalamin (B_{12})	0.003
Thiamine.HCl (B_1)	0.3
Tris (hydroxymethyl) aminomethane	1 g
Trace metals (see Guillard and Ryther 1962)	
Seawater	500 ml
Demineralized water	to 1 liter

2. *Containers.* For preparation of media for inocula and for maintaining carboy cultures, media is dispensed into the following:

200 ml in 500 ml Erlenmeyer flasks, wide mouth, cotton plugged.

1500 ml in Fernbach flasks with glass siphon inserted in cotton plug (siphon tube closed with Hoffman clamp and cotton plugged 20 mm aseptic filling bell inserted at end of siphon).

4 or 9 l aspirator bottles with 24 in amber latex rubber tubing on tubulation outlet (tubing closed with Hoffman clamp, ending in 20 mm aseptic filling bell plugged with cotton).

c. *Sterilization*

Following assembly, the components are prepared for sterilization with all parts of each section attached. All parts terminating in rubber tubing or glass connecting tubes (to be attached following sterilization) are inserted in shell vials packed with cotton wool. Cotton plugs are wrapped with brown wrapping paper. Valves in empty containers are left open and a small quantity of water placed in each section to ensure wet sterilization. A plastic 'sterilization bag' is useful for covering carboy tops and holding tubing and other parts during sterilization. After sterilization the bag may be left on the carboy top between manipulations as a protection against dust, drafts, and loose tubing. All parts of the assembly are autoclaved for 20 min at 15 lb/in². This includes: growth chamber, medium reservoir, culture harvest vessel, culture collecting bottles, bottles containing medium, covered containers for hemostats and serum stoppers, Fernbach flasks and Erlenmeyer flasks with medium.

D. *Assembly*

Following sterilization the growth chamber and other vessels are placed in position as indicated in Figs. 15-1 or 15-2. Wrappings are removed and the following aseptic connections made: in apparatus no. 2 the medium reservoir outlet tube is aseptically attached by the glass connector to the medium filling tube of the growth chamber with the aid of liberal amounts of ethyl alcohol. Similarly, the culture harvest tube is attached to the culture harvest bottle via the stopcock connector. In both no. 1 and no. 2 the culture collecting bottle is put into position on the floor shelf, the cotton plugs quickly removed, and the 70 mm aseptic filling bell placed over the culture collecting bottle. Media bottles are placed on the top shelf and added to the culture carboys as needed by rapidly replacing the serum stopper on the medium filling port with the 20 mm aseptic filling bell.

E. *Inoculation*

1. Preparation. Stock cultures (200 ml) are allowed to attain high densities in wide mouth 500 ml Erlenmeyer flasks in an illuminated incubator at 18–20 °C. The entire contents of the flask are used to inoculate a Fernbach flask and similarly incubated until dense growth appears. This usually is ready for use in 10–15 days, but may be incubated until needed for an inoculum, which may be as long as 3 months without adverse effects.

2. Inoculation. Cultures in Fernbach flasks are placed on the top shelf directly over the growth carboy, the serum stopper on the inoculating port is quickly exchanged for the aseptic filling bell on the siphon of the Fernbach flask, the clamp on the siphon partially opened and the culture allowed to gravity feed slowly into the carboy. After the entire culture has drained into the carboy, a sterile serum stopper taken from a covered sterile container with the aid of a hemostat is quickly placed on the media filling port as the aseptic filling bell is removed. Each carboy is inoculated with 2 Fernbach flask cultures (*ca.* 3 l) and 4 l of sterile medium.

3. Initial growth phase. After exponential growth starts (*ca.* 3–5 days) an additional 4 l of sterile medium is added. The culture is then again allowed to increase in density before additional medium is added. This procedure is continued until the total capacity of the growth chamber and the desired culture density are reached. The total carboy is filled with medium and the culture density is increased before vigorous aeration is started. During early stages of growth, bubbling is kept very low, with actual rates of flow being adjusted by reducing valves. As the culture increases in density, the amount of bubbling is also increased. This must not be premature or it will result in culture failure. Agitation, as such, is not necessary since bubbling prevents stratification of cells, provides for gas and heat transfer, light intermittency, dispersion of dissolved materials, and inhibition of the adherence of cells to the carboy wall.

F. *Maintenance*

In practice the original inoculum is allowed to grow until the required population density is reached before harvesting is started. A constant population density is maintained by adding fresh medium as the culture grows establishing an equilibrium between growth rate and dilution rate. In apparatus no. 1 a sample is withdrawn and the culture diluted at a fixed time each day; in apparatus no. 2 the culture is continuously diluted.

The harvest siphon tube is primed initially by clamping off the effluent air outlet, closing the inoculating port and turning on the air/CO_2 system so that aeration is vigorous. Once the siphon is started the air pressure is reduced and the clamps removed.

In apparatus no. 1 the siphon stopcock is opened and the predetermined culture volume (e.g., 3–4 l of a dense culture) is allowed to drain into the collecting bottle. The stopcock is then closed and an empty sterile collecting bottle rapidly substituted for the bottle containing the harvested culture. This is done by exchanging the sterile cotton plug on the empty bottle for the aseptic filling bell. With the air/CO_2 system closed, sterile medium equal in volume to culture harvested is added to the culture carboy. The serum stopper from the medium addition port on the carboy is removed while simultaneously pulling the cotton plug from the aseptic filling bell of the medium bottle and placing it on the port. The Hoffman clamp is partially opened and sterile medium (e.g., 3 l) enters the carboy in a slow stream. The clamp is closed and the filling bell is removed and the port quickly covered with a sterile serum stopper or Bacti-Cap (removed from a covered container with the aid of a sterile hemostat).

In apparatus no. 2 medium is fed continuously from the medium reservoir to the culture carboy by means of a single or multi-channel peristaltic pump. The rate of flow is controlled by adjustment of the stopcock on the tube from the medium reservoir to the culture carboy and by controlling the micrometer flow knobs on the pump. The rate is adjusted to about 4 ml/min and an interval timer regulates the pump to operate 30 min in each hour for 24 h. Culture overflow in the harvest vessel is primed by a brief vacuum and then regulated by adjusting the opening of the stopcock in the harvest tube and the extent of bubbling. The rates of medium inflow and culture outflow are made equal so that a constant volume is maintained.

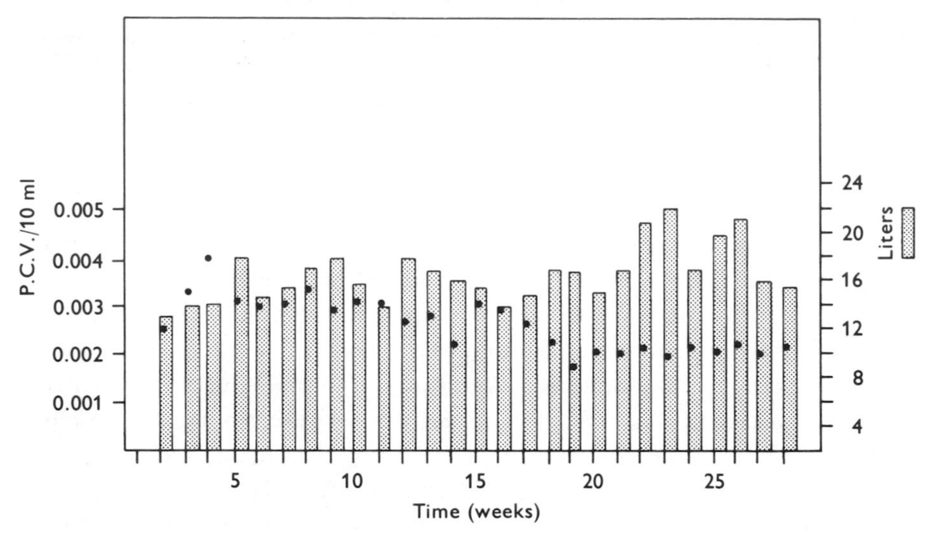

Fig. 15-6. Weekly average of daily packed cell volume (dots) and volume of culture harvested (strips) from carboy culture of *Monochrysis lutheri* (apparatus no. 1).

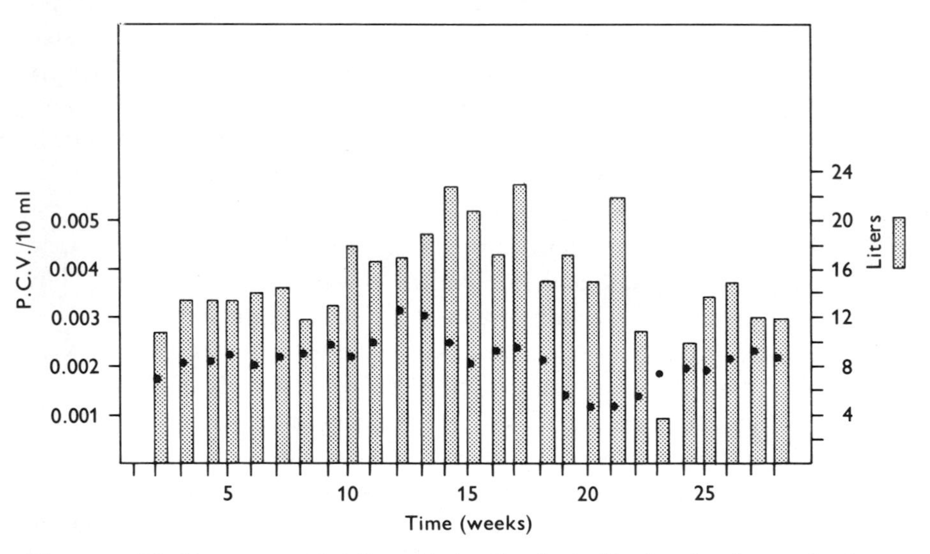

Fig. 15-7. Weekly average of daily packed cell volume (dots) and volume of culture harvested (strips) from carboy culture of *Isochrysis galbana* (apparatus no. 1).

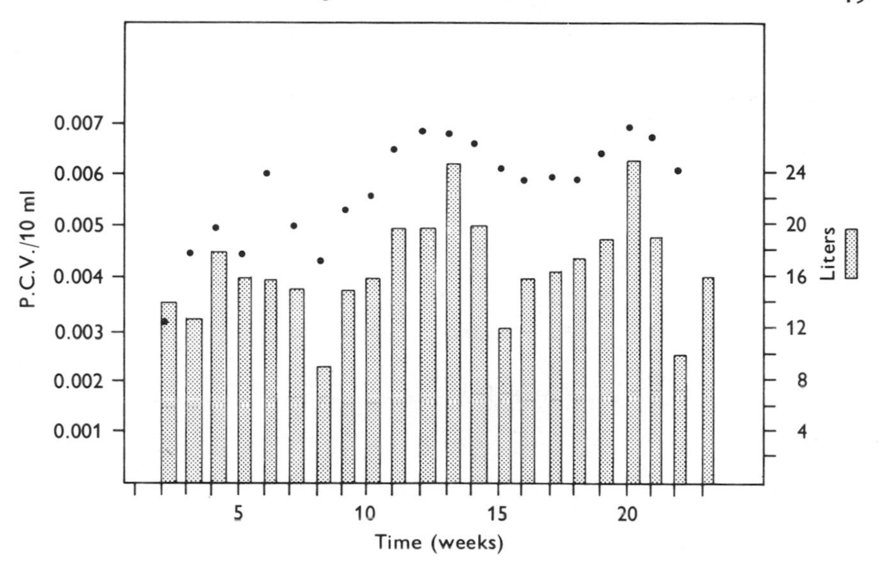

Fig. 15-8. Weekly average of daily packed cell volume (dots) and volume of culture harvested (strips) from carboy culture of *Phaeodactylum tricornutum* (apparatus no. 1).

VI. SAMPLE DATA

Determinations of cell densities of the harvested culture are made by cell counts on a representative sample or determining the centrifuged packed cell volume of a 10 ml aliquot. This is determined in a modified Hopkins centrifuge tube with the capillary portion extended and calibrated to read 1×10^{-3} ml (Fig. 15-5*b*).

A flexibility in harvesting schedule of cultures used for feeding purposes is useful since volume requirements will vary depending upon the age of the animal and variations in culture density. Figs. 15-6, 15-7, 15-8 show the weekly average packed cell volumes and liters harvested for *Monochrysis lutheri*, *Isochrysis galbana* and *Phaeodactylum tricornutum*. Cultures were incubated with overhead lighting (Fig. 15-11). The daily cell counts for each 24 h where 3 l of culture/24 h were harvested are shown for *M. lutheri* and *I. galbana* in Fig. 15-9 and for *P. tricornutum* in Fig. 15-10. These cultures were incubated by vertical lighting, shown in Fig. 15-12.

VII. PROBLEM AREAS

One of the potential problems in construction of a culture apparatus is toxicity in formulations of some materials (Dyer and Richardson 1962). New materials should be evaluated for effects on individual algal species

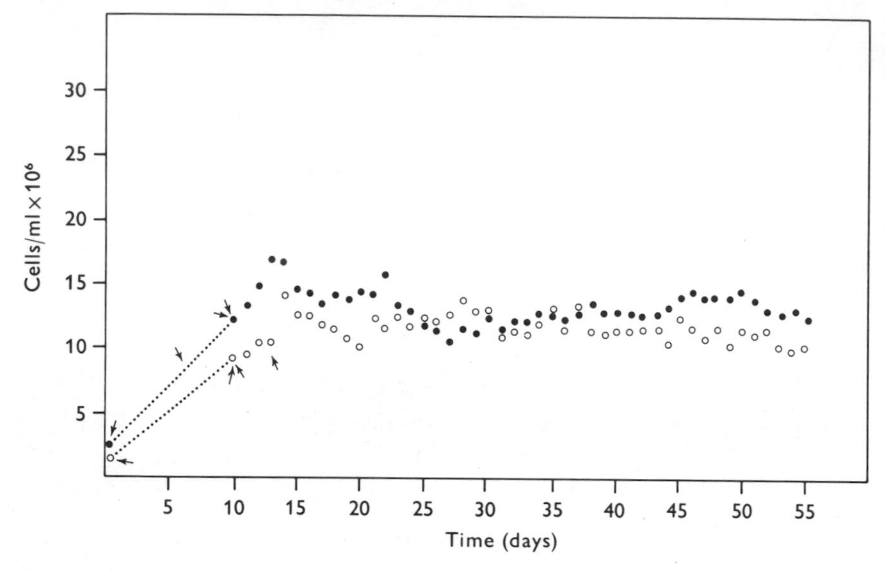

Fig. 15-9. Cell counts of cultures (apparatus no. 2) harvested daily at 3 l/24 h; arrows indicate addition of media (3 l), harvesting started at 14th day; ○ *Isochrysis galbana*; ● *Monochrysis lutheri*.

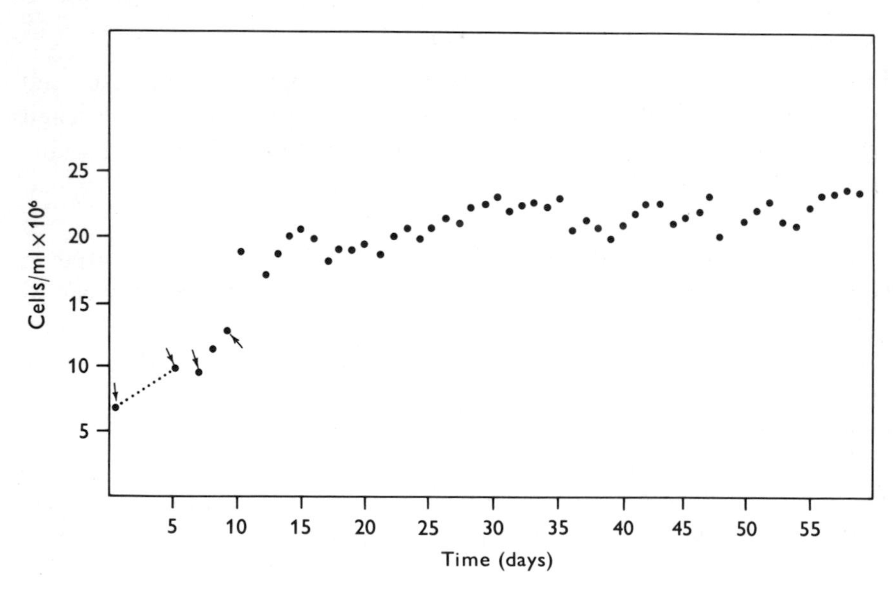

Fig. 15-10. Cell counts of *Phaeodactylum tricornutum* (apparatus no. 2) harvested daily at 3 l/24 h; arrows indicate addition of media (3 l), harvesting started at 14th day.

and for toxicity to the species feeding on the algae (also, see Chp. 14). Seawater is brought into the laboratory through polyvinyl chloride or lead delivery lines and hard rubber stopcocks. Seawater or media must be filtered in the absence of stainless steel since contact with certain types of stainless steel makes the media toxic to young larvae. Oil in the air line is another source of toxicity that can adversely affect algal growth. Filters should be used with compressed air supplies to deliver clean air free from oil, condensed water, and other impurities.

If adhesion of cultures to glass walls of carboys is excessive, light may be occluded. Generally this does not occur if cultures remain free of contaminants. Certain media, however, may enhance development of this phenomenon since the mechanism of adhesion is dependent upon the ionic constituents of the medium (Nordin, Tsuchiya and Fredrickson 1967).

A significant difficulty in operation of continuous culture is the susceptibility to contamination. Cotton wool is usually an effective filter, but in some situations may be a source of contamination. Random porosity, uneven packing, rapid pressure changes, and excessively high air flow may produce passages large enough for contaminants to enter. Moisture in the filter fibers also promotes channel formation and growth of contaminants within the filter. Another possible source of contamination is an unreliable seal of plastic tubing stretched over glass, and cracked or distorted rubber tubing. Slow growing contaminants are excluded from competition and would wash out, but contaminants with a high growth rate may compete with the algae. The area chosen for the culture apparatus should be evaluated for density of airborne contaminants by exposure of nutrient agar plates for short periods of time. Other potential sources of contamination, such as the medium, inoculum, and the filtered air/CO_2 gas mixture, should be ascertained for purity by inoculation of fluid thioglycollate media made up in seawater.

Early stages of carboy growth are most critical. Special attention should be paid to the rate at which media is added and the gas introduced. An excessive concentration of CO_2 will result in a pH less than optimum (for marine species less than pH 7.3). A CO_2 concentration of 1–2% is generally adequate for culturing algae at low light intensities but this depends on a variety of factors as light intensity, culture density, growth rate, gas dispersion, and size and shape of growth chamber. If pH does begin to decrease due to excessive CO_2, turn off the CO_2 supply for 1–3 days until the pH is more favorable (if the culture is not contaminated).

VIII. REFERENCES

Ammann, E. C. and Fraser-Smith, A. 1968. Gas exchange of algae. IV. Reliability of *Chlorella pyrenoidosa*. *Appl. Microbiol.* **16**, 669–72.

Bach, M. K. 1960. Mass culture of *Euglena gracilis*. *J. Protozool.* **7**, 50–2.

Benoit, R. J. 1964. Mass culture of microalgae for photosynthetic gas exchange. In Jackson, D. F., ed., *Algae and Man*, pp. 413–25. Plenum Press, New York.

Bohlin, K. 1897. Zur Morphologie und Biologie einzelliger Algen. *Öfvers, Kongl. Svenska Vetenskapsakad. Forhandl.* **9**, 507–29.

Butcher, R. W. 1952. Contributions to our knowledge of the smaller marine algae. *J. Mar. Biol. Ass. U.K.* **31**, 175–91.

Butcher, R. W. 1959. Introduction and Chlorophyceae. *An Introductory Account of the Smaller Algae of British Coastal Waters*. Part I, pp. 1–74. Fishery Investigations Series IV, London.

Carpenter, E. J. 1968. A simple, inexpensive algal chemostat. *Limnol. Oceanogr.* **13**, 720–1.

Conger, A. D. 1969. A simple gas-washing bottle. *BioScience* **19**, 54–5.

Cook, J. R. 1968. A continuous culture device for protozoan cells. *J. Protozool.* **15**, 452–5.

Davis, H. C. 1969. Shellfish hatcheries – present and future. *Trans. Amer. Fish. Soc.* **98**, 743–50.

Droop, M. R. 1953. On the ecology of flagellates from some brackish and fresh water rockpools of Finland. *Acta Bot. Fennica* **51**, 1–52.

Droop, M. R. 1966. Vitamin B_{12} and marine ecology. III. An experiment with a chemostat. *J. Mar. Biol. Ass. U.K.* **46**, 659–71.

Dyer, D. L. and Richardson, D. E. 1962. Materials of construction in algal culture. *Appl. Microbiol.* **10**, 129–31.

Eppley, R. W. and Dyer, D. L. 1965. Predicting production in light-limited continuous cultures of algae. *Appl. Microbiol.* **13**, 833–7.

Fogg, G. E. 1965. *Algal Cultures and Phytoplankton Ecology*. University Wisconsin Press, Milwaukee. 126 pp.

Fuhs, G. W. 1969. Phosphorus content and rate of growth in the diatoms *Cyclotella nana* and *Thalassiosira fluviatilis*. *J. Phycol.* **5**, 312–21.

Guillard, R. R. L. and Ryther, J. H. 1962. Studies of marine planktonic diatoms. I. *Cyclotella nana* Hustedt, and *Detonula confervacea* (Cleve) Gran. *Can. J. Microbiol.* **8**, 229–39.

Hannan, P. J. and Patouillet, C. 1963. Gas exchange with mass cultures of algae. I. Effects of light intensity and rate of carbon dioxide input on oxygen production. II. Reliability of photosynthetic gas exchanger. *Appl. Microbiol.* **11**, 446–52.

Hare, T. A. and Schmidt, R. R. 1968. Continuous-dilution method for the mass culture of synchronized cells. *Appl. Microbiol.* **16**, 496–9.

Howell, J. A., Tsuchiya, H. M. and Fredrickson, A. G. 1967. Continuous synchronous culture of photosynthetic microorganisms. *Nature* **214**, 582–4.

James, T. W. 1961. Continuous culture of microorganisms. *Ann. Rev. Microbiol.* **15**, 27–46.

Lerche, W. 1937. Untersuchungen über Entwicklung und Fortpflanzung in der Gattung *Dunaliella. Arch. Protistenk.* **88**, 236–68.

Maddux, W. S. and Jones, R. F. 1964. Some interactions of temperature, light intensity, and nutrient concentration during the continuous culture of *Nitzschia closterium* and *Tetraselmis* sp. *Limnol. Oceanogr.* **9**, 79–86.

Málek, I. and Fencl, Z., eds. 1966. *Theoretical and Methodological Basis of Continuous Culture of Microorganisms.* Academic Press, New York. 654 pp.

Matthern, R. O. and Koch, R. B. 1964. Developing an unconventional food, algae, by continuous culture under high light intensity. *Food Technol.* **18**, 58–65.

McLachlan, J. 1960. The culture of *Dunaliella tertiolecta* Butcher – a euryhaline organism. *Can. J. Microbiol.* **6**, 367–79.

Monod, J. 1950. La technique de culture continue; théorie et applications. *Ann. Inst. Pasteur* **79**, 390–410.

Myers, J. and Clark, L. B. 1944. Culture conditions and the development of the photosynthetic mechanism. II. An apparatus for the continuous culture of *Chlorella. J. Gen. Physiol.* **98**, 103–12.

Nordin, J. S., Tsuchiya, H. M. and Fredrickson, A. G. 1967. Interfacial phenomena governing adhesion of *Chlorella* to glass surfaces. *Biotech. Bioengineer.* **9**, 545–58.

Parke, M. 1949. Studies on marine flagellates. *J. Mar. Biol. Ass. U.K.* **28**, 255–85.

Pipes, W. O. and Koutsoyannis, S. P. 1962. Light-limited growth of *Chlorella* in continuous cultures. *Appl. Microbiol.* **10**, 1–5.

Shihira, I. and Krauss, R. W. 1965. *Chlorella: Physiology and Taxonomy of Forty-one Isolates.* Univ. Maryland, College Park. 97 pp.

Spoehr, H. A. 1953. The need for a new source of food. In Burlew, J. S., ed., *Algal Culture from Laboratory to Pilot Plant*, pp. 24–8. Carnegie Inst. Washington Publ. 600.

Tamiya, H. 1960. Role of algae as food. In Kachroo, P., ed., *Proceedings, Symposium on Algology*, pp. 379–89. Indian Council Agric. Res. and UNESCO South Asia Sci. Coop. Office, New Delhi.

Taub, F. B. and Dollar, A. M. 1968. Improvement of a continuous-culture apparatus for long-term use. *Appl. Microbiol.* **16**, 232–5.

NOTES ADDED IN PROOF

Contamination of culture carboys is less likely to occur if the following modification in the 20 mm aseptic filling bell is made: the tubular projection on inside of the bell is inserted into a hole drilled in a serum bottle rubber stopper, and the stopper pushed into the upper part of the bell so that it fits securely. The tight fit of the stopper in the medium addition port, and coverage by the filling bell is protection against airborne contamination during medium and culture addition to the carboy. A 200 mm drying tube packed with cotton should be attached at Fig. 15-4*a*.

Packed cell volumes three and four times as high as shown in the figures (Figs. 15-6 to 15-8) can be achieved by allowing the cultures to reach higher densities before harvesting is begun and by harvesting cultures at intervals greater than 24 hours.

Fig. 15-11. Culture vessels incubated in water bath with overhead lights.

Fig. 15-12. Culture carboys incubated in cold room with vertical light bank.

16: Mass culture

GLEN HEMERICK

Route 5, Box 5606, Gig Harbor, Washington 98335

CONTENTS

I. INTRODUCTION

Mass algal culture is contemplated as a future food or animal feed because of the high protein content of some microalgae and because their high yield of photosynthetic products per unit time and area of solar illumination (Tamiya 1962). Other proposed applications of large scale culture include purification of water, production of oxygen (Hemerick and Benoit 1962), phycocolloids (such as algin), and pigments for converting solar energy into electrical energy (Hemerick 1968).

Different objectives determine the species and environmental conditions controlled. Among objectives can be listed: (a) production of large amounts of a particular algal species; (b) production of an algal product obtainable from any one of a number of species; and (c) production of large samples of algae from a number of species.

II. EQUIPMENT

Choice of equipment varies with major source of illumination and growth habits of organism: solar *vs* artificial illumination; and planktonic *vs* attached, or benthic algae.

A. *Vessels*

1. Planktonic algae. Algae such as *Chlorella* (Chlorophyceae), *Chlamydomonas* (Chlorophyceae), *Porphyridium* (Rhodophyceae), and *Anacystis* (Cyanophyceae), when grown in sunlight, are maintained on agar test tube slants, and then the populations increased by growth in CO_2/aerated liquid medium in flasks (1 l), jars (5 l battery jars, 10 l wide mouth bottles, or 1 gal jars) or pools. When these algae are grown indoors, space is saved by illuminating dense cultures internally with high-intensity lamps. Culture may be grown in internally illuminated vertical jars or horizontal or vertical vessels made from glass or plastic cylinders (see B following, and Hemerick and Benoit 1962). These vessels are adaptable to continuous culture (see Chp. 15, II.A).

2. *Attached algae.* Algae, such as filamentous Cyanophyceae, are also maintained on agar test tube slants. The population is increased in CO_2/aerated liquid cultures in polyethylene plastic bags inside flasks or jars, and in pools lined with plastic sheets. Convenient pools for mass culture consist of wooden trays 1 × 3 m by 200 mm deep. These trays are supported on concrete blocks 300 mm above the ground. In use, the trays are lined with two sheets of polyethylene. The rolls of polyethylene are 500 mm wider than pool. They are cut at the fold so that the fold-creases are not under water as they may leak. The trays are housed in a 'greenhouse' constructed of plastic sheet (Mylar) stretched over wire netting on a wooden frame. Ventilation is provided by an exhaust fan with a dust filter for incoming air. The greenhouse has thermostatically-controlled heat, electrical outlets and cold tap water. Indoor pools can be made by lining the bottom of an aquarium or a box with double sheet plastic. They are illuminated by overhead lamps.

An internally-illuminated culture has three concentric cylinders: the inner cylinder has an air-cooled lamp, the middle has coolant water, and finally the outer contains the culture. These cylinders may consist of: long, thin test tube for a quartz–iodine lamp and air (inner cylinder); larger test tube for water (middle cylinder); and jar for the culture. The lamp is contacted and supported by copper wires insulated with borosilicate (Pyrex) glass tubing. The copper wires extend through borosilicate glass tubing bent to form spring contacts for a 125 V, 500 W quartz–iodine lamp. The inner cylinder is inserted into one hole of 3-hole rubber stopper that fits the middle tube. The other two holes are for water circulation in the middle tube. The middle tube is inserted through a metal cover with several openings of a 10 l wide mouth jar containing the culture. CO_2/air mixture can be bubbled through a tube inserted in one opening on the cover and O_2-enriched air through another tube. If desired, fresh medium can be added through a third tube; whereas a fourth opening serves as air-exit or overflow tube. A thermometer should also be inserted in one opening.

B. *Lights*

Algal cultures can be grown with external or overhead illumination if the algae are either planktonic or benthic. If the algae are planktonic, they can be also illuminated internally by a 500 W quartz–iodine lamp.

The cultures may be lighted horizontally by 40 W fluorescent lamps. These lamps are available in various shades (also see Chp. 11, II.C.1, concerning lamps). Incandescent tungsten lamps require further temperature control systems.

Overhead illumination of pool cultures can be by clear heat lamps, but outdoor flood lamps are more resistant to shattering, if water drops should strike the bulb.

C. *Carbon dioxide*

The laboratory and pool areas should be provided with cylinders (40 lb) of CO_2 and systems for mixing and distribution of air and CO_2. Various systems under high or low pressure are possible, but a satisfactory one is as follows. A pressure regulator with valve is attached to the CO_2 cylinder, and plastic tubing conducts the CO_2 to a flowmeter. A supply of air can be provided by an aquarium pump. The air passes through a flow-meter at 50–100 times the flow rate of CO_2 and the two gas streams are mixed at a tee. This gives a mixture of 1–2% CO_2 in air. The mixture is distributed through plastic tubing to the cultures. A 3-way brass valve is provided for each culture; the valves are connected by plastic tubing. A plastic tube extends from each valve to a culture, terminating in a glass tube (or inverted pipette) in the culture. At the point farthest from the gas source, a long tube with no valve is immersed in a cylinder of water whose depth is greater than that of the cultures. If bubbling occurs in the deep water, all of the cultures will be aerated. The brass valves are adjusted so that limited bubbling occurs in all cultures. Excessive bubbling can cause excessive evaporation of water from the medium. The mixture of air and CO_2 can be passed through activated charcoal to remove organic contaminants, and through a bacterial filter to remove bacteria. The entire system can be autoclaved if the tubing is of the Tygon type. If not, it can be sterilized by connecting a bottle with 2-hole stopper, at the head of the system, forcing liquid through the system with air pressure, first a bleach solution, followed by sterile water, and finally thiosulfate solution (see II.F, following).

D. *Temperature*

Aquaria (50 gal) serve as controlled temperature tanks for algal cultures in glass vessels, immersed in water in the aquaria. Temperature of water in the aquarium (tank) can be controlled by circulating water through a coil of copper tubing immersed in the tank. The circulated water can come from a tap where hot and cold water are blended to maintain a desired temperature; or only cold water can be circulated through the coils in the tanks and then each tank is heated by a thermostatically-regulated immersion heater.

Internally illuminated cultures can have a tin or stainless coil in the culture itself, or the culture vessel can be immersed in a larger basin of water of controlled temperature. The coolant water for the lamp should be recirculated distilled water because tap water contains organisms and particulate matter which adhere to the glass and obstruct the light.

Pools are heated by adding overhead lamps, or by turning up the thermostat of the greenhouse when no one is in it (cooling it with the exhaust fan when work is to be done). Pools are cooled by a thermostat-regulated exhaust fan, and by flat coils of copper tubing, placed under the double plastic liner of the pool, and through which cold or constant temperature water is circulated.

Minimum–maximum thermometers are placed in the water baths or in a sterile beaker of water in a pool to record temperature extremes. This is because cooling systems may fail, or thermostats stick, temporarily, allowing a temperature rise that is deleterious to the algae.

E. *Turbulence, circulation*

Turbulence and circulation in the temperature tanks and culture medium are important in maintaining uniform temperature. In addition, they help provide light, CO_2, and nutrients; distribute oxygen and other metabolic products. The medium in pools of both planktonic and attached algae can be circulated by an immersion pump. Care must be exercised that large clumps of inoculum do not clog the intake screen of the pump. The pump can be placed inside a sterile wire basket which will screen out the larger clumps. The air/CO_2 supply tube can be placed at the intake of the pump with the result that very small bubbles of gas are distributed throughout the medium.

F. *Media*

In my work with about 200 species of Chlorophyceae, Cyanophyceae and Rhodophyceae, only two media have been used. These are the freshwater BGM (Allen 1963) and seawater-based SWM (see Chp. 2, IV.A.5). For agar slants or plates use either 1 or 2% agar.

1. BGM. BGM contains 13 compounds prepared as individual concentrated solutions (200 times) in individual containers. All are used at 5 ml/l water. The preparation is as follows:

a. Major stocks (for each use 5 ml of stock solution/liter water)

i. KNO_3	200 g/l
ii. K_2HPO_4	70
Should be stored in Pyrex or plastic bag.	
iii. $MgSO_4.7H_2O$	50
iv. NaCl	34
v. FeEDTA	10

('Versene' or 'Sequestrene' supplied by Carolina Biological Supply Co., 2700 York Rd, Burlington, North Carolina 27215.) Store in refrigerator; anaerobically, if possible.

vi. $CaCl_2$	5 g/l

b. Boric acid (use 5 ml/l water of 200 × solution)

H_3BO_3	20 g/l

Dilute 20 ml/l; this is 200 ×.

c. Minor stocks (for each, dilute the original solution at 5 ml/l to get 200 × concentration; then, for each, use 5 ml of 200 × solution/liter water).

i. $MnCl_2.4H_2O$	72 g/l
ii. $(NH_4)_2MoO_4$	5.2
iii. $ZnCl_2$	4.0
iv. $CuSO_4.5H_2O$	3.2
v. $CoSo_4$	2.0
vi. NH_4VO_3	0.8

2. *SWM.* This seawater medium contains 14 compounds. Iron (FeEDTA), boron (H_3BO_3) and the six minor element compounds (1.c, preceding) are used as in BGM. The phosphate was used at half the concentration in BGM (use 2.5 ml stock/liter water). Other compounds are:

i. NaCl	29 g/l
ii. $MgSO_4.7H_2O$	12.3
iii. $NaNO_3$	1.7
iv. $CaCl_2.2H_2O$	1.47
v. KCl	0.75

The NaCl and $MgSO_4$ are made new each time. They are dissolved separately in a blender, adding water to the undissolved residue until completely dissolved, more water is added, leaving room for the other solutions, and finally bringing to correct volume.

For pools, the major compounds are weighed dry, dissolved and added to sterile water in the pool (see 3, following); minor compounds are added from stock solutions. The level of medium is marked in the pool and maintained at this level by adding sterile tap water.

3. Water. In most areas tap water is not sterile. In addition to bacteria it may contain living algae as well as zooplankton. Furthermore, public water supplies are treated with chlorine, copper sulfate, or other algicides at times of algal blooms.

Water can be tested for toxicity by bioassay with an alga of known sensitivity. Motile cells of some algal species (such as *Chlamydomonas*) are sensitive to toxic substances and stop swimming if the water is toxic. If the only water supply is toxic, wait a few days for the toxicity to decline as at most seasons of the year public water supplies are not very toxic.

For small cultures, distilled water can be used; but for larger cultures, large volumes of pure water are needed. Hot tap water may be used by storing in large plastic tanks until cool. Or water can be heated by immersion heaters to kill nuisance organisms.

If cold water is used for mass culture, unwanted organisms can be eliminated by the following procedure.

Fill the pool 100 mm with water. Add 5% chlorine (commercial bleach, 3 ml/l = 150 ppm Cl_2); let stand overnight. Then add sodium thiosulfate (150 mg/l); let stand overnight again. Take a sample for bioassay with the test organism.

Equipment, such as the circulation pump, is sterilized at higher concentrations of bleach or boiled in water while running.

III. TEST ORGANISMS

Porphyridium cruentum (Allen 1963, R-1.1.1) is a unicellular red alga (Rhodophyceae) which grows well in SWM at 25 °C.
Fremyella diplosiphon (IUCC 590) is an attached filamentous cyanophyte (Cyanophyceae) growing well in BGM at 35 °C. Both species move on agar surfaces.

IV. METHOD

A. *Mass culture*

1. Single species. Liquid cultures are grown similarily to that described in Chp. 15, V.E.1, 2. Growth rates of planktonic species can be determined turbidimetrically; whereas, yield of filamentous, attached species by fresh or dry weight. Cultures can be tested for growth with different light sources, and temperatures. Temperature significantly affects yield at high light intensity (Hemerick and Benoit 1962). Sometimes it is necessary for the algal culture to adjust to high levels of light, temperature, CO_2 and nutrients.

2. Algal product. If an algal product is desired that is available from any one of a number of cultures, it should be tested for comparative yield. This should be either at the ambient conditions of the mass culture system or at high levels of light and temperature.

3. Large mixed cultures. When mixed samples are desired, the various species can be tested for conditions permitting high productivity. It may be convenient to grow high temperature algae during summer, and low temperature algae in winter.

B. *Axenic cultures*

The probability of bacterial contamination of cultures increases with the size of culture units used; however, from a practical standpoint, a few bacterial cells with a large mass of algae is of little significance.

The size of axenic culture units is limited to means of sterilization of the apparatus and media. Planktonic algae can be grown in a number of 1 gal sterile jugs with 2-hole silicone rubber stoppers containing two glass tubes. One long tube is for aeration and a short one for effluent air. These jugs are connected in parallel, or series, by valves and sterile rubber tubing. Bacterial filters are placed at the inlet and outlet ends. Several sets of jugs containing liquid medium can be autoclaved, cooled and carefully inoculated with axenic cultures of algae. Keeping bacteria out of the mass culture during and after harvest is difficult.

Attached algae may be grown in liquid media in clear plastic (Mylar) bags, in sterile jugs prepared as for planktonic algae. The plastic bags should be reinforced with autoclave tape (see C.2, following).

C. *Xenic cultures*

In xenic cultures rapid growth of the algae is most important. Agar cultures should be transferred frequently and grown at nearly optimum light and temperature. Usually 2–4 days gives abundant growth.

1. Sterilization, materials. For mass culture of either planktonic or benthic algae, small volumes of medium can be sterilized by autoclaving, while large culture units are sterilized by the chlorine and thiosulfate method (see II.F.3, preceding). The CO_2/air tubing and valve system is also sterilized and sterile air filters installed. Inocula can be grown in glass containers lined with clear plastic bags as the bags can be discarded after use. The glass vessel remains clean and need not be washed. New plastic bags are usually free of contaminating organisms, but they can also be rinsed with bleach, sterile water, and thiosulfate (see II.F.3, preceding).

2. Inoculation. Pieces of agar culture are placed in the bag and the closed bag rubbed with the fingers to disperse the algae. The bags are placed in the glass containers and sterile medium added. Glass aeration tubes are placed in the bags and elastic bands placed around the mouth of the bags. The cultures are illuminated and aerated at favorable temperature and light.

When algal growth is visible the cultures are transferred to larger and/or additional culture units. Planktonic algae are poured into new media. Transfer of attached algae is more complicated. Pour off the liquid in the bag. Then rub the bag to loosen the algae. Sterile medium is added to the bag, resuspending the algae which are then put into new media.

An algal crop should be harvested in a few days, or less than a week, because bacteria grow considerably in culture, and algal growth rates slow down.

3. Harvesting

a. Planktonic algae are harvested with a continuous centrifuge, the used medium going to an empty pool. A dairy cream separator is a good centrifuge if the inner disks are removed and it is operated at a very low flow rate. The algae collect inside the head or bowl of the separator, and are removed periodically. Other types of continuous centrifuges collect the algae as a thick suspension which is further concentrated in a tube or bottle-type centrifuge.

b. Non-motile planktonic algae can be harvested by a combination of physical methods. If the culture is shaded, cooled, and the circulation pump shut off, the algae settle. The upper layer of medium can be removed by siphoning or pumping. Then only the bottom portion is centrifuged.

c. For attached forms, the aeration tubes are removed from the plastic bag and the liquid medium poured off. The bag collapses or may be collapsed by hitting the surrounding container against the hand. The bag is removed from the container and the algae are loosened by rubbing the bag between the hands. Sterile medium is added to the bag, shaken some, and the algae are poured into a collection vessel. The sample may be treated as with nonmotile planktonic forms.

Attached algae may also be harvested after the medium is removed by cutting the bag, laying it flat and scraping the algae with a sterile spatula.

Algae attached to a plastic sheet at the bottom of a pool are harvested by removing the medium. Then the algae on the sheet are scraped together and scooped into a container. They may be further concentrated by centrifuging in large plastic tubes, and dried (at 50 °C) or frozen.

V. SAMPLE DATA

TABLE 16-1. *Fresh weight in relation to temperature of Fremyella diplosiphon (Cyanophyceae)*

Temperature °C	20	25	30	35	40
Algae (g/10 l)	12	28	33	42	31

TABLE 16-2. *Fresh weight in relation to iron in medium of Fremyella diplosiphon (Cyanophyceae)*

FeDETA (mg/l)	12.5	25	50	100	200
Algae weight (g/10 l)	24	31	35	34	29

TABLE 16-3. *Dry weight in relation to CO_2/air concentration of Porphyridium cruentum (Rhodophyceae)*

% CO_2	0.5	1	2	4	8
Algae (mg/10 l)	629	752	750	755	321

VI. POSSIBLE PROBLEMS

A. *Overheating*

Cultures may become too warm due to a sticking thermostat or coolant pump; this may happen in the night and cultures may be the right temperature in the morning, but dead. Use minimum–maximum thermometers to monitor temperature fluctuations and check and reset them every morning.

B. *Electric shock*

Fluorescent lamps often leak electricity; ground them with copper wire to water pipes. Also, the temperature tanks, both the metal frame and the water inside should be grounded. A pool of medium with an electric immersion pump, all in a plastic liner, is hazardous and it should be grounded with a sterile stainless steel wire clamped to a water pipe.

C. *Aeration*

Carbon dioxide cylinders are usually uniformly filled and their length of service can be predicted if a uniform flow rate is used. They should be labeled 'empty' and returned before becoming empty.

The aeration systems somctimes develop a leak and stop bubbling. The leak can be located by clamping off sections of the system, and by placing the valves in water, with normal gas pressure inside of them.

D. *Toxicity to algae*

If the algae fail to grow, try comparative flask cultures using the questionable medium, distilled water and fresh tap water medium. Keep a supply of sterile water of known quality available for this purpose. Sometimes people have difficulty getting algae cultures to grow, especially in new laboratories. The culture environment may have any number of possibly toxic conditions which are new to the algae cultures, such as toxic rubber tubing, rubber stoppers, air, CO_2, glass, water, distilled water, etc. New cultures transferred to large agar slants, kept cool (20 °C) or at a series of temperatures, and illuminated with fluorescent light, will usually survive and grow. (Also see Chp. 14 concerning toxic materials.)

VII. ALTERNATIVE TECHNIQUES

Some planktonic algae such as *Chlorella* have been grown with fluorescent lamps immersed in the culture medium. This method is not very satisfactory because the algae tend to stick to the lamp, reducing illumination. The low light intensity does not permit rapid growth nor dense cultures.

Some filamentous, attached algae are grown as planktonic algae, with violent agitation. Even so, they usually stick to the vessel. However, some filamentous species vary between the attached and planktonic habits, and can be selected for whichever habit is preferred. Some algae grow as planktonic clumps which can be harvested by straining from the medium, which may be convenient, but may not be highly productive.

VIII. ACKNOWLEDGMENTS

The work mentioned in this chapter was supported by the United States Air Force, the Office of Naval Research, the National Aeronautics and Space Administration, and the U.S. Department of Agriculture.

I am grateful to Drs Mary Belle Allen, Jack Myers, Frank Trainor, Constantine Sorokin and Thomas Brown for cultures.

I am also grateful to my daughter, Kathleen, for typing this manuscript.

IX. REFERENCES

Allen, M. B. 1963. List of cultures maintained by the Laboratory of Comparative Biology, Kaiser Foundation Research Institute, Richmond, California.

Hemerick, G. 1968. Action spectra and mass culture of variously pigmented algae. Air Force Cambridge Research Laboratories Report No. AFCRL-68-0417. (See Clearing House for Federal Scientific and Technical Information, U.S. Dept. of Commerce.)

Hemerick, G. and Benoit, R. 1962. Engineering research on a photosynthetic gas exchanger. Report No. U-413-62-018. General Dynamics/Electric Boat. Groton, Connecticut.

Tamiya, H. 1962. Chemical composition and applicability as food and feed of mass-cultured unicellular algae. U.S. Army Research and Development Group (9852) (Far East) Contract No. DA-92-557-FEC.

17: Light–temperature gradient plate

CHASE VAN BAALEN

University of Texas Marine Science Institute, Port Aransas, Texas 78373

and

PETER EDWARDS

Nelson Biological Laboratories, Rutgers University, New Brunswick, New Jersey 08903

CONTENTS

I. OBJECTIVE

A complete investigation of growth responses of an alga to light intensity and temperature by conventional algal culture techniques, would be a lengthy undertaking. Halldal and French (1956, 1958) suggested the principle of growing an alga on a uniformly seeded, large agar block under continuous gradients of light intensity and temperature. Thus in one experiment an alga maps its growth responses to these important variables. Jitts *et al.* (1964) using the same principle adapted the system to test tube cultures and lighting conditions approaching those found in nature. The apparatus described originally by Edwards and Van Baalen (1970) embodies features of both the above systems with simplification of design as the basic rationale.

The apparatus can be used to study different algal cultures. When used with the glass–plastic insert and agar medium, quantitative growth can be tested. By use of small petri dishes organisms can be examined in liquid culture. The apparatus also has predictive value for natural systems. Dishes inoculated with an untreated natural water sample will reveal not only the kinds of algae present in the sample but also growth preferences towards light intensity and temperature. A further advantage is that a desired organism may be isolated as unialgal.

Although not discussed here other gradients could be substituted for either the light or temperature gradient. Critical medium components such as the iron-chelator system, Ca^{++} to Mg^{++} ratio, Na^{+} to K^{+} ratio, salinity, etc. can then be examined for their effects on growth.

II. EQUIPMENT

A. Aluminium plate (alloy 2024-T4) $14 \times 16 \times 0.75$ in thick. Aluminium strips 0.25 in thick $\times \frac{5}{8}$ in high.

B. Circulating cooling bath (suggested unit, Blue M Constant Flow Portable Cooling Unit No. 1241, Blue M Electric Co., 138th and Chatham St., Blue Island, Illinois 60406).

C. Circulating heating bath. A number of commercial units are available; one may be constructed using an adequate pump (such as, No. 8512, Gorman-Rupp Industries, Bellville, Ohio).

D. Copper-constant thermocouples (3–4), external resistance 10 ohms (Barber-Coleman Model, 801A-6 Thermocouple selector, Barber-Coleman, Industrial Instruments Division, Rockford, Illinois). Readout meter (Weston 1941, Weston Instruments, 614 Frelinghuysen Ave., Newark, New Jersey 07114).

E. Fluorescent lamps and ballasts with interval timer (see IV.B, following).

F. Pyrex Infrared Reflecting Glass, 14 × 16 in (Corning Glass Works, Corning, New York 14830).

G. Compression fittings, 0.25 in pipe, $\frac{3}{8}$ in tube; copper tubing, $\frac{3}{8}$ in.

H. Styrofoam insulation for plate (bottom, 14 × 16 in; sides, 2 strips 14 in, 2 strips 16 in).

I. Glassware and media as necessary for cultures (see V.A, following).

J. Dish with glass bottom $13\frac{3}{8} \times 12\frac{5}{8}$ in. Sides of Plexiglass strips 0.25 in wide × 0.5 in high (Rohm and Haas Co., Independence Mall West, Philadelphia, Pennsylvania 19105). Glue the plastic strips to the glass (Acryloid B-7 glue, Cadillac Plastics, Houston, Texas).

K. Electrical conducting coating (Silpaint, Fansteel Electronic Materials Laboratory, Compton, California) to which *ca.* 10 V is applied using Variac and alligator clips.

III. TEST ORGANISM

Agmenellum quadruplicatum (Cyanophyceae). Strain PR-6 (Van Baalen 1967).

IV. PROCEDURE

A. *The plate* (see Fig. 17-1)

Drill two $\frac{7}{16}$ in holes centered 0.5 in from each end on long sides of the aluminium plate. Tap ends of holes for 0.25 in pipe thread. Install 0.25 in pipe to $\frac{3}{8}$ in compression fittings. Install short piece of $\frac{3}{8}$ in O.D. copper tubing to each compression fitting to provide rubber tubing connection to water baths. It is convenient to install $\frac{1}{4}$ in thick × $\frac{5}{8}$ in high aluminium strips centered $1\frac{1}{8}$ in in from edges of plate along the back and sides of the plate as a permanent centering device. The plate should be insulated with styrofoam or equivalent on the bottom and along the edges. Connect the cooling and heating baths to either side of the plate via rubber tubing. The tubing should be thermally insulated if room temperature changes are large. Maintain a flow rate of 2–3 l/min.

Three or four thermocouples are fastened to the front edge of the plate with Scotch Tape and covered with a small piece of cotton to lag thermal effects. The thermocouples are connected through the selector switch to the readout meter. It is advisable to verify the temperature gradient along the edge of the plate several times a day. More accurate values of the temperature in the petri dishes or in the agar block can be made by temporarily inserting a thermocouple into the desired location.

B. *Lighting*

This can be varied over a limited range in intensity and wavelength distribution by the choice of the fluorescent lamp phosphor and its distance from the front edge of the plate. A convenient set-up is one 20 W, deluxe cool-white fluorescent lamp positioned 3.75 in above the front edge of the plate. A linear light intensity gradient exists from the front edge of the plate back *ca.* 8 in. The gradient can be checked with a Weston Footcandle Meter or with a thermopile (such as Kipp and Zonen CA-1 pile with readout on a Keithley 150B Microvolt-Ammeter). Daylength is programmed with the interval timer.

Tungsten lamps present a heat problem not entirely solved by the infrared reflecting character of the Pyrex glass cover. If tungsten illumination is desired, a water filter (heat filter) must be used as in the original design (Halldal and French 1958). Light intensities and wavelength distributions approaching sunlight present special design problems (Jitts *et al.* 1964).

V. OPERATION

A. *Liquid cultures*

Petri dishes, 60 × 15 mm containing 15–20 ml of medium are inoculated as uniformly as possible. They are placed directly on the aluminium block and covered with the Pyrex plate.

B. *Seeded-agar block inlay*

1. Sterilize the glass–plastic dish (see II.J, preceding) by rinsing several times with warm, sterile distilled water.

2. Pour a base layer of 550 ml agar medium at *ca.* 42 °C into the glass dish.

3. Cool this layer until firm; then reheat on a warming tray until the agar is *ca.* 42 °C.

4. 300 ml of agar medium at 42 °C, containing the desired number of algal cells is poured on the base agar layer.

5. Mix quickly and uniformly by tilting.

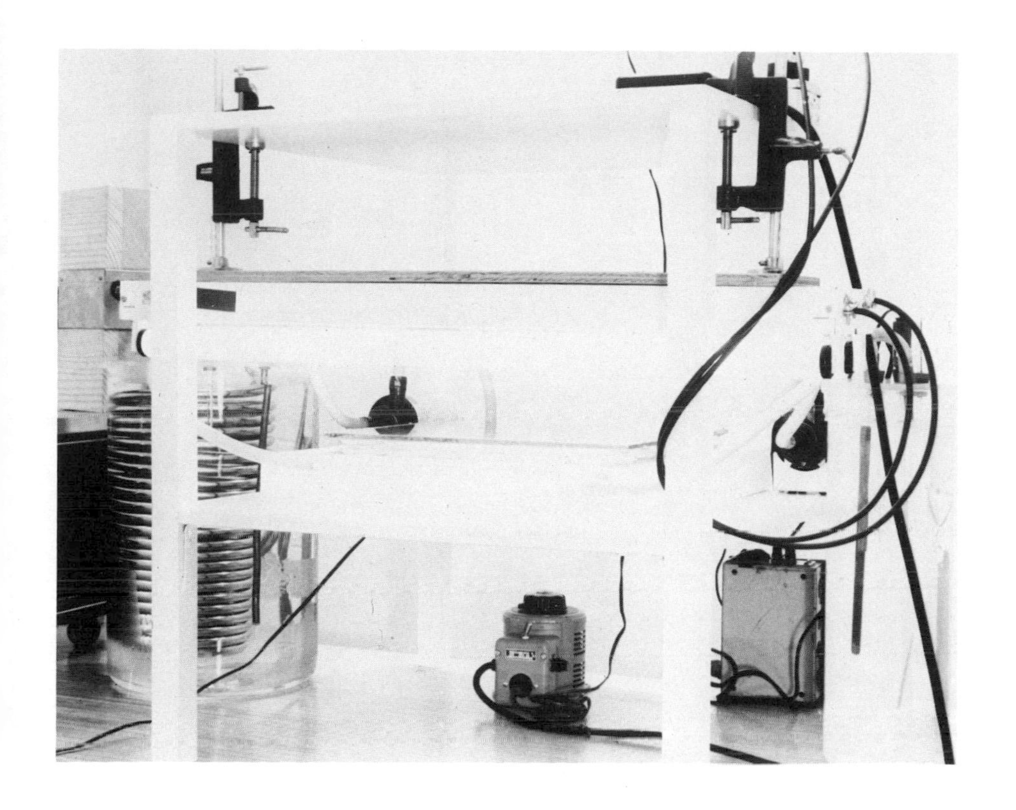

Fig. 17-1. Front view of light–temperature gradient plate.

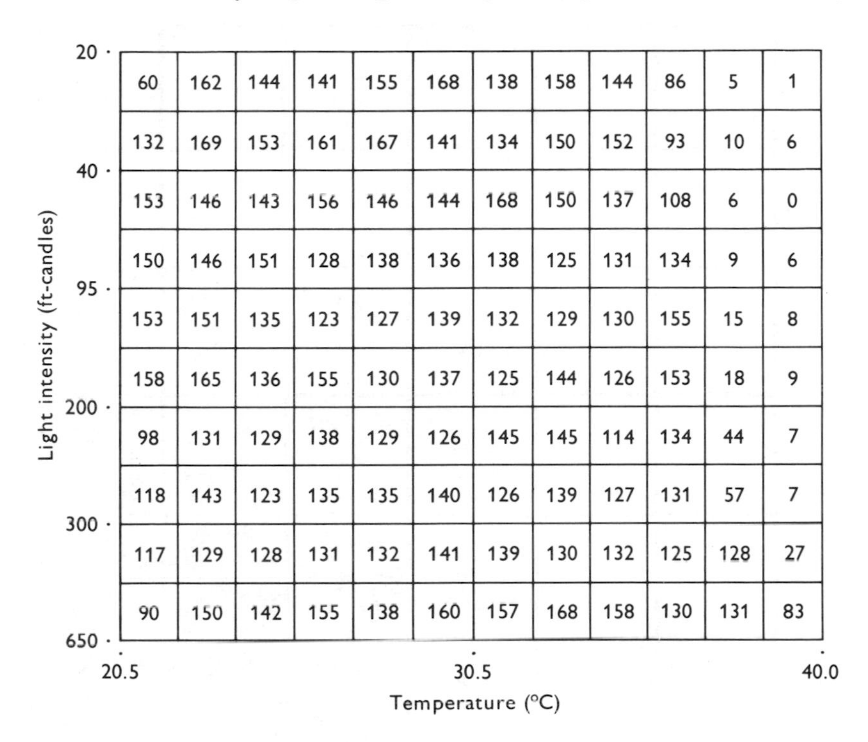

60	162	144	141	155	168	138	158	144	86	5	1
132	169	153	161	167	141	134	150	152	93	10	6
153	146	143	156	146	144	168	150	137	108	6	0
150	146	151	128	138	136	138	125	131	134	9	6
153	151	135	123	127	139	132	129	130	155	15	8
158	165	136	155	130	137	125	144	126	153	18	9
98	131	129	138	129	126	145	145	114	134	44	7
118	143	123	135	135	140	126	139	127	131	57	7
117	129	128	131	132	141	139	130	132	125	128	27
90	150	142	155	138	160	157	168	158	130	131	83

Fig. 17-2. Quantitative plate count of colonies of *Agmenellum quadruplicatum* (Strain PR-6, Van Baalen 1967). Incubation 5 days, Medium ASP-2 + 2 μg/l cyanocobalamin (Provasoli, McLaughlin and Droop, 1957; Van Baalen 1967). Agar overlay contained approximately 155 cells/in² as determined by count of inoculum with Petroff–Hauser Bacterial Counter (see Chp. 19, II.B.5). Number is colonies/in².

C. *Evaporation and temperature control*

The agar block or the petri dishes are directly covered by the plate of Pyrex Infrared Reflecting Glass. Along the two edges of the glass plate parallel to the holes in the aluminium block, a 0.25 in strip of electrically conducting coating is applied. A low AC voltage of *ca.* 10 V is applied (see II.K, preceding). The voltage is varied until the temperature of the glass plate is just sufficient to overcome too rapid evaporation of water from the petri dishes or the agar block at the high temperature end of the plate.

In the petri dishes some evaporation occurs at the high temperature end of the plate. With large algae in xenic culture, the medium can be renewed at approximately weekly intervals. With axenic cultures, evaporation loss is compensated by adding sterile distilled water. The agar block can be usefully incubated up to *ca.* 7 days which is sufficient for most experiments.

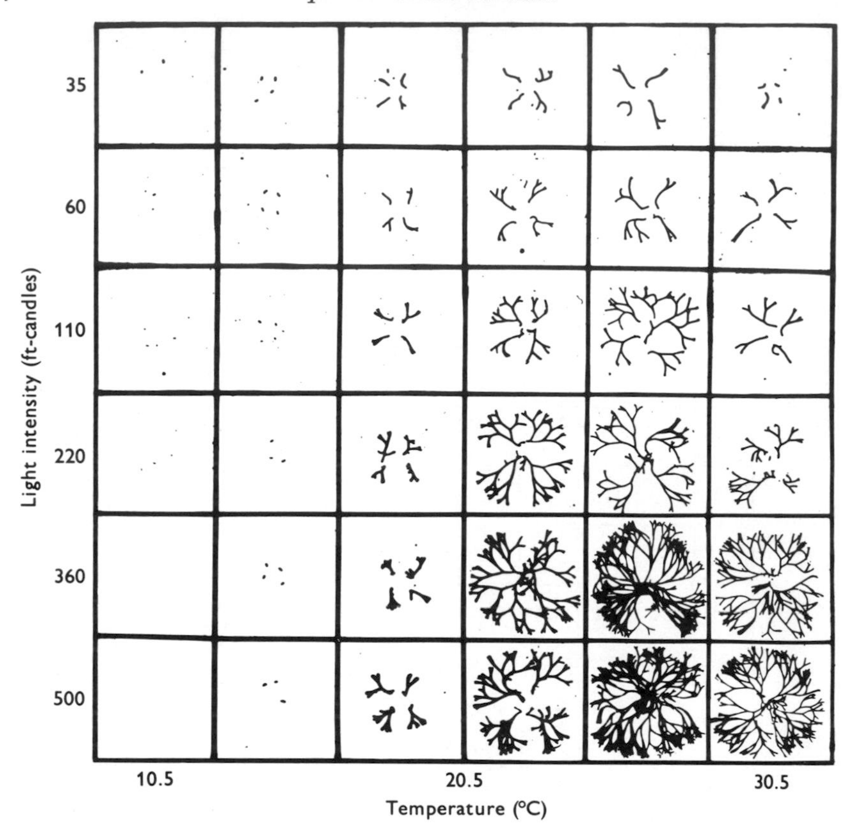

Fig. 17-3. The growth responses (in 19 days) of a unialgal culture of *Centroceras clavulatum* on the gradient plate. The medium was von Stosch's enriched seawater (Ott 1966) with 33‰ salinity changed every 7 days, daylength 14 h. Each square represents a petri dish. Further details in Edwards (1969).

D. *Recording results*

There can be severe limitations on growth imposed by the various operating conditions, for example: (1) the nutrient supply may become limiting in the case of mildly enriched or unenriched natural waters; (2) CO_2 limitation or light limitation may result as cultures become dense; and (3) competitive effects may occur with a mixed inoculum. These factors make frequent visual or microscopic observations of growth imperative if a true picture is to be obtained. A variety of final readouts is possible: dry weight, turbidity, colony count, photography. Two methods are represented in Figs. 17-2, 17-3. In Fig. 17-2 the colonies were counted microscopically over a grid after an incubation period long enough to insure visible colonies in plate positions of low growth rate. In Fig. 17-3 the algae were mounted on a herbarium sheet and photographed at the end of the experiment.

VI. SAMPLE DATA

Results (Fig. 17-2) with *Agmenellum quadruplicatum*, Strain PR-6 (Van Baalen 1967), a marine coccoid Cyanophyceae, show the value of the gradient plate in judging the adequacy of growth conditions for initiation of growth from single cells. The results also demonstrate the wide latitude of temperature and light intensity characteristically tolerated by coccoid bluegreen algae.

Fig. 17-3 shows the usefulness of the technique for unialgal cultures with an undefined medium of enriched seawater using the Rhodophyceae *Centroceras clavulatum*.

VII. REFERENCES

Edwards, P. 1969. Field and cultural studies on the seasonal periodicity of growth and reproduction of selected Texas benthic marine algae. *Contrib. Mar. Sci. Univ. Texas* **14**, 59–114.

Edwards, P. and Van Baalen, C. 1970. An apparatus for the culture of benthic marine algae under varying regimes of temperature and light intensity. *Bot. Marina* **13**, 42–3.

Halldal, P. and French, C. S. 1956. The growth of algae in crossed gradients of light intensity and temperature. *Carnegie Instit. Washington Year Book* **55**, 261–5.

Halldal, P. and French, C. S. 1958. Algal growth in crossed gradients of light intensity and temperature. *Plant. Physiol.* **33**, 249–52.

Jitts, H. R., McAllister, C. D., Stephens, K. and Strickland, J. D. H. 1964. The cell division rates of some marine phytoplankters as a function of light and temperature. *J. Fish. Res. Bd Canada* **21**, 139–57.

Ott, F. D. 1966. A selected listing of xenic algal cultures. *Systematics Ecology Program*, Mar. Biol. Lab., Woods Hole, Massachusetts, Contrib. 72, 1–45 (Mimeo).

Provasoli, L., McLaughlin, J. J. A. and Droop, M. R. 1957. The development of artificial media for marine algae. *Arch. Mikrobiol.* **25**, 392–428.

Van Baalen, C. 1967. Further observations on growth of single cells of coccoid blue-green algae. *J. Phycol.* **3**, 154–7.

18: Growth with organic carbon and energy sources

ALASDAIR H. NEILSON,* WILLIAM F. BLANKLEY†
AND RALPH A. LEWIN

Scripps Institution of Oceanography, University of California at San Diego, La Jolla, California 92037

CONTENTS

* Present address: Botaniska Institutionen Stockholms Universitet Stockholm 50, Sweden.

† Present address: Duke University Marine Laboratory, Beaufort, North Carolina 28516.

I. INTRODUCTION

A. *Objectives*

This section is concerned with methods for growing algae in media supplemented with organic substrates. Its purpose is to outline procedures used, to note some problems which may arise, and to give examples which have demonstrated an obligate requirement for, or facultative utilization of, an organic compound in more than trace amounts. The compound is used for growth, by which we mean an increase in mass with sustained cellular differentiation and continued synthesis of new cell material.

We shall discuss growth of: (a) algae which, having an impaired photosynthetic capacity, are able to grow in the light and in presence of CO_2, only when there is an additional source or organic carbon; and (b) those algae which are able to utilize an exogenous substrate as a source of both carbon and energy, and can thereby grow in the dark. These algae are distinguished from those which grow normally with only light and CO_2 but can apparently use an additional organic carbon source for faster or more luxuriant growth.

B. *Definitions*

1. Photo-autotrophy. Growth in the light, using CO_2 as the sole source of carbon for synthesis of cell material. Energy for growth is derived only from the light.

2. Auxotrophy. Growth requiring, in very low concentrations, at least one organic growth factor not used primarily as a source of carbon or energy. Perhaps most algae with the exception of the Cyanophyceae are auxotrophic; auxotrophy and heterotrophy seem to be independent.

3. Mixotrophy. Growth in the light in the presence of CO_2, but with an additional and obligate requirement for at least one major organic substrate which is photometabolized during growth. This is the situation in algae such as *Ochromonas malhamensis* (Chrysophyceae; Myers and Graham 1956) and *Chlamydobotrys stellata* (Chlorophyceae; Pringsheim and Wiessner 1961).

4. Heterotrophy. Utilization of one or more organic substrates for the provision of energy and cell carbon. CO_2 may or may not be required. Usually this will enable the cells to grow in total darkness, e.g., *Chlorella pyrenoidosa* (Chlorophyceae; Samejima and Myers 1958), *Tolypothrix tenuis* (Cyanophyceae; Kiyohara *et al.* 1960). However, in some instances (e.g., the Cyanophyceae, *Agmenellum quadruplicatum* and *Lyngbya lagerheimii*; Van Baalen, Hoare and Brandt 1971) light is required for assimilation of the substrate, though only at intensities far below those able to support photo-autotrophic growth. This must be clearly distinguished from the situation in which addition of an organic substrate to the medium results in increased yields or accelerated growth in the light. There may be explanations for this phenomenon other than direct utilization of the substrate.

Unlike higher plants, most algae that can grow in darkness retain their normal photosynthetic pigments. There are, however, a number of morphologically more or less closely related microorganisms which completely lack photosynthetic pigments and are therefore obligate heterotrophs. Some are unquestionably apochlorotic algae, such as three species of *Nitzschia* (Bacillariophyceae) described by Lewin and Lewin (1967). Others bear a more remote resemblance to their pigmented counterparts, e.g., *Prototheca* (Chlorophyceae, possibly near *Chlorella*; see Poynton 1970).

C. *Topics excluded*

Auxotrophic growth will not be discussed, nor phagotrophy, in which particulate food is ingested and digested intracellularly. Also excluded from this section are: (1) accounts of the physiology of symbionts and parasites; (2) methods for the elimination or bleaching of photosynthetic pigmented organelles; (3) methods for the initial isolation of mixotrophic or heterotrophic algae; and (4) methods for detecting uptake or assimilation of the substrate (see Fay 1965; Van Baalen, Hoare and Brandt 1971). As mentioned in I.B, preceding, those algae utilizing a carbon substrate in light for faster or more luxuriant growth, are excluded.

II. EQUIPMENT

This is essentially the same as that required for growth in the light, and has been covered in Chps. 1–4. The major differences arise from the need for exclusion of light in heterotrophic growth and for periodic sampling in darkness or in subdued light. These topics are dealt with in section IV.B.2, 3, following.

III. TEST ORGANISMS

A. *Mixotrophic growth*

1. Chlamydobotrys stellata (Chlorophyceae; Göttingen collection, strain 10/le). This freshwater strain has impaired photosynthetic capacity and grows poorly in the light unless the mineral medium, containing vitamins thiamine and cyanocobalamin, is further supplemented with sodium acetate (4 mM) and asparagine (10 mM) (Pringsheim and Wiessner 1961).

2. *Diplostauron elegans* (Chlorophyceae; IUCC 1896). This freshwater species is physiologically similar to the preceding with a requirement for the same two vitamins and for a carbon source such as acetate (1 mM), pyruvate (2 mM), or ethanol (1 mM) (Lynn and Starr 1970).

B. *Facultative heterotrophic growth*

1. Chlorella pyrenoidosa Emerson strain (Chlorophyceae; possibly IUCC 251, 252 or 395). This freshwater clone will grow in the dark in a mineral medium supplemented with glucose or galactose (1% w/v or 55 mM), or sodium acetate trihydrate (0.1% w/v or 7 mM) (Samejima and Myers 1958). Ammonia or urea is a rather better source of nitrogen than nitrate (see section IV.C.4). No vitamin requirement has been demonstrated.

2. *Agmenellum quadruplicatum* strain PR-6 (Cyanophyceae; Van Baalen, Hoare and Brandt 1971). This marine bluegreen alga grows heterotrophically, but only in low light intensities at which photo-autotrophic growth does not occur. It may be cultured on synthetic seawater medium (see Chp. 2, IV.B) containing cyanocobalamin supplemented with glucose (1% w/v or 55 mM), sodium nitrate and ammonium chloride (50 mg/l for each) (Van Baalen, Hoare and Brandt 1971).

C. *Obligate heterotrophy in apochlorotic algae*

1. *Euglena gracilis* var. *bacillaris* (Euglenophyceae). Though the wild type is normally pigmented, and grows heterotrophically in the dark, preferably on acetate, this colorless mutant grows on a mineral medium supplemented with thiamine and cyanocobalamin and glucose (1% w/v) (Hurlbert and Rittenberg 1962).

2. *Nitzschia* spp. (Bacillariophyceae; Lewin and Lewin 1967). There are three naturally occurring apochlorotic diatom species which grow on an

artificial seawater medium supplemented with thiamine and cyanocobalamin and either sodium lactate (0.5% w/v or 45 mM) or sodium succinate (0.5% w/v or 18.5 mM) (Lewin and Lewin 1967).

IV. METHODS

A. *Design of experiments*

1. Axenic cultures. Since many experiments may extend over long periods, it is essential that the culture be axenic and this be maintained throughout the experiment. The results should be discarded if bacteria or molds can be shown to be present, either by microscopic examination using phase-contrast illumination at high magnification (1000 ×), or by streaking on a nutrient-rich medium, such as the mineral base supplemented by addition of 0.2% each of yeast extract and casamino acids. For further information refer to Chp. 3, VII.

2. Medium

a. Defined media – initial isolations may be done in a complex medium containing, for example, yeast extract, beef extract, or peptone. For determining specific requirements and for critical studies of heterotrophy, a defined medium containing a limited number of known substrates must be used.

b. Preliminary screening – it is sometimes advantageous to screen isolates for mixotrophic or heterotrophic potential by streaking on agar plates prepared with a suitable concentration of the substrate (C.2, following). It is thus possible to observe directly the effects of the various treatments on cell morphology and multiplication. Cultures are set up on petri plates prepared with or without substrate, and incubated in the light or in darkness. Comparison of these permits an assessment of the viability and growth of the inoculum in light; of stimulatory or inhibitory effects of the substrate on growth in light; and, of stimulatory effects of the substrate on growth in darkness. A slight stimulatory effect of the substrate on multiplication in darkness may suggest that the substrate would be more effective at a different concentration. Since many algae continue to divide for a limited time after being placed in darkness in the absence of substrate, it is essential to demonstrate sustained growth at the expense of the substrate before asserting an alga is capable of heterotrophy. As growth in darkness is usually much slower than growth in light, incubation periods should extend several weeks. Preliminary results obtained using solid media should be confirmed by testing for growth in a defined liquid medium, to eliminate possible effects of unknown substances derived from the agar.

3. Effects of substrates

a. Uptake and incorporation of a substrate – experiments on the uptake and incorporation of labelled substrates in the dark, though not treated here provide unequivocal proof of heterotrophic growth (see Fay 1965; Van Baalen, Hoare and Brandt 1971).

b. Utilization of a substrate – the effect of a substrate on growth, in the light or in the dark, may be determined by a comparison of growth rates or yields, or both, with those of control cultures from which the substrate has been omitted (see Chps. 19–23 for discussion of measurements).

c. Osmotic effects – if a substrate is added at a relatively high concentration, its osmotic effect must be considered when assessing growth by comparison with that in the control medium. Additional controls, with appropriate concentrations of a different compound which is neither utilizable nor inhibitory should be included. In seawater media, the deleterious osmotic effects of high concentrations of an added substrate may be mitigated by lowering the salinity of the basal medium (Blankley 1971).

4. Adaptation to substrate. After isolation, or during initial experiments designed to test for heterotrophy, growth may be slow, especially in the dark. In some instances this is due merely to a lag in synthesis of enzymes required for substrate utilization, but in others no such simple explanation suffices. In some situations, growth through several sub-cultures may eliminate or reduce the lag period and increase the growth rate, e.g., in *Chlorogloea fritschii* (Cyanophyceae; Fay 1965). For this reason, if cells are to be used for physiological or biochemical studies, the inoculum should be taken from a culture already growing in a medium supplemented with the substrate.

5. Use of photosynthetic inhibitors. Generally, tests for heterotrophic growth are carried out in the dark. It is, however, also possible to carry out experiments on the utilization of organic carbon substrates in the light, in the presence of a specific inhibitor of photosynthesis such as DCMU (3–(3:4 dichlorophenyl)1:1 dimethyl urea); a concentration of 10^{-5} M usually inhibits CO_2 uptake completely. This technique seems to have been little exploited, though it should be noted that a positive result does not necessarily mean that the compound can be utilized in total darkness.

B. *Culture methods*

Methods for studying growth in the light are discussed in several chapters (Chps. 3, 11, 15, 17, 21). Included here are some additional problems which may arise in investigation of mixotrophic or heterotrophic growth.

1. Gas phase

a. Growth in air – generally oxidation of the substrate takes place in cells growing mixotrophically in the light or heterotrophically in the dark. As in studies on the growth of bacteria, therefore, care must be taken that oxygen is not a limiting factor. If experiments are to be conducted in the presence of CO_2, aeration may be accomplished simply by bubbling a rapid stream of sterile air (see Chps. 15, 16) through the culture, or by otherwise agitating the medium. Since many experiments last for considerable periods, evaporation of the medium should be minimized by humidifying the air by passing it through one or more wash bottles containing sterile water.

b. Growth in absence of CO_2 testing for growth in the absence of CO_2 may be accomplished by enclosing agar plates or non-aerated liquid cultures in bell jars or desiccator jars furnished with shallow vessels containing 40% KOH. To facilitate diffusion of CO_2, liquid cultures should be in shallow layers. To avoid oxygen starvation, the bell jar or desiccator jar may be connected to the atmosphere through a CO_2 trap containing 'Ascarite' (asbestos coated with sodium hydroxide; J. T. Baker Chemical Co., 222 Red School Lane, Phillipsburg, New Jersey 08865).

An alternative method for eliminating CO_2 is to pass a stream of CO_2-free air through liquid cultures. However, it must be borne in mind that removal of the last traces of CO_2 may be deleterious.

c. Anaerobic growth – energy for heterotrophic growth is normally provided by oxidation of the substrate; though fermentation has been demonstrated in some algae (e.g., in *Ochromonas malhamensis*, Reazin 1956), there has apparently been no conclusive demonstration of heterotrophic growth of any alga under anaerobic conditions. This should certainly be attempted, using standard methods employed for bacteria (see Aaronson 1970).

2. Growth in darkness. Experiments on growth in darkness may be carried out in light-tight incubators for larger flasks, in small light-tight boxes for tubes or plates, or in flasks either individually wrapped in black paper, black plastic sheet or black tape, or painted with black epoxy paint on vapor-blasted surfaces. When paint or tape is to be used, it should be first ascertained that inhibitory fumes are not imparted to the medium during sterilization (see Chp. 14). If gas is added to cultures, care must be taken that stray light does not enter through the tubes used as gas inlets and outlets; their exposed sections should be wrapped with black tape or enclosed in black 'shrink-fit' insulating tubing. Alternatively, the entire gas unit may be enclosed in a dark incubator provided with gas inlets. Agitation of the cultures may be provided by magnetic stirring, shaking, or by gassing.

3. Sampling

a. Cultures in atmospheres other than air – standing cultures in which the gas phase is not air must be protected during sampling against introduction of either CO_2 or O_2. Though appropriate traps on sampling tubes may sometimes suffice, it will often be easier to prepare many replicate cultures and to sacrifice one for each sample.

b. Cultures in darkness – for cultures growing in the dark, sampling tubes should be opaque as suggested above for gas inlet tubes. When the entire culture must be exposed during the course of repetitive sampling, the best expedient is to work rapidly in light of minimal physiological effect. The light should be as dim as possible, while still allowing vision, and of a wavelength near an absorption minimum and thus only minimally utilizable by the cell.

Sacrifice of whole cultures for each sample is the best alternative sometimes, especially for filamentous algae and those which tend to clump and cannot easily be kept in suspension (e.g., *Tribonema aequale*; Xanthophyceae, Belcher and Miller 1960).

C. *Media*

Media for cultivation of photo-autotrophs have been discussed (Chps. 1, 2). Only special problems concerned with media containing an organic substrate are considered here. For an extended discussion of medium components and preparation see Bridson and Brecker (1970).

1. General preparation. Usually, conditions for mixotrophic growth are provided by adding a substrate to the basal mineral medium in which photo-autotrophic growth of the same or related organisms is known to take place, albeit poorly.

For liquid media it is best to sterilize the mineral base separately, by autoclaving; when it has cooled add a sterile solution of the organic substrate (see 2, following). For the preparation of solid media the safest procedure is to autoclave separately, in equal volumes, twice-concentrated solutions of the agar and of the mineral base; these are then mixed after cooling to about 60 °C. Lastly a sterile solution of the substrate is added. Final concentration of agar should be 1–1.5%, and the medium should be poured fairly thick (about 5 mm deep) to reduce the effects of drying out during the lengthy incubations which may be needed.

2. *Preparation of substrates.* Solutions of the sodium salts of most organic acids may be sterilized by autoclaving without appreciable loss or decomposition. Sugar solutions, most of which decompose or polymerize during autoclaving, or volatile substrates like ethanol, are best sterilized by filtration through sterile membrane filters of 0.22 or 0.45 μm porosity (see Chp. 3, II.B; and 12, II.E). Substrates can be prepared as concentrated stock solutions, so only small volumes are added to the mineral base. This is especially important in preparation of marine media, where it may be undesirable to lower the salinity of the medium by addition of large volumes of other aqueous solutions. Conversely if a substrate such as glycerol has to be used in high concentration, it may be necessary to lower the salinity of the medium accordingly (Blankley 1971). If the substrate must be added in a non-aqueous solution (e.g., in ethanol) it is necessary to employ a solvent control.

It is difficult to generalize about concentrations of substrates. High concentrations (up to 1–10% w/v) may be needed to provide adequate amounts for sustained growth, but they may also be toxic (see also 3, following, on pH). The optimum concentration may be critical, and it may be necessary to measure the growth rate or yield in a range of substrate concentrations. Undissociated molecules of organic acids are generally toxic at high concentrations, but since they penetrate the cell membranes more readily than the anions, they are therefore more effectively utilized. Acetic acid has a pH of 4.74, and in solutions of pH 6 and 7 there is about 5% and 0.5% of the free acid, respectively. Compounds of this type must therefore be tested at several pH values, and at low concentrations (e.g., 0.01–0.1% w/v, or 0.7–7 mM, sodium acetate). By contrast, most common carbohydrates are non-toxic and may be used at much higher concentrations (e.g., 0.2–5% w/v glucose or 10–275 mM). For heterotrophic growth of certain algae, extremely high substrate concentrations are needed (e.g. 0.5 M is the optimum concentration of glycerol for growth of *Prymnesium parvum*, Haptophyceae, in the dark; Rahat and Jahn 1965).

3. Control of pH. Due to removal of CO_2, or formation of precipitates such as calcium phosphate or magnesium ammonium phosphate, appreciable changes in pH may take place as a consequence of autoclaving. Such effects can often be minimized by adjusting the pH prior to heat sterilization or by including buffers such as tris or glycylglycine. Note, however, that these substances are inhibitory to some algae.

Changes of pH during heterotrophic or mixotrophic growth of a culture are generally smaller with neutral substrates such as carbohydrates or polyols than with acidic or basic compounds. Utilization of a substrate may cause changes in the pH (e.g., due to concomitant CO_2 production) which cannot readily be simulated in the control cultures where the substrate is lacking. The only way to deal with this difficulty is to monitor pH, and if necessary to carry out experiments over a range of pH values.

4. Nitrogen sources. During growth in light, many algae are able to use nitrate as a nitrogen source as well as, or even better than, ammonia. For algae growing in the dark, however, nitrate may be an inferior source, perhaps because nitrate reductase is light-induced, and the reduction process itself requires more energy than the direct incorporation of ammonia into amino acids. Other sources of reduced nitrogen (e.g., urea) may sometimes be more readily utilized than nitrate during heterotrophic growth. For at least one alga (*Tolypothrix tenuis*), dark growth is stimulated by the addition of casamino acids (Kiyohara *et al.* 1960).

V. ACKNOWLEDGMENTS

We gratefully acknowledge support for A.H.N. from N.S.F. Grant No. GV 27110 (to Dr O. Holm-Hansen, Institute of Marine Resources) and for W.F.B. from a N.S.F. predoctoral fellowship.

This is a contribution from the Scripps Institution of Oceanography.

VI. REFERENCES

Aaronson, S. 1970. *Experimental Microbial Ecology*, pp. 34–6. Academic Press, New York.

Belcher, J. H. and Miller, J. D. A. 1960. Studies on the growth of Xanthophyceae in pure culture. *Tribonema aequale* Pascher. *Arch. Mikrobiol.* **36**, 219–28.

Blankley, W. F. 1971. Auxotrophic and heterotrophic growth and calcification in coccolithophorids. *Ph.D. Dissertation*, pp. 1–186. University of California, San Diego.

Bridson, E. Y. and Brecker, A. 1970. Design and formulation of microbial culture media. In Norris, J. R. and Ribbons, D. W., eds., *Methods in Microbiology* **3A**, 251–95. Academic Press, New York.

Fay, P. 1965. Heterotrophy and nitrogen fixation in *Chlorogloea fritschii. J. Gen. Microbiol.* **39**, 11–20.

Hurlbert, R. E. and Rittenberg, S. C. 1962. Glucose metabolism of *Euglena gracilis* var. *bacillaris*; growth and enzymatic studies. *J. Protozool.* **9**, 170–82.

Kiyohara, T., Fujita, Y., Hattori, A. and Watanabe, A. 1960. Heterotrophic culture of a blue-green alga, *Tolypothrix tenuis* I. *J. Gen. Appl. Microbiol.* **6**, 176–82.

Lewin, J. and Lewin, R. A. 1967. Culture and nutrition of some apochlorotic diatoms of the genus *Nitzschia. J. Gen. Microbiol.* **46**, 361–7.

Lynn, R. I. and Starr, R. C. 1970. The biology of the acetate flagellate *Diplostauron elegans* Skuja. *Arch. Protistenk.* **112**, 283–302.

Myers, J. and Graham, J. R. 1956. The role of photosynthesis in the physiology of *Ochromonas. J. Cell. Comp. Physiol.* **47**, 397–414.

Poynton, R. O. 1970. The characterization of *Hyalochlorella marina* gen. et sp. nov. a new colourless counterpart of *Chlorella. J. Gen. Microbiol.* **62**, 171–88.

Pringsheim, E. G. and Wiessner, W. 1961. Ernährung und Stoffwechsel von *Chlamydobotrys* (Volvocales). *Arch. Mikrobiol.* **40**, 231–46.

Rahat, M. and Jahn, T. L. 1965. Growth of *Prymnesium parvum* in the dark; note on ichthyotoxin formation. *J. Protozool.* **12**, 246–50.

Reazin, G. H., Jr. 1956. The metabolism of glucose by the alga *Ochromonas malhamensis. Plant Physiol.* **31**, 299–303.

Samejima, H. and Myers, J. 1958. On the heterotrophic growth of *Chlorella pyrenoidosa. J. Gen. Microbiol.* **18**, 107–17.

Van Baalen, C., Hoare, D. S. and Brandt, E. 1971. Heterotrophic growth of blue-green algae in dim light. *J. Bact.* **105**, 685–9.

Section IV

Growth measurements

19: Division rates*

ROBERT R. L. GUILLARD
Woods Hole Oceanographic Institution, Woods Hole, Massachusetts 02543

CONTENTS

* Contribution Number 2533 from the Woods Hole Oceanographic Institution. This work was supported by National Science Foundation grant GB 20488.

I. OBJECTIVE

The objective is to determine division rate from the time course of increase in cell numbers. Division rate may be expressed in a number of different ways, and can be computed or derived graphically from cell count data. In turn, division rates may be used to calculate rates of increase of biomass, cell volume, plasma volume, or of other properties, if appropriate conversion factors are known. (Also see discussion of growth measurements for macroscopic (benthic) algae in Chp. 5, IV.)

Counting methods described are those suited to algal cultures and are not necessarily useful for counting algae in natural situations. Methods for enumerating natural populations generally involve sedimentation of plankton. These methods may of course be used to count cultures, but the delay between sampling and counting is often disadvantageous. It may be necessary to use them if a culture is too dilute to count optically without concentration. The procedures standard for use of the inverted microscope (Lund, Kipling and LeCren 1958; Utermöhl 1958), or for separable settling chambers with the ordinary microscope (Lund 1951; Utermöhl 1958) can be followed. An alternative is to concentrate the sample by sedimentation, then count in an appropriate chamber as described in this chapter.

The use of electronic counting devices is treated by Parsons in Chp. 22.

II. EQUIPMENT

A. *Microscope and accessories*

1. Microscope. Standard or phase contrast with mechanical stage.

2. Optical accessories. The microscope should be equipped with an ocular containing a Whipple disc for counting fields or strips. (For counting strips an ocular with two parallel lines is satisfactory.) An ocular micrometer and stage micrometer should be available.

3. Hand tally. A mechanical hand tally is useful but not essential.

TABLE 19-1. *Counting devices and uses*

Counting device	Sizes handled easily (in μm)	Culture densities handled easily (cells/ml)
Sedgwick–Rafter	50–500	$30–10^4$
Palmer–Maloney	5–150	$10^2–10^5$
Hemacytometer (0.2 mm deep)	5– 75	$10^3–10^7$
Hemacytometer (0.1 mm deep)	2– 30	$5 \times 10^4–10^7$
Petroff–Hausser	<1– 5	$10^5–>10^8$

Spines and sometimes extracellular threads or sheaths increase the effective size of an organism. If in doubt about this or the effective size of colonial species, use the chamber appropriate to the next larger size. Experiment will usually reveal exclusion of cells from the counting chamber and obviously irregular distribution. The chi-square test (e.g., Lund, Kipling and LeCren 1958) can be used to test for irregular distribution.

B. *Counting devices* (see Tables 19-1, 19-2)

Several different devices may be used depending on the materials to be enumerated (Table 19-1).

1. Sedgwick–Rafter counting chamber with cover glass as provided by manufacturer (A-2431, Clay-Adams, Division Becton, Dickinson and Co., 299 Webro Rd, Parsippany, New Jersey 07054).

2. Palmer–Maloney chamber, with no. 1½ or no. 2 cover glasses (Palmer and Maloney 1954).

3. Hemacytometer, 0.2 mm deep. The Speirs–Levy type (Eosinophil counter) is recommended, with cover glass as provided, *ca.* 0.4 mm thick. (No. 2 cover glasses can also be used.)

4. Hemacytometer, 0.1 mm deep, having Improved Neubauer ruling, with cover glasses as provided or no. 1½ cover glasses (A. O. Spencer 'Bright Line').

5. Petroff-Hausser bacteria counter (Improved Neubauer ruling) with Hausser reinforced cover glass 0.18–0.25 mm thick (thinner preferred) (A-2424, Clay Adams).

C. *Accessories*

1. Pipettes

a. Large bore 1 ml diluting pipettes, for filling Sedgwick–Rafter chambers.

b. Pasteur-type pipettes (see Chp. 3, II.C), selected for uniformity and large diameter of bore and with tips fire polished or smoothed by rubbing on emery paper. These are for filling Palmer–Maloney chambers, hemacytometers and bacteria counters. Certain medicine droppers will also serve.

2. *Wire loops*. Platinum or nichrome wire loops 0.2–0.5 mm in diameter can be used to fill hemacytometers.

3. *Cleaning supplies*

a. Wash bottle with distilled water.
b. Dropping bottle of ethanol.
c. Cheesecloth, 100 mm squares.
d. Waste dish for water and ethanol poured over counting chambers.
e. Staining dish, small, for holding cover glasses upright in ethanol.

4. *Settling tubes*. These are for counting dilute cultures. The best are 15 ml flat-bottomed, screw-capped test tubes or small volume graduated cylinders with stoppers.

5. *Fixatives*. These are necessary to immobilize flagellates for counting and to preserve cells being concentrated by sedimentation.

a. Iodine in particular is helpful because it speeds the settling of cells in counting chambers as well as in sedimentation tubes. Utermöhl's (1958) solution made with sodium acetate as follows is the most useful: KI, 10 g; H_2O, 20 ml; I_2, 5 g; H_2O, 50 ml; $NaC_2H_3O_2 . 3H_2O$, 5 g. Less than 1% need be used, thus correction for the volume increase on adding this preservative is unnecessary.

b. Neutral formalin, 5%, is also used. Correction for volume increase is indicated.

Some flagellates do not preserve well with any fixative, even glutaraldehyde or osmium. These algae are best counted as soon as possible, whatever the fixative.

III. METHODS

A. *Preliminary considerations*

Most of the sampling problems involved in enumeration of natural populations do not arise in counting cultures. However, sampling technique cannot be ignored. The main point is to homogenize the culture and select an appropriate counting chamber in which the organisms will assume a random distribution for counting.

1. Attaching algae. To begin with the most difficult situation, if an alga grows largely attached to the culture vessel walls, a sample can only be taken by scraping the cells loose, then homogenizing the resulting suspension. In practice, this generally means sacrificing that particular culture. The resulting suspension may or may not be countable, depending on the separation of the cells. Some algae grow in suspension, but in such large and irregular clumps that individual cells cannot be distinguished. Sometimes the clumps can be dispersed by brief ultrasonic treatment, or by the use of 3–5% formalin or a detergent followed by agitation or ultrasonic treatment. If not, growth must be estimated by some means other than counting devices (also see Chp. 5, IV).

2. Colonial algae. Colonial growth poses no unmanageable problems provided the cells can be distinguished for counting and the number of cells per colony does not vary too widely. This problem is considered by Lund, Kipling and LeCren (1958). The basic point is that colonies, not individual cells, are the units distributed over the surface of the counting area, hopefully in a Poisson distribution. Therefore colonies must be tallied, and the average number of cells per colony determined separately. In some instances, notably that of chain-forming diatoms (Bacillariophyceae), ultrasonic or other treatment as described before will separate colonies into such small fragments that little accuracy is lost if cells rather than colonies are tallied directly. Reducing the number of cells per colony is generally an advantage. The error involved in counting colonial forms is the product of the errors involved in counting the colonies and in determining the average number of cells per colony, which Lund, Kipling and LeCren (1958) mention as generally negligible.

3. Random distribution. Lund, Kipling and LeCren (1958) also consider statistics applicable to counting algal populations. For practical purposes the important point is to test, for randomness, the distribution of cells or colonies in the counting chamber being used. This can be done by the chi-square test (outlined by Lund, Kipling and LeCren 1958). This permits use of standard confidence limits. If the distribution is not random, then it seems reasonable to assume some degree of contagion or clumping of individuals, which may reflect such factors as association of cells after division. Holmes and Widrig (1956) considered this and show, for some, how the width of the confidence interval is increased when the distribution follows the negative binomial rather than the Poisson distribution.

Inasmuch as the reliability that can be obtained by counting increases only with the square root of the number of individual objects counted, the level of diminishing returns is quickly reached. At the 95% confidence level, the accuracy is roughly ±20% if 100 objects are counted, ±10% if 400 are counted, and ±5% if 1600 are counted. For most purposes the 10% level is adequate.

In general, the counts made of cells in replicate Whipple fields, hemacytometer areas, chambers, etc., should be recorded individually to permit statistical treatment if necessary.

It is not possible to determine by eye if a given distribution of cells in a chamber is random, but it is often possible to tell that it is not. For example, some chain-forming algae will cluster at the edges or near the entry slot of the Palmer–Maloney chamber, or along the edges of a thin hemacytometer chamber. If an alga extrudes mucilage and the culture has not been mixed adequately, long strands of cells may be seen in the chamber fields. Cells that are too large for a counting chamber may cluster near the entry port. In these situations a change of mixing technique, sample treatment, or of counting chamber is required. However, with some colonial forms, no treatment or chamber seems to give random distributions, so that all individuals in the chamber must be counted. If the number of individuals is too large, the sample can be diluted without introducing appreciable error, provided pipetting errors are minimized by the use of large bore pipettes of suitable accuracy.

The rejection of a suspect count (a deviant in a replicate series) can be made by the Chauvenet criterion. This is illustrated by Calvin *et al.* (1949, Appendix II).

4. Dividing cells. Counting dividing cells can be problematical. In the simplest situation (e.g., a flagellate dividing by fission or a diatom elongated and about to divide) the individual can be counted as two because it has about twice the amount of cell material. There are usually few enough such specimens so that the culture need not be considered to consist of 'colonies' of one or two cells. If there are many such instances, it implies synchrony of division, which suggests simply changing the time of sampling. Species that divide by producing several daughter cells (e.g., autospores) pose a more severe problem. If the proportion of cells that are at various stages of the life cycle remain similar in the population during the growth period, it makes no difference how they are scored, provided it is consistent and that the considerations applying to colonies need not be applied. If, on the other hand, the proportions change, then it is necessary to estimate the number of cells present at each count. Note that the property under consideration

TABLE 19-2. *Dimensions, rulings of counting devices and microscope objectives for each*

Sedgwick–Rafter. The chamber is 50 × 20 × 1 mm in size, with area of 1000 mm² and volume of 1.0 ml. There are no rulings on commercial models. Objectives of 2.5–10 × (initial magnifications) are most useful, but certain 20 × long working distance objectives can be used to examine small organisms.

Palmer–Maloney. The chamber is 17.9 mm in diameter and 0.4 mm deep, with area of 250 mm² and volume of 0.1 ml. There are no rulings. Standard objectives of 10–45 × can be used.

Speirs–Levy hemacytometer (Eosinophil counter). This hemacytometer is 0.2 mm deep and has four separate chambers each consisting of 10 squares 1 mm on a side. (The squares are further divided.) Total volume in all 4 chambers (40 squares) is thus 0.008 ml. Ruling is a modified Fuchs–Rosenthal type. Count with 10 × or at most 20 × objectives.

Hemacytometer with Improved Neubauer ruling. The model designed for use with phase contrast microscope is recommended. (This yields best resolution, which is needed for counting small algae, particularly in mixed cultures.) This hemacytometer is 0.1 mm deep and has 2 chambers each consisting of 9 squares 1 mm on a side. (The squares are further divided.) The total volume in both chambers (18 squares) is thus 0.0018 ml. Counting is best done with 10 × or 20 × objectives. Standard high dry objectives (40–45 ×) can be used, but it is absolutely necessary to employ no. 1½ or no. 2 cover glasses for adequate resolution.

Petroff–Hauser (bacteria). This has one chamber, with the Improved Neubauer ruling of 9 squares each 1 mm on a side (further divided), but with depth of only 0.02 mm. The total volume under the 9 squares is thus 0.00018 ml. High-dry (40–45 ×) and oil immersion (50–100 ×) objectives can be used.

here is division rate, *not* the rate of synthesis of protoplasm or rate of increase of cell colume or plasma volume. If the latter are desired, then cells must be measured as well as counted.

5. Choice of counting device. The first practical consideration is choice of counting apparatus, which depends upon culture density, the size and shape of the cells or colonies being counted, and the presence and amount of extra-cellular threads, sheaths, or dissolved mucilage, which influence the filling of the counting chamber. With the apparatus chosen properly, experience shows that cleanliness of the counting chamber and cover slip is the most important factor in attaining proper distribution of the cells or colonies. Care should be taken to distribute the count over a wide area of the counting apparatus, or to count more than one complete chamber.

It may be necessary to dilute the sample if more than one chamber is counted.

Table 19-1 lists the various devices and circumstances indicating application of each. Table 19-2 gives essential dimensions of the chambers, rulings if present, and magnification of microscope objectives useful with each. (Microscope oculars are selected to give an appropriate field of view and total magnification. Magnification should not exceed 1000 × the numerical aperture of the objective.)

B. *Procedure*

1. Culture density

a. If the culture density is *too low* to be counted in a chamber available, concentrate a sample as follows:

i. Homogenize the culture by shaking or swirling by hand (taking care to reverse directions) or by using a mechanical shaker. Note that some species are sensitive to violent agitation. (At some point, the effectiveness of the shaking process should be tested by repeated sampling.)

ii. Sample the culture with a large bore pipette, taking a volume calculated to contain enough cells or colonies. Transfer quantitatively to a settling cylinder and add *ca.* 0.5% iodine solution (to the color of light tea).

iii. Allow to settle at least one hour for each 10 mm height of the sample in the cylinder (in practice, usually overnight).

iv. With tubing, attach a Pasteur-type pipette to a reservoir (small suction flask) then to an aspirator. Gently aspirate 75–90% of the supernatant fluid, thus concentrating the sample 4–10 fold. The reservoir permits both recovery of a sample lost by error and a check to be certain all cells had settled.

v. Homogenize the cells in the remaining liquid and continue with the counting procedure.

vi. The concentration factor is the initial volume/final volume.

b. If the culture is *too dense* to be counted conveniently, take a sample by pipette as described in a.i, ii, preceding, transfer to a cylinder, and dilute as needed with particle-free distilled or seawater. Mix and proceed with the counting.

The dilution factor is the final volume/initial volume. Dilution is needed only for samples to be counted in the Sedgwick–Rafter or Palmer–Maloney chambers, and only when: (i) the density is too low to permit counting by ocular fields (to be described following) and just too high for convenient counting of the whole chamber; or (ii) when the culture consists of long colonies of cells (e.g., *Skeletonema*, Bacillariophyceae) that cross the ocular fields, hence the entire chamber must be counted.

2. *Cleaning counting devices.* The chambers should be washed initially with mild detergent, water, ethanol (or clean acetone), then dried with clean cheesecloth. Between samples the chamber need only be held over the waste container and flushed with distilled water, then ethanol or acetone, then wiped dry. A square of cheesecloth three or four layers thick can be used perhaps 10 times, then should be discarded. The cover glass should be cleaned and dried in the same way, then set on the counting chamber or rested on a pair of clean glass or metal rods. Wash hands free of grease before working with counting chambers. A surprisingly small amount of grease can make it almost impossible to load the chamber uniformly, and the precision and accuracy of counting drop significantly. Examine the empty chamber occasionally to see that particles and organisms have been removed by the cleaning process.

3. *Filling counting device, settling.* Allow the algal cells to settle for 3–5 min in the counting device before counting. This is chiefly important when they are small and contained in relatively deep chambers. It is common to use two counting devices, allowing the organisms to settle in one while the other is being counted. If a filled chamber is to stand for any length of time before counting, prevent evaporation by putting it into a petri dish with a glass triangle to hold it off the bottom (see Chp. 3, II.C), or else in a divided bottom petri plate. Add some distilled water to the bottom of the petri plate and cover.

Before counting, examine the contents of the chamber under low magnification to detect any obviously unsatisfactory distribution patterns. High power of a stereomicroscope will do.

a. Sedgwick–Rafter chamber – set the cover glass diagonally across the chamber so that a space is left open at opposite corners. Use a large bore measuring pipette or a milk dilution pipette to deliver 1.0 ml of suspension to the chamber through one opening; air will exhaust by the other. Slide the cover glass into position. If the cover glass interferes with dispersal of very large organisms, it may be better not to use it, even though visibility becomes poor at the edges of the chamber.

b. Palmer–Maloney chamber – place cover glass in position and tilt the slide slightly. Firm contact between the clean chamber ring and cover

TABLE 19-3. *Number of Chaeoceros simplex cells/1 mm² in a hemacytometer with improved Neubauer ruling, 0.1 mm deep*

Chamber no.	Filled by	Cells per 1 mm square									Sum
1	pipette	39	12	17	26	26	27	30	38	36	251
2	pipette	26	16	27	44	26	38	27	23	19	246
3	loop	30	33	25	24	32	27	18	23	24	236
4	loop	22	26	30	20	21	22	26	29	23	219
5	pipette	19	31	33	30	45	27	20	30	32	267
6	pipette	30	28	18	36	22	22	35	33	28	252
7	loop	24	21	30	30	14	23	34	25	22	223
8	loop	34	17	19	40	16	23	18	25	25	217
									Total		1911

$9 \times 8 = 72$ 1 mm squares were counted; the average number of cells per square is $1911/72 = 26.5$; hence the population density is estimated to be 26.5×10^4 cells per ml (see text, IV.A).

glass is necessary to prevent significant loss of liquid between their surfaces. Add 0.1 ml *via* the lower entry port. A large bore measuring pipette is best, but with experience, a large bore Pasteur-type pipette can also be used with acceptable precision.

c. Hemacytometers – place the cover glass in position, seating it firmly on the support pillars. Fill a smooth tipped pipette of adequate bore diameter with algal suspension and touch-off any liquid remaining on the tip. Hold the pipette at an angle of 45° for small bore pipettes; to about half this for large diameter ones. Place the tip of the pipette next to the entry slit or groove of the hemacytometer chamber. Release the liquid flow and almost simultaneously remove the pipette from contact, so that liquid flows quickly and evenly into the chamber, barely filling it, with no overflow, or even bulging of the liquid surface into the canals surrounding the chamber. Refill if this is not achieved.

Hemacytometers were designed to be filled with a pipette, and studies of hemacytometer precision have been carried out with pipette-filled chambers. However, a wire loop can also be used to fill hemacytometers; the shallow ones more easily than those 0.2 mm deep. Note carefully, however, that the counts obtained with chambers filled in these two ways are not necessarily comparable, as illustrated in Tables 19-3, 19-4, and sample data (IV.A, following). This difference has no influence on the value of the growth rate obtained from the counts, as long as all counts were made by the same system.

TABLE 19-4. *Comparison of algal numbers in chambers filled by pipette and loop (see text,* IV.A)

	Cells/chamber (9 1 mm squares)	
	Pipette filled	Loop filled
First trial	251	—
	246	—
Second trial	—	236
	—	219
Third trial	267	—
	252	—
Fourth trial	—	223
	—	217
Total	1016	895
Average ($\overline{X}$)	254	223.8
Standard deviation (s)	9.05	8.54
95 % confidence limit	243.3–264.6	213.8–233.9

Size of the loop is critical, and it must be lifted out of the culture medium so that its plane is parallel to the surface of the liquid. The lens of liquid in the loop is touched to the loading slot of the hemacytometer, with the plane of the loop still horizontal. The loop must be withdrawn as the liquid flows, so that the chamber barely fills, as before.

d. Petroff–Hausser chamber – the construction makes it necessary to load the chamber with a pipette. Treat it as a thin hemacytometer (c, preceding).

C. *Counting*

The suspension contained in the counting chambers may be a concentrate or diluate of the original culture. If so, the densities (symbolized by d in Eqs. 3, 5, 6, 7, 8, 9 of the following sections) must be divided by the concentration factor (see III.B.1.a.vi) or multiplied by the dilution factor (see III.B.1.b) to give density of the original culture.

In dilute cultures, tally all organisms contained in the counting chamber, then refill and count as often as necessary to get the desired level of reliability.

To count a dense culture of a non-randomly distributing species (usually, one that clumps), dilute the culture appropriately as described in 1.b, preceding, and count the entire contents of several filled chambers. Tally enough cells to obtain the level of reliability desired. Record separate chamber counts for statistical treatment if desired.

To count a dense culture (or concentrate) of a randomly distributing species, the algal cells in representative areas of the chamber floor can be counted. Hemacytometers and Petroff–Hausser counting devices are ruled into areas of known size, thus no further problems arise. The Sedgwick–Rafter and Palmer–Maloney devices have no ruled division lines, hence the 'area of known size' must be contained in the microscope's field of view, which is then moved about the counting chamber in random fashion and cells tallied. See the section concerning use of Sedgwick–Rafter cell for description of method to be used.

1. Sedgwick–Rafter chamber. In principle, it is possible to use the whole field of view of the ocular as the known area, but in practice it is better to fit the ocular with a Whipple disc, which has a square field. (The square is further divided, but this is not significant for our purposes.) Whipple discs are too big for some oculars. However, the diameter can be reduced by grinding the edge by hand with carborundum paper or a stone, with the faces of the disc carefully protected.

Preparation for counting by areas is as follows: place the Whipple disc in the ocular. Determine the area covered by the disc by measuring with a stage micrometer with each objective. (Take into account the draw tube or interpupillary distance settings of the microscope, because in some models these change the magnification.) As an example, in a microscope with 12.5 × oculars and a 10 × objective, a Whipple square will measure *ca.* 700 μm and its area therefore 0.49 mm^2.

When sampling the chamber with the Whipple disc do not look through the microscope while changing fields. The eye will stop the hand when there is something in the field to count, which eliminates randomness of choice. While in theory the area of the counting chamber should be sampled entirely at random, it may be well in practice to sample each quadrant about equally. Count enough fields to get the precision desired.

a. Counting the whole chamber

i. To count the whole chamber, position it so the Whipple square is in one corner of the counting chamber, e.g., the upper left as seen through the microscope. Slowly move the stage horizontally, tallying the organisms as they pass the leading boundary (right side) of the square, until the Whipple square is in the upper right corner. (Count the organisms that are cut by the lower boundary; they will be ignored on the next sweep when they will lie on the upper boundary.)

ii. Move the stage vertically so that the Whipple square has (apparently) moved down by its own width. (A particle, or mark on the glass of the chamber on or near the bottom line of the Whipple square, can be used as an index.)

iii. Then sweep horizontally from right to left, again tallying organisms. Neglect organisms that lie on the upper boundary of the square.

iv. Repeat until the whole chamber has been covered. The total count is the algal density in either, cells/ml or colonies/ml.

If the number of sweeps required to cover the chamber is known, then the number of algae tallied can be reduced. When this method is indicated, count only those in every other sweep, or some other fraction. The concentration of organisms (cells or colonies/ml) in the sample placed in the chamber will then be the number of individuals counted multiplied by the reciprocal of the fraction of the chamber scanned.

b. Counting Whipple fields – decide first on a system for selecting the positions to be counted. (See McAlice 1971 for detailed discussion of the statistics of sampling with the Sedgwick–Rafter chamber.) Count every organism within the boundaries of the square and every other organism cut by a boundary. The area of the Sedgwick–Rafter chamber is 1000 mm²; the chamber holds 1.0 ml. The area of the Whipple field is A mm². Obtain the average number of cells (or colonies) per Whipple field, N. The density (d) of the algal suspension in the chamber (cells or colonies per ml) will be:

$$d = \frac{\text{cells}}{\text{field}} \times \frac{\text{Whipple fields}}{\text{chamber}} \times \frac{\text{chambers}}{\text{ml}}, \tag{1}$$

$$d = N \times \frac{\text{area of Sedgwick–Rafter chamber}}{\text{area of Whipple field}} \times \frac{1}{1.0}, \tag{2}$$

$$d = N \times \frac{1000}{A}. \tag{3}$$

In the example given previously, where the Whipple disc area was 0.49 mm², Eq. (3) becomes: $d = 2040 \times N$ per ml.

2. Palmer–Maloney chamber. See the general remarks and directions for the Sedgwick–Rafter device (1, preceding). There are two points of difference: (*a*) the Palmer–Maloney chamber is round, hence the sweeps must begin at the top or bottom inner tangent (there is no upper right-hand corner); and (b) the right-to-left sweeps differ in length. It is therefore necessary to estimate the fraction of the area of the chamber that will be surveyed if alternate sweeps (or some other fraction) are tallied.

Assume that fields are counted, that there are M organisms per field, and that the area of the Whipple field is A mm². The area of the Palmer–

Maloney chamber is 250 mm², and it holds 0.1 ml. Then the density (d) of the suspension in the chamber is:

$$d = \text{Eq. (1), preceding,}$$

$$d = M \times \frac{250}{A} \times \frac{1}{0.1}, \tag{4}$$

$$d = M \times \frac{2500}{A}. \tag{5}$$

For $A = 0.49$ mm² as before, $d = 5100 \times M$ (cells or colonies) per ml.

3. Speirs–Levy hemacytometer (0.2 mm deep). This counting device has four separate chambers each with 10 squares 1 mm on a side – the word 'square' in reference to all hemacytometers will mean such a 1 mm square, regardless of depth. The 10 squares of the Speirs–Levy chamber are in a 5×2 array and each is further divided into sixteen smaller squares (250 μm on a side and referred to *only* as '250 μm squares').

The aim is to determine the average number of organisms per square, P, which can be done by counting algae in all forty squares (or more), or in any smaller number, or in any fraction of a square if the suspension is dense. By whatever means the total count desired is accumulated, it is better to sample as many chambers as possible and more than one filling of the whole device if practical.

The density of the algal suspension in the hemacytometer is then given by:

$$d = 5 \times 10^3 \times P. \tag{6}$$

4. Hemacytometer (0.1 mm deep). The type recommended has two chambers each with 9 squares 1 mm on a side. The various squares are again divided, but not identically. The only subdivision of practical importance to us is that of the central square, which is divided into twenty-five smaller squares 200 μm on a side, called '200 μm squares'. (These are divided into sixteen 50 μm squares, but this does not concern us.)

Again the aim is to determine the average number of algae per square, Q, which can be done in many ways depending on the suspension density. Spread the total accumulated count desired over both chambers and at least two fillings of the hemacytometer, if possible. As an example, if there are in fact 100 algae per square, one could count the organisms in the upper half of the top left square in each chamber, then repeat the filling and counting. The finding would be 200 organisms in four half squares, or 100/square. Given the average Q, the density of the suspension in the hemacytometer is:

$$d = 10^4 \times Q. \tag{7}$$

For very dense suspensions it is sometimes enough to get the average number of cells, R, in the 200 μm central squares, in which case:

$$d = 10^6 \times \tfrac{1}{4}R. \tag{8}$$

5. *Petroff–Hausser chamber*. The ruling pattern is like that of the 0.1 mm deep hemacytometer described above, but the chamber is only 0.02 mm deep. It is normally used only for dense suspensions of very small algae. Let the average numbers of organisms (see 4, preceding) be as follows: S per square, T per 200 μm square, and V per 50 μm square. Then density of the suspension in the counting chamber is:

$$d = 5 \times 10^4 \times S = \tfrac{5}{4} \times 10^6 \times T = 2 \times 10^7 \times V \text{ organisms/ml.} \tag{9}$$

D. *Computation of division rates*

1. *Growth constant.* Under suitable circumstances a culture of micro-organisms grows so that the rate of addition of cells is proportional to the number present – 'exponential growth'. Cells, on the average, divide in a characteristic time called the division time (also generation or doubling time). Population development follows the solution of the equation:

$$dN/dt = K_e N, \tag{10}$$

where N is number or concentration of cells in the culture, t is time, and K_e (the growth constant) is a quantity with dimensions of t^{-1}. The solution is:

$$K_e = \frac{\ln(N_1/N_0)}{t_1 - t_0}, \tag{11}$$

where subscripts denote values at two times and ln indicates natural (base e) logarithms. Counts are ordinarily made at more than two times to establish that the population is in fact growing exponentially.

The growth constant K_e (Eq. (11)) is the number of 'logarithm-to-base-e' units of increase per day. Growth rate is sometimes expressed as 'logarithm-to-base-10' units of increase per day, K_{10}; or as 'logarithm-to-base-2' units of increase per day, k; where:

$$K_{10} = \frac{\log(N_1/N_0)}{t_1 - t_0} \text{ and } k = \frac{\log_2(N_1/N_0)}{t_1 - t_0}. \tag{12}$$

For algal cultures, it is usually most convenient to use base 2 logarithms, which yields k as divisions per day.

Different expressions for the growth rate can be interconverted by using the relationship between logarithms of any number to two bases. Conversions are given in Eq. (13), following, in which K_e, K_{10}, and k are as defined in Eqs. (11) and (12), with time in days; $0.6931 = \ln 2$, and $2.3026 = 1/\log e$:

$$k \text{ (div./day)} = \frac{K_e}{0.6931} = \frac{2.3026}{0.6931} K_{10} = 3.322 K_{10}. \qquad (13)$$

Time is sometimes expressed as hours for organisms (mostly bacteria) with short generation times. If growth rates (k, K_e, or K_{10}) are thus given in logarithmic increases per hour (h^{-1}), multiply by 24 to obtain the corresponding values expressed as units of increase per day (day^{-1}).

From k, compute the division (generation) time by:

$$T_d = 1/k \text{ days per division} = 24/k \text{ hours per division}. \qquad (14)$$

In the following (2–5) are given four methods of finding k from the same data. An example is treated by all four methods in IV.B of this chapter.

2. *Computation of k from two counts.* If N_1 and N_0 are the cell concentrations at the end and beginning of a period of time t days, then:

$$k = \ln (N_1/N_0)(1.443/t) = \log (N_1/N_0)(3.322/t). \qquad (15)$$

3. *Graphical estimate of k from two counts.* Define the Ns and t as above. Form the ratio N_1/N_0 (ordinarily a number between 1 and 16). Read the divisions per time t from Fig. 19-1 corresponding to the value N_1/N_0. If $t \neq 1$, then $k = $ (divisions in time t)/t.

4. *Graphical estimate from semi-log plot of successive counts.* This method is relatively fast and permits ready identification of the period of exponential growth, which appears as a straight line on the semi-log plot.

a. Plot the cell concentrations $N_0, N_1, \ldots, N_i$ as ordinates on semi-log (base 10) paper, with the times $t_0, t_1, \ldots, t_i$ on the linear abscissa. (Both linear scale 10 and 12 paper are available. Paper with three or four decades usually covers the range of population increase encountered.) It is best to plot each count as made.

b. Identify the portion of the graph corresponding to exponential growth, and fit, by eye, a straight line through the appropriate points.

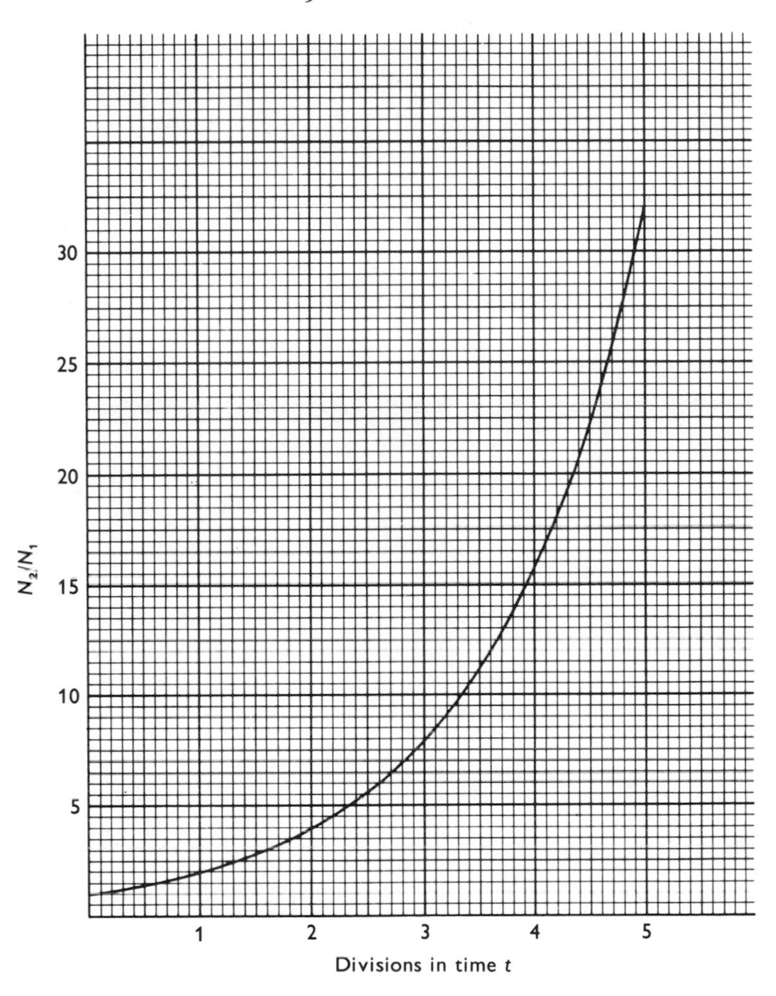

Fig. 19-1. Graph for estimating division rate k from two counts of a culture, assumed to be in exponential growth.

The ratio N_2/N_1 of the counts at times t_2 and t_1 ($t = t_2 - t_1$) is located on the ordinate. The corresponding abcissa then yields the number of divisions in time t needed to produce the increment in N. Divide by t to obtain k as divisions per day. (From a computer plot of $y = 2^x$, $x \leqslant 5$). See text, III.D.3.

c. Using the line drawn above, estimate the time T_{10} (in days and decimal fractions of days) required for a tenfold increase in N.

$$\text{Then } k = \frac{3.322}{T_{10}} \text{ divisions per day.} \tag{16}$$

5. *Computation of k from successive counts.* This method is the most rigorous; besides permitting ready identification of the exponential phase of growth, it allows use of standard statistical procedures.

a. Choose the points corresponding to exponential growth by forming the semi-log plot as 4, preceding.

b. Define $Y = \log N$ (or $Y = \ln N$) and form a table of Y *vs* t for the points chosen. (Note: the numbers $N_0, N_1, \ldots, N_i$ can be written $M_0 \times 10^a$, $M_1 \times 10^a$, ... $M_i \times 10^a$, and only the Ms need be carried through the calculations.)

c. By least squares, find the regression of Y on t and compute the slope, which is the growth constant K_{10} or K_e (depending on the base chosen). Convert to divisions/day by Eq. (13).

IV. SAMPLE DATA

A. *Hemacytometer counts*

The hemacytometer had Improved Neubauer ruling with chambers 0.1 mm deep. A culture of *Chaetoceros simplex*, a solitary-growing centric diatom (Bacillariophyceae), was mixed and both chambers of the hemacytometer were filled and counted four times. Both chambers were filled with a pipette the first and third times, and with a wire loop the second and fourth times. Record was kept of the counts in individual 1 mm squares (Table 19-3), and the number of algal cells in each chamber (9 squares) was considered as a separate estimate of the population. The estimate is given in Table 19-3, based on all counts.

In Table 19-4, the four estimates of the population density derived from counts of chambers filled by pipette are compared with the estimates derived from loop filled chambers. A t-test, for 6 degrees of freedom, was applied to test the hypothesis that the 8 counts shown in Table 19-4 were from the same population of estimates (i.e., that the method of filling had no effect on the count). The test is one for comparison of two sample means, with unpaired observations, equal variances. The computed value of t, 4.854, indicates that there is less than one chance in 100 that the 8 counts would derive from the same population. The method of filling the chambers therefore very probably influenced the results. This is only an argument for consistancy in method. Additional study would be required to determine which method yields better estimates of the culture population.

TABLE 19-5. *Concentration of Cyclotella cryptica, as* 10^4*/ml*

Day	0	1	2	3	4	5	6	7
10^4 cells/ml	0.5	2.14	4.62	11.8	20.5	33.8	35.7	35.3

Cells were counted in a Speirs–Levy hemacytometer until day 2, then in a phase-contrast hemacytometer (see text, IV.B, and Fig. 19-2).

B. *Calculation of growth rate*

The data, taken from an experiment on growth of the centric diatom *Cyclotella cryptica*, are shown in Table 19-5 and in semi-log plot in Fig. 19-2.

1. Computation of k from successive counts. The data and computations are shown in Table 19-6. The growth rate obviously falls after day 4, as shown best in Fig. 19-2. (This example does not show a lag phase at the beginning of the period of growth; the cells were apparently dividing faster before inoculation into the experimental conditions.) The period of exponential growth reflecting the experimental conditions is taken to extend from day 1 to day 4; the high count at day 3 causes k_{2-3} to be high and k_{3-4} to be low. The computation ($k_{1-4} = 1.05$) using the counts from days 1 through 4 seems the most reasonable estimate. Note that averaging the separate values of k over the same 3 day interval does not yield the same value as k_{1-4}; the former is 1.21, *vs* 1.05 divisions per day.

2. Graphical estimate of k. This is much faster than the computation illustrated before. Using the values for N_2/N_1 as given in Table 19-6, read off values of k from Fig. 19-1. The estimates are shown in Table 19-7 and can be compared with the computed values in Table 19-6. The same considerations apply as if the values were computed; the value 1.09 (over the period days 1–4) would be taken.

3. Graphical estimate of k from semi-log plot. Examine Fig. 19-2, the semi-log plot of the data. It is obvious that the first point and the last three must be excluded because exponential growth characteristic of the conditions being tested was not attained until day 1 and did not last beyond day 4. (It will not always be the case that the end points are to be excluded, and knowledge of things not directly in the data must help determine which data are to be considered.) Here, the points for days 1 through 4 will be used, though there might be grounds for rejecting even the fourth.

TABLE 19-6. *Computation of k (divisions/day) from:* $k = [3.322/(t_2 - t_1)].(\log N_2/N_1)$

Interval, days	0–1	1–2	2–3	3–4	4–5	5–6	6–7	1–4
N_2/N_1	2.14/0.5	4.62/2.14	11.8/4.62	20.5/11.8	33.8/20.5	35.7/33.8	35.3/35.7	20.5/2.14
	4.28	2.16	2.55	1.74	1.65	1.06	<1	9.58
$\log N_2/N_1$	0.6314	0.3345	0.4065	0.2405	0.2175	0.0253	—	0.981
$t_2 - t_1$	1	1	1	1	1	1	—	3
k	2.11	1.11	1.35	0.80	0.72	0.08	—	1.05

Using data from Table 19-5 for various time intervals (also see text, IV.B.1, 2, and Fig. 19-2).

TABLE 19-7. *Graphical estimates of k (divisions/day) from Fig. 19-1*

Interval, days	0–1	1–2	2–3	3–4	4–5	5–6	6–7	1–4
N_2/N_1	4.28	2.16	2.55	1.74	1.65	1.06	<1	9.58
k (from Fig. 19-1)	2.08	1.05	1.36	0.77	0.70	0.05	—	1.09[a]

Data are taken from Table 19-6 (also see text, IV.B.2.) [a] 1.09=3.26/3.

The line AA' was drawn by eye as an estimate of the slope of the line representing exponential growth. The two arrows show where AA' intercepts horizontal lines an order of magnitude apart (here, the three line in each decade). The time interval T_{10} is estimated on the horizontal axis as 3 days 3 h = 3.125 days, and k is given from:

$$k = \frac{3.322}{T_{10}} = 1.07 \text{ divisions per day,} \tag{17}$$

in good agreement with previous estimates.

4. *By least-squares fit of a straight line to the data, logarithmically transformed.* It is best to use the semi-log plot shown in Fig. 19-2 to decide which points are to be fitted, as explained in 3, preceding. Again, the points for days 1 through 4 will be considered. A logarithmic transformation is required. The equation assumed to hold is:

$$N = N_0 \, e^{K_e t} \tag{18}$$

where the Ns refer to cell concentrations, K_e is the growth constant in base e units, t is time, and $N = N_0$ at $t = 0$. It is usual to use common (base 10) logarithms for calculations, and K is wanted in units of divisions per day (base 2). The transformation and conversion of units are accomplished by first taking logarithms to the base 10:

$$\begin{aligned} \log N &= \log N_0 + K_e t \log(e) \\ &= \log N_0 + (0.4343) K_e t \end{aligned} \tag{19}$$

but according to Eq. (13) $K_e = (0.6931)\, k$ (k expressed as divisions per day), hence:

$$\log N = \log N_0 + (0.301)\, kt. \tag{20}$$

This has the form:

$$Y = a_i + a_k\, t \tag{21}$$

where $Y = \log N$ and $k = 3.322 a_k$. (22)

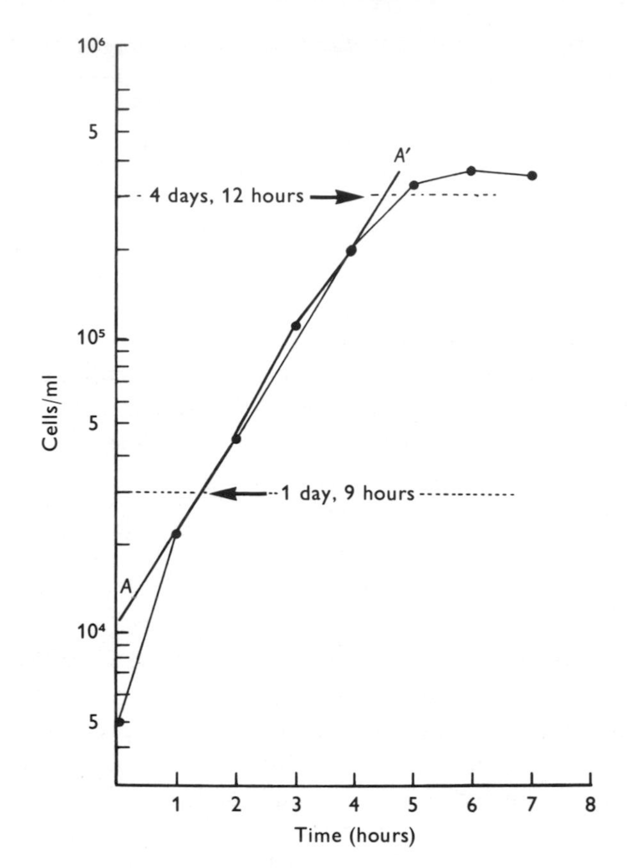

Fig. 19-2. Semi-log plot of an experiment with *Cyclotella cryptica*. The straight line AA' was fitted by eye. See text IV.B.3.

The numbers a_i and a_k are produced by the least squares procedure.

A calculation (by one process) is shown in Table 19-8. This yields a value of 0.33565 for a_k, which, from Eq. (22) gives $k = 1.12$, in good agreement with previous estimates.

The antilog of the number a_i is the intercept of the regression line on the vertical axis; it corresponds roughly to the theoretical inoculum density. The antilog of a_i (0.063) is 1.156, which, multiplied by 10^4, yields 11560 cells/ml. The intercept of the line drawn by eye in Fig. 19-2 is just above 10^4 cells/ml.

TABLE 19-8. *Computation of k (divisions/day) from slope of a line fitted to the data by the method of least squares*

	t	N^a	$Y = \log N$	t^2	$t.Y$
	1	2.14	0.3304	1	0.3304
	2	4.62	0.6646	4	1.3292
	3	11.8	1.0719	9	3.2157
	4	20.5	1.3118	16	5.2472
Sum	10	—	3.3787	30	10.1225

[a] With 10^4 omitted. This makes no difference except in computation of a_i later (see text). M = number of observations = 4.

In the following expressions, all sums are over all data pairs (4 in this example):

$$a_k = \frac{M.\Sigma(t.Y) - (\Sigma t).(\Sigma Y)}{M.\Sigma(t^2) - (\Sigma t)^2} = \frac{4(10.1225) - 10(3.3787)}{4(30) - (10)^2} = 0.33565$$

$$k = 3.322\,(a_k) = 1.12$$

$$a_i = \frac{(\Sigma Y).(\Sigma t^2) - (\Sigma t).[\Sigma(t.Y)]}{M.\Sigma(t^2) - (\Sigma t)^2} = \frac{(3.3787)(30) - 10(10.1225)}{4(30) - (10)^2}$$

$$a_i = 0.063.$$

V. REFERENCES

Calvin, M., Heidelberger, C., Reid, J. C., Tolbert, B. M. and Yankwick, P. F. 1949. *Isotopic Carbon. Techniques in Its Measurement and Chemical Manipulation.* John Wiley and Sons, Inc., New York.

Holmes, R. W. and Widrig, T. M. 1956. The enumeration and collection of marine phytoplankton. *J. Cons. Int. Expl. Mer* **22**, 21–32.

Lund, J. W. G. 1951. A sedimentation technique for counting algae and other organisms. *Hydrobiologia* **3**, 390–4.

Lund, J. W. G., Kipling, C. and LeCren, E. D. 1958. The inverted microscope method of estimating algal numbers and the statistical basis of estimations by counting. *Hydrobiologia* **11**, 143–70.

McAlice, B. J. 1971. Phytoplankton sampling with the Sedgwick-Rafter cell. *Limnol. Oceanogr.* **16**, 19–28.

Palmer, C. M. and Maloney, T. E. 1954. A new counting slide for nannoplankton. *Amer. Soc. Limnol. Oceanogr. Special Publ.* **21**, 1–6.

Utermöhl, H. 1958. Zur Vervollkommung der Quantitativen Phytoplankton-Methodik. *Mitt. Int. Ver. Limnol.* No. 9. 38 pp.

20: Growth measurements by analysis of carbon*

DAVID W. MENZEL† AND
WILLIAM M. DUNSTAN†
Woods Hole Oceanographic Institution, Woods Hole, Massachusetts 02543

CONTENTS

* Contribution no. 2481 from the Woods Hole Oceanographic Institution. Supported by National Science Foundation Grant BG 15103 and by Atomic Energy Commission Contract AT(30–1)–3862, Ref. NYO 3862–29.

† Present address: Skidway Institute of Oceanography, University System of Georgia, 55 West Bluff Drive, Savannah, Georgia 31406.

I. OBJECTIVE

Changes in biomass of cultures may be quickly and easily determined by analytical determination of cell carbon. The method provides a direct measure of net growth within single culture vessels over an extended period of time (i.e., in excess of 24 h) when cultures are serially sampled. Growth estimates are integrated between given sampling periods and are cumulative over the length of time of observations. The method does not provide an instantaneous measure of photosynthetic activity such as can be obtained with carbon-14 uptake or oxygen production measurements. These methods are preferable for short-term estimates of growth (*ca.* 6 h maximum) though both are indirect measures of carbon fixation and are based on many interpretative assumptions (these techniques are described by Strickland and Parsons 1968).

II. EQUIPMENT

Many carbon, and carbon:nitrogen:hydrogen analyzers are available commercially. All use high temperature induction furnaces to pyrolize organic matter to carbon dioxide. Most are automated except for sample changing. Quantitative detection of the resulting carbon dioxide is achieved by: (a) gravimetry after adsorption onto 'Ascarite' (Model 33 Carbon/Hydrogen analyzer, Coleman Instruments Division, Perkin Elmer Corp., 42 Madison St., Maywood, Illinois 60153; 8250 Mountain Sights Ave., Montreal 308, Quebec); (b) measurement of gas volume (Leco Carbon Determinator, Laboratory Equipment Corp., 3000 Lakeview Ave., St Joseph, Michigan 49085); (c) nondispersive infrared detection (Menzel and Vaccaro 1964; Strickland and Parsons 1968); (d) gas chromatography (240 Elemental Analyzer, Perkin Elmer Corp., Instrument Division, 800 Main Ave., Norwalk, Connecticut 06852: Model 185 C:H:N Analyzer, F and M Scientific Corp.); or (e) adsorption in base accompanied by coulometric titration, Dal Pont and Newell (1963) (Leco Carbon Determinator). In general the last three techniques (c to e) are capable of detecting smaller quantities of carbon dioxide with greater precision than the first two, thus minimizing the size of sample required for accurate analysis.

Also required for treatment of sample material are suitable filters (III.A, following); a source of vacuum and in-line reservoir for the filtrate; storage vials; and, drying chambers. Reagents required vary with the recommendations of the manufacturer of analytical instruments.

III. CONCENTRATION OF SAMPLE MATERIAL

A. *Filter choice, preparation*

Although centrifugation can be used to concentrate samples for carbon analysis, total recovery of the cells by this technique is difficult. For this reason selected volumes of culture are usually filtered onto one of two types of filters containing little organic material.

All filters must be combusted in a muffle furnace prior to use to remove extraneous carbon. Generally this procedure reduces included carbon to about 3 μg C/filter. We are not aware of any microcarbon analyzers which can accommodate filters larger than 24 mm diameter.

1. Metal. Selas Flotronics (P. O. Box 300, Spring House, Pennsylvania 19477); available in a range of pore sizes from 0.25 to 5.0 μm. Metal filters are combusted at 400 °C in an O_2-atmosphere.

2. Glass fiber. Whatman GFC or Gelman Type A (Gelman Instrument Co., 600 S. Wagner Rd, Ann Arbor, Michigan 48106) are also satisfactory and have the added advantage of rapid filtration rates. Glass fiber filters are combusted for 1 h at 600 °C.

B. *Filter pore size, vacuum*

Metal filters of all pore sizes tested and those of glass fiber retain phytoplankton cells with approximately equal efficiency when the applied vacuum does not exceed 100 mm Hg (Table 20-1). At lower vacuums it is extremely time consuming to use pore sizes smaller than 1.2 μm. Within the size ranges and external morphology of the species tested (Table 20-1), the choice of pore size and type of filter used therefore is arbitrary as long as the applied vacuum does not exceed 100 mm Hg.

C. *Filtering*

The quantity of material which must be filtered for accurate carbon measurements is dependent upon the concentration of cells present and the analytical precision required. As with any chemical technique, precision

TABLE 20-1. *Effect of filter pore size and vacuum on the retention of cell carbon in selected phytoplankton cultures (values in μg C) (figures in parenthesis by clone are approximate sizes in μm)*

	Filter Type-12 cm vacuum						Vacuum (cm Hg)–Glass filters					
	Metal					Glass						
Clone[a] ▼ Pore size (μm) ▶	0.45	0.80	1.20	3.00	5.00		5	12	25	38	51	64
Dun (11.0 × 7.6)	123	99	102	101	110	118	185	187	184	173	162	150
3H (5.9 × 5.5)	104	82	83	97	90	95	—	—	—	—	—	—
Cocco 2	154	161	159	160	159	170	—	—	—	—	—	—
—												
BT 6 (5.0)	90	92	89	86	92	94	843	842	847	802	806	761
Skel (7.1 × 6.4)	98	93	92	97	99	94	778	780	754	685	652	605
BBsm (6.6 × 5.1)	256	288	265	295	284	285	—	—	—	—	—	—
Pyr 1 (10.9 × 8.1)	310	295	306	291	302	305	—	—	—	—	—	—
Mil Nanno (3.0)	106	107	98	98	96	100	260	258	235	228	210	185

[a] Cultures – Woods Hole Oceanographic Institution, Woods Hole, Massachusetts 02543

Clone–WHOI	Species
Dun	*Dunaliella tertiolecta* – (Chlorophyceae)
3H	*Cyclotella nana* – (Bacillariophyceae)
Cocco 2	*Cricosphaera carterae* – (Haptophyceae)
BT 6	*Coccolithus huxylei* – (Haptophyceae)
Skel	*Skeletonema costatum* – (Bacillariophyceae)
BBsm	*Chaetoceros simplex* (?) – (Bacillariophyceae)
Pyr 1	*Pyramimonas* sp. – (Prasinophyceae)
Mil Nanno	*Nannochloris atomus* (?) – (Chlorophyceae)

is a function of care in handling to avoid external contamination and inherent analytical error. In our experience, using the method of Menzel and Vaccaro (1964), the total error of determination at carbon levels above 100 μg is $\pm 4\%$. Below 100 μg C the relative error increases to $\pm 100\%$ at 5 μg C. It is usually not necessary to concentrate more than 10 ml from actively growing cultures. If color is visually detectable on the filter, sufficient material is usually present to allow analyses at the $\pm 4\%$ level.

D. *Filter treatment for analysis*

If inorganic precipitates or coccoliths (calcium scales of Coccolithophoraceae, Haptophyceae) are present, it is necessary to remove the associated carbonate prior to drying the filter for analysis. One method is to rinse the filter with isotonic 0.01 N HCl or H_3PO_4. Another method is to place the filter over fuming HCl for 5 min. Experience has shown that the latter technique is preferable in most instances since some lysis of cells occurs at lowered pH. Ruptured cell material may then pass through filters with the acid solution.

Following filtration and fuming the samples are thoroughly dried in wide mouth glass vials in a desiccator over silica gel or in a drying oven at 60 °C. They may be stored indefinitely if kept dry or frozen.

IV. ANALYTICAL PROCEDURES

A. *Combustion techniques*

Step by step analytical procedures vary depending on the type and manufacture of carbon analyzer available. All commercially available models include adequate operating instructions for standardization and determination of carbon in unknowns.

B. *Determination of error; blank corrections*

Once proper standardization is achieved it is useful to validate the consistency of results by filtering various quantities of cells and plotting carbon as a function of volume filtered. When extrapolated to zero volume, the intercept on the carbon axis is a measure of carbon contained in the filter. The value so obtained is subtracted from each other value to obtain the 'real' carbon concentration of individual samples. This procedure, however, is not a practical means of determining 'blanks' on a day to day basis. Filters through which no media has been passed give false corrections (too low) since most filter surfaces absorb some dissolved organic material from solution (Menzel 1966). To circumvent this problem, a barrier of two filters can be used; the upper representing particulate and adsorbed carbon, and the lower adsorbed carbon only. Proper blank corrections are obtained by this method even when high levels of soluble organic compounds are present in culture media.

V. INTERPRETATION OF DATA

A. *Interferences*

Increase in cell carbon is a straightforward assay of growth that is not subject to serious interference or misinterpretation of results except in two situations: (1) if the organisms in culture produce solid excretory by-products; and, (2) if dead cells are present and their concentration changes with time. In each instance living cells cannot practically be separated from the total organic matter present as filterable material and growth estimates will be biased toward the high side.

B. *Sample data*

An example of the type of information obtained by carbon analyses of a growing axenic culture of *Skeletonema costatum* (Bacillariophyceae) (Woods Hole Oceanographic Institution, Woods Hole, Massachusetts 02543, Clone-'Skel') is shown in Fig. 20-1. The cultures were grown in 2 l Fernbach flasks containing 'f/2' media (Guillard and Ryther 1962; Chp. 2, IV.A.4) and their growth monitored by carbon analysis and cell counts for 13 days. The two methods are not entirely comparable since the C/cell varies depending upon the growth phase. In this culture, as in others tested, the C/cell is highest at the time of initial transfer of 5 day cultures and decreases during exponential growth. After exponential growth, at the upper asymptote of the growth curve, cell division slows down more rapidly than carbon is synthesized and the C/cell increases. Only between the second and fifth day of the experiment did both methods give comparable estimates of growth.

A comparison of short-term growth estimates by carbon-14 uptake and changes in cell carbon have been discussed by Ryther and Menzel (1965).

VI. ALTERNATIVE METHOD

The most economical method of carbon analysis is based on the wet oxidation of organic matter with chromic acid. The amount of the chromate reduced has been measured by titration (Fox, Isaacs and Corcoran

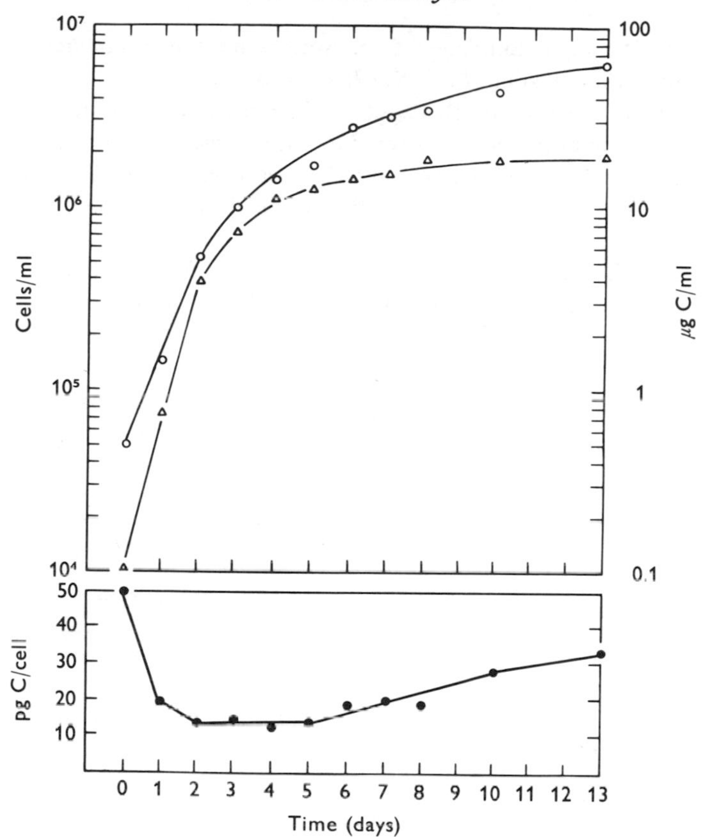

Fig. 20-1. The top half of the figure shows the relation between growth measured by carbon analysis (○) and cell division (△) both as a function of time. The lower half shows changes in C/cell during the course of the same experiment.

1952) and spectrophotometry (Strickland and Parsons 1968). Both techniques are subject to relatively large error (> 50 μg C). They are not recommended here because of relative insensitivity and incomplete knowledge of the extent of oxidation achieved.

VII. REFERENCES

Dal Pont, G. and Newell, B. 1963. Suspended organic matter in the Tasman Sea. *Aust. J. Mar. Freshwat. Res.* **14**, 155–65.

Fox, D. L., Isaacs, J. D. and Corcoran, E. F. 1952. Marine leptopel, its recovery, measurement and distribution. *J. Mar. Res.* **11**, 29–46.

Guillard, R. R. L. and Ryther, J. H. 1962. Studies on marine planktonic diatoms. I. *Cyclotella nana* Hustedt and *Detonula confervacae* (Cleve) Gran. *Can. J. Microbiol.* **8**, 229–39.

Menzel, D. W. 1966. Bubbling of sea water and the production of organic particles: a re-evaluation. *Deep-Sea Res.* **13**, 963–6.

Menzel, D. W. and Vaccaro, R. F. 1964. The measurement of dissolved organic and particulate carbon in sea water. *Limnol. Oceanogr.* **9**, 139–42.

Ryther, J. H. and Menzel, D. W. 1965. Comparison of the ^{14}C-technique with direct measurement of photosynthetic carbon fixation. *Limnol. Oceanogr.* **10**, 490–2.

Strickland, J. D. H. and Parsons, T. R. 1968. A practical handbook of seawater analysis. *Fish. Res. Bd Canada*, Ottawa, Bull. 167. 311 pp.

21: Dry weight, packed cell volume and optical density

CONSTANTINE SOROKIN
Department of Botany, University of Maryland, College Park, Maryland 20742

CONTENTS

I. INDICES OF GROWTH

Physiological and biochemical investigations of microscopic algae are generally conducted on cultures of algal cells as experimental objects. The obtained data reflect the activities of a multitude of cells. Quantitatively, they are usually referred to a unit of cell mass. Cell mass can be expressed as dry weight of cells or as packed volume of cells in a volume unit of a given microbial culture. In a growing culture, dry weight and packed volume of cells per volume of cell suspension (of medium) increase with time, and this increase, termed growth, can be the specific subject of observation. Another characteristic, widely used as a measurement of growth, is an increase in optical density (turbidity) of a suspension of algal cells. Other indices of growth, such as increase in cell number, accumulation of carbon, nitrogen, protein, or some products of cell metabolism (starch, acids, etc.), are used in growth measurements (see Chps. 19, 20).

II. YIELD AS A GROWTH INDICATOR

Growth can be expressed as yield or as growth rate. Yield, as an expression of organic production, is usually given in terms of dry or fresh weight of the organic mass produced over a period of time per unit of volume or unit of area occupied by a given organism. Thus,

$$Y = \frac{X_1 - X_0}{A \text{ (or } V)} \tag{1}$$

where X_0 and X_1 are quantitative expressions of the mass of cells given usually in terms of dry or fresh weight of cells at the beginning (X_0) and at end (X_1) of the growth period under observation and A (or V) are correspondingly the area or volume occupied by a given population of microbial cells. Since illumination is one of the most critical factors determining the productivity of photoautotrophic organisms, the yield for these organisms is often expressed in terms of production per unit of illuminated area.

Yield, as defined in the preceding paragraph, is equal to what has been previously designated as total growth, G (Monod 1949). Since the term 'growth' generally applies not only to the product of the process but also, and even more frequently, to the process itself, it is expedient to use the term 'yield' for the designation of the product and to reserve the term 'growth' only for the process of the accumulation of the product.

The adjective 'total', in the term 'total growth', is also misleading, because total growth, or what is here called yield, does not represent total production as the outcome of all growth processes, but only the balance between the anabolic and catabolic processes. In addition to losses of organic mass due to catabolism, there may be losses caused by other factors – such as death of some cells. These losses may or may not have any relation to productivity of the organism. If an attempt is made to estimate these losses, the distinction can be made between net yield (designating the actually-harvested cell mass) and the corrected or potential yield, which would also include losses due to external factors and, depending on the purposes of the investigation, also losses due to catabolic processes.

The limitations, in using 'yield' as an indicator of growth, must be clearly understood. In estimations of yield, time period is either chosen arbitrarily or is determined by the cessation of growth. Cessation of growth is a valid consideration for determinations of yield of organisms with limited growth. Cessation of growth comes naturally, for instance, in annual and biennial plants at the end of the vegetative period, and thus indicates the moment of harvesting when the yield must be determined.

In populations of microorganisms, growth is generally unlimited unless external factors change to preclude further growth. By manipulating external factors, the moment of growth cessation can be moved or eliminated altogether. The decision to interrupt growth and to determine the yield or to wait until the complete cessation of growth is, therefore, arbitrary and may contradict the basic purposes of the investigation.

A serious limitation of yield, as an indicator of growth, is that the course of growth, within the period for which yield is estimated, remains unknown. During some portions of that period, growth rate may be lower, net growth may be absent, or it may even give place to loss of organic substance due to a predominance of catabolic over anabolic processes. For all of the above reasons, in all possible cases, estimations of yield must be accompanied by evaluations of the kinetics of growth as a process within the overall period for which yield is determined. The kinetics of growth are described by examination of the rate of growth.

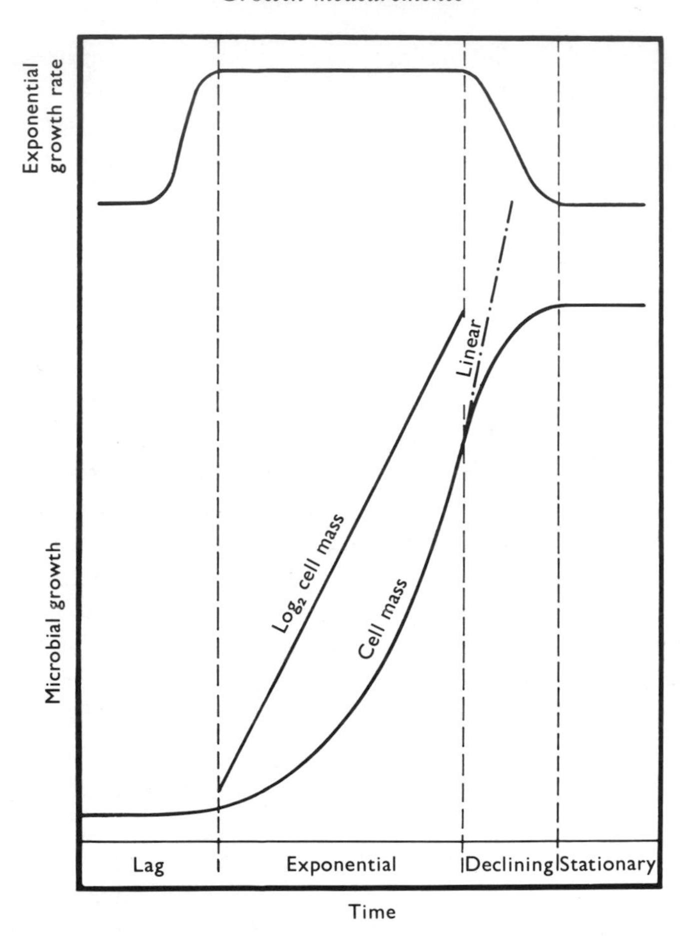

Fig. 21-1. Growth phases in batch cultures of microbial organisms.

III. GROWTH RATES, PHASES OF GROWTH CYCLE

As a measure of velocity of growth, growth rate is an index relating the process of multiplication of organic mass to time. The concept of growth rate and the techniques of its measurements, as applied to populations of microbial organisms, has been based on observations originally made on the so-called batch cultures. Growth of a microbial culture, confined to a constant volume of unchanged medium, is usually described by a sigmoid growth curve (Fig. 21-1). Special significance has been attached to the universality of the sigmoid curve and the whole period of growth of a batch culture, for which the sigmoid curve stands, has been denoted as the 'growth cycle' (Monod 1949). Historically, the term 'growth cycle' is an adaptation to microbial cultures of an earlier term, 'grand period of growth', introduced by Sachs (1873).

Based largely on the way in which the growth rate behaves, that is on whether it is increasing, constant or falling, the growth cycle has been divided into phases of growth (Fig. 21-1). For the purposes of the following discussion, the phases of growth to be mentioned here are:

lag phase
exponential phase
declining phase, a portion of which may become a linear phase
stationary phase.

The lag phase of growth, or its subdivisions, has also been labeled by different authors as the increased logarithmic phase (Buchanan 1918), the phase of adjustment (Winslow and Walker 1939) or the acceleration phase (Monod 1942). During the lag phase, growth rate is changing – usually increasing with time. However, during a portion of the lag phase, net growth may be absent or there may even be a decline in cell mass (dry weight of cells, turbidity of cell suspension, etc.) per unit volume of cell suspension.

During the exponential (Monod 1942) or logarithmic (Buchanan 1918) growth phase, the mass of microbial cells doubles over each of the successive and equal time intervals. The doubling time and, therefore, the growth rate during the entire exponential phase is thus constant. Inasmuch as the stability of the rate, within the time period for which it is measured, is a basic condition for meaningful determinations of growth rate, the exponential phase of growth is uniquely suitable for growth rate determinations. In fact, when growth rate is mentioned without further specifications, it generally means an exponential growth rate.

The declining phase of growth has also been called the retardation phase (Monod 1942) and the decreased logarithmic growth phase (Buchanan 1918). The doubling time for the cell mass is increasing during the declining phase of growth and this phase is generally unfit for growth rate measurements. However, under certain conditions, the decline in the exponential growth rate with time may assume such a pace that increments in cell mass over the successive time intervals may, in absolute figures, be uniform. The mass of cells in a given and unchanged volume of cell suspension increases linearly with time though the (exponential) rate related to the increasing mass of cells is falling. Under these specific conditions, growth phase and growth rate are called linear phase of growth and linear growth rate respectively.

During the stationary growth phase, a stable concentration of cell mass per unit volume of cell suspension is maintained. This is, of course, possible only if losses due to catabolism are exactly compensated by anabolic processes. If a microbial culture is maintained beyond the stationary phase, then the preponderance of catabolic processes causes a decline in cell mass and eventual death of the culture. This phase of decline can hardly be considered as growth.

IV. EXPONENTIAL GROWTH RATE

The most comprehensive way to introduce the technique and significance of determinations of exponential growth rate is to present it as the number of doublings of cell material over a specified time interval – commonly over a one day (24 h) period. The mathematical expression for the exponential growth rate is:

$$R_E = \frac{\log_2 X_2 - \log_2 X_1}{t_2 - t_1} \qquad (2)$$

where

R_E = exponential growth rate,

X_1, X_2 = the numerical values for cell mass (given as dry weight, packed cell volume, optical density, etc.) at the beginning and end of the period during which exponential growth rate is constant,

t_1, t_2 = corresponding times, at which X_1 and X_2 values are determined, in terms of the unit (or its fractions) of time (day = 24 h) used in calculations.

Despite the convincing reasoning advanced more than twenty years ago (Monod 1949) in favor of usage of logarithms to the base 2, the practice persists to express exponential growth rate in terms of common or natural logarithms. The usefulness of the exponential growth rate in biological research lies in its direct application as an index of the number of doublings of cell material per specified period of time. From this, the generation (doubling) time is easily obtainable by dividing time period, for which the rate has been determined, by R_E – the exponential growth rate. References to the exponential growth rate based on logarithms other than those to the base 2 involve, in practical applications, the usage of conversion factors ($\log_{10} \times 3.32 = \log_2$ and $\log_e \times 1.44 = \log_2$) and thus additional efforts and possible confusion.

The ambiguity of growth rate calculations using natural logarithms as well as of attaching to the exponential growth rate the term 'specific growth rate' with the symbol 'k' or 'μ' designated as a 'first-order reaction rate constant' involves implications that 'k' is an intrinsic characteristic of a given organism with a theoretical significance well above that of other cellular characteristics. Actually, growth rate is not characteristic of any specific reaction and may not be characteristic of any individual cell.

For a particular cell, growth (as an increase in cell mass) is an outcome of a multitude of anabolic and catabolic reactions. For an organism of a given genetic constitution, the direction and rate of each of these reactions depends on the intracellular environment which, in turn, is affected by extracellular conditions operating during and prior to observations. An important factor affecting the growth rate is the developmental stage of the cell. A multitude of factors affects the exponential growth rate as they do any other cellular characteristic.

One outstanding feature of the exponential growth rate is its easily obtainable reproducibility. However, the reproducibility of the exponential growth rate for a given organism depends on the reproducibility of external conditions under which the rate is observed. Individual cells in a microbial culture are at any given moment in different developmental stages characterized by different rates of metabolic activities and different growth rates. The exponential growth rate, as any other cellular characteristic, is actually a statistical average which may not be characteristic of any particular cell. The stability and reproducibility of the exponential growth rate depends more than anything else on the stability and reproducibility of the age composition of individual cells comprising a particular population of cells.

An understanding of the concept of the growth rate and of factors affecting its reproducibility is essential in an evaluation of the significance and limitations of growth rate measurements. By using the abbreviation '*R*' (some authors prefer a small *r*) for the rate, by attaching the subscript 'E' to distinguish the exponential rate from the linear rate, R_L, and by expressing the exponential rate as number of doublings of cell material per day, the concept and usefulness of the exponential growth rate are defined in simple and unequivocal terms.

V. DRY WEIGHT

Measurements of growth as increase in dry weight of cells involve taking aliquots (samples) of an algal suspension, drying samples to a constant weight, and expressing the dry weight of cells per unit of volume or of illuminated area of the algal culture. Taking a sample, which is representative of a given culture, is one of the crucial conditions for reliable estimation of dry weight of cells. Adequate stirring of the algal suspension and fast pipetting, to prevent cell settling in the process of sampling, are routine requirements for proper sampling. Each measurement of dry weight of cells is usually done on at least two parallel samples (in duplicate). Size of the sample generally depends on the population density of the culture – the aliquot increasing with the decrease in population density.

After taking the aliquot of algal suspension, cells are separated from culture medium by centrifugation. The deposit of cells is then resuspended in distilled water and centrifuged again. This washing of cells in distilled water removes salts (present in the nutrient medium) which can affect measurements of dry weight. After being washed once in distilled water, cells are transferred with a pipette in a small volume of distilled water into a tared weighing dish. Washing in distilled water must be omitted for marine algae if transferring cells from high salinity medium into distilled water causes plasmolysis resulting in bursting of cells and release of their contents. The techniques employed to circumvent this handicap have not been subjected to a rigorous verification to enjoy a general acceptance.

The weighing dishes may be of disposable or reusable type, glass or metal, open or covered with a lid. The aluminum, light-weight, open, disposable dishes are much in use for quick determinations. For precise measurements on small samples, glass covered dishes are preferable. In our laboratory, drying at 100 °C is continued until two successive weighings of the dish, done at intervals, give a constant weight. It is essential to avoid excessive temperatures and the extension of drying time beyond that necessary to achieve constant weight, since an excessive drying causes changes (oxidation) in dry weight of cells other than loss of water. Weighing is done after the dish is cooled in a desiccator (in our laboratory for 15 min) to room temperature. Dry weight of cells is determined for each of the duplicates and the average weight from these duplicates is used to calculate dry weight per unit of volume, or of surface, of the original algal culture. The sample protocol of dry weight determinations is given in Table 21-1.

Instead of centrifugation, cells can be separated from the medium by filtration on a membrane filter, washed on the filter with distilled water and dried on the filter to a constant weight. The dry weight of cells is determined by subtraction of the dry weight of the filter.

Among factors most commonly affecting the accuracy of dry weight determinations the following are the most important.

1. Poor sampling resulting in an aliquot of cells which is not representative of the original culture.

2. Loss of cells during separation of cells from the medium or after washing the cells in distilled water.

TABLE 21-1. *Dry weight measurements. Sample protocol*

Weighing dish, no.	Suspension volume, ml	Dish weight, g	Successive weighings			Net weight, g	Dry weight, g/l
			1	2	3		
(1)	(2)	(3)	(4)	(5)	(6)	(7)	(8)
19	25	22.6989	22.7417	22.7417	—	0.0428	1.712
95	25	23.6019	23.6448	23.6445	23.6445	0.0426	1.704
						Average	1.708

3. Improper drying – either insufficient drying to bring the sample to constant weight or excessive drying resulting in changes other than removal of water.

4. Improper cooling of the sample before weighing it.

VI. PACKED CELL VOLUME

Determination of growth, as increase in packed volume of cells, involves sampling, centrifugation of the sample until the deposit of cells is compressed to a constant volume, and calculations of packed volume of cells per volume or surface unit of algal culture. Sampling does not generally differ from that involved in dry weight measurements. However, depending on the population density of an algal suspension, it may become necessary to condense (or to dilute) the original suspension before using it for packed volume measurements. To avoid changes in volume of individual cells, due to osmotic effects, measurements of packed cell volume are done in the same nutrient medium in which cells have been grown.

Centrifugation to a constant packed volume is done in calibrated capillary tubes. The tube consists of an enlarged upper portion, the receiver, and a lower calibrated portion which is a capillary tube. The capacity of the calibrated section is usually 0.05 ml. It is calibrated to 0.001 ml and, with an approximation, the volume of packed cells can be read to 0.0001 ml. The capacity of the upper section, in two different makes employed in our laboratory, is either 4 or 7 ml. A sample of the suspension of algal cells pipetted into the receiver must contain an amount of cells which, when compressed to a constant volume, will occupy not more than 0.05 ml (the capacity of the calibrated section). On the other hand, to achieve higher accuracy, the packed volume of cells should be large enough to occupy at least half of the calibrated section of the packed cell volume tube.

The volume of the algal suspension placed into the receiver, and/or

TABLE 21-2. *Packed cell volume determinations. Sample protocol*

Cell volume tube, no.	Volume suspension, ml	Volume packed cells, cm^3	Volume packed cells, cm^3/l
(1)	(2)	(3)	(4)
1	20	0.0425	2.13
2	20	0.0430	2.15
		Average	2.14

the amount (concentration) of cells in this suspension, can be manipulated by dilution or condensation. Thus, in case of an algal suspension of low population density, it may become necessary to take a large sample (e.g., 100 ml), to condense it by centrifugation, and to transfer (with a pipette) this condensed suspension into the receiver. The dilutions (or condensations) must be, of course, taken into consideration in final calculations of the packed volume of cells in a given algal suspension.

The conditions for centrifugation of a particular organism must be found by experimentation. For *Chlorella*, centrifugation at 1500 g for 30 min is sufficient to pack cells to a constant volume. This can be proven by reading the packed volume after 30 min of centrifugation and then again after additional centrifugation for extended periods of time. The packed volume of cells in the capillary tubes must be read immediately after centrifugation, since the postponement of readings results in swelling of cells and thus in exaggerated values for packed cell volume. Determinations of packed cell volume are usually done in duplicate and the average from two measurements is expressed as packed volume of cells in cm^3 per volume (liter) or surface unit of the original algal culture. The sample protocol of packed cell volume determinations is given in Table 21-2.

In the given example (Table 21-2), volume of the sample of the suspension of *Chlorella* cells was 20 ml. Inasmuch as this volume was too large to be placed in the receiver section (4 ml capacity) of the packed cell volume tube, it was first condensed by centrifugation, then resuspended in a small volume of medium and transferred with a pipette into the receiver. After centrifugation for 30 min at 1500 g, the volume of packed cells in the calibrated section of the tube was read (column 3 of Table 21-2) and then expressed in terms of a liter of cell suspension by dividing the figures in column 3 by 20 and multiplying them by 1000.

The sources of errors during packed cell volume determinations are:

1. Poor sampling producing a sample which is not representative of the original culture.

2. Inadequate centrifugation (insufficient time and/or insufficient speed).

3. The delay in readings of packed cell volume causing swelling of cells and thus inflated values of packed cell volume.

VII. OPTICAL DENSITY

Of all indices of growth, measurements of optical density (turbidity technique) are particularly suitable for determinations of growth rate. The basic advantage of the turbidity technique in growth rate measurements, in addition to its efficiency, is the possibility of taking repeated readings on the increase in turbidity of the same batch of the suspension of microbial cells. The same growth vessel, such as a test tube, containing a given volume of cell suspension, can be repeatedly taken out of the constant temperature bath (at specified intervals), subjected to turbidity measurements, and then returned to the bath. The time during which the growth vessel is out of the bath and, therefore, the interruption in conditions under which growth is observed, can be made minimal, and its effect on growth observations can be considered negligible.

In estimations of other indices of growth, the cells subjected to observations are either killed, as in dry weight determinations, or affected unfavorably as to their capacity for subsequent growth – as in packed cell volume measurements. The destruction or damage to cells, used for measurements, makes it imperative either to limit observations on growth to just one determination of yield; or, if measurements of the growth rate are intended, to rely in subsequent determinations on samples repeatedly withdrawn at intervals from the growth vessel. Sampling is an integral and a crucial element of statistical investigations. The possibility of avoiding sampling by the turbidity technique saves time and increases the accuracy of growth measurements.

To assure dependability of the turbidity measurements, they can be checked against some other index of growth – such as the increase in dry weight or in packed cell volume. However, a technique is available which makes such comparison superfluous. This technique, described in a later section, makes turbidity measurements self-sufficient and as dependable as the determination of growth rate by any other index of growth.

In measurements of exponential rate as increase in optical density (O.D.), percent transmission for a growing cell suspension is determined at intervals with a measuring device (spectrophotometer, colorimeter, or

TABLE 21-3. *Measurements of the exponential growth rate. Sample protocol*

Time, h	T	O.D.	log O.D.	$\log_2$ O.D. (log O.D. × 3.32)	$\log_2$ O.D. + 10	Δ
(1)	(2)	(3)	(4)	(5)	(6)	(7)
0	70.00	0.155	−0.810	−2.689	7.311	0
1	63.75	0.196	−0.709	−2.353	7.647	0.336
2	55.25	0.258	−0.589	−1.955	8.045	0.734
3	45.50	0.342	−0.466	−1.547	8.453	1.142
4	35.25	0.453	−0.344	−1.142	8.858	1.547
5	24.75	0.606	−0.217	−0.721	9.279	1.968
6	15.25	0.817	−0.088	−0.292	9.708	2.397
7	9.00	1.046	0.020	0.065	10.065	2.754

similar instrument suitable for measurements of the increase in turbidity of microbial suspensions) as:

$$T = I/I_0, \tag{3}$$

where

I = the transmission of the sample in percent of the transmission of the blank (similar vessel, same medium, no cells),

I_0 = latter adjusted to read 100%.

Transmission, T, is then converted into optical density:

$$\text{O.D.} = \log (I_0/I). \tag{4}$$

Since tables of logarithms to the base two are not readily available, the $\log_2$ O.D. can be obtained by using tables of common logarithms and the conversion factor ($\log_{10} \times 3.32 = \log_2$).

These steps are illustrated in detail in a sample protocol of a growth experiment given in Table 21-3. Suspensions of the high-temperature green alga, *Chlorella sorokiniana* Shihira and Krauss (Shihira and Krauss 1965) strain 7-11-05 (IUCC 1230) were grown in a culture medium (Sorokin and Krauss 1958), in 18 × 150 mm test tubes placed in an illuminated constant temperature bath. The test tubes were fitted with cotton plugs through which cotton-plugged bubbling tubes were passed to supply the cultures with a 5% CO_2/air mixture. The bubbling of gas served also to stir the cell suspensions. Test tubes, in groups of four at a time, were taken from the bath at regular (1 h) intervals (column 1 in Table 21-3) and wiped dry with clean dry cheese cloth. The cotton plugs (together with the bubbling tubes) were removed. The tubes were stoppered with rubber stoppers and shaken to insure a uniform distribution of cells throughout the volume of the medium prior to turbidity measurement.

It may be noticed that turbidity measurements of the exponential growth rate are generally conducted under nonsterile conditions. In our practice, sterile conditions are maintained in stock agar cultures. From these agar cultures, inoculations are made into sterile culture medium several days in advance of actual growth measurements. This preculture of algae in sterile liquid medium serves the purpose of bringing algal cells into a good physiological condition, thus eliminating or at least shortening the lag phase during subsequent growth measurements. Sterile conditions are maintained during preculture until the morning of the growth experiment. The abandonment of sterile conditions during growth measurement is justified on two grounds (but see following):

1. During the course of a growth experiment, some algal cells may settle to the bottom and/or stick to the walls of the growth vessel despite the stirring by gas bubbling. By vigorous shaking of the growth vessel before it is inserted into colorimeter, these cells are returned into the suspension and the accuracy of measurements of optical density is greatly improved.
2. Under conditions of adequate illumination necessary for good algal growth in inorganic media, bacterial contamination is usually at a low level and generally does not affect the algal growth or measurements of optical density of the algal cultures. If observations are conducted in organic media and particularly in darkness (heterotrophic growth), the maintenance of sterile conditions during growth measurements becomes imperative. Under these conditions, bubbling tubes are lifted (but not removed) enough to permit light during turbidity measurements to pass through the algal suspension unimpeded by the bubbling tube. Shaking of the growth vessel before inserting it into the colorimeter is omitted with the consequence that the growth measurement may be expected to be less precise.

In the sample experiment (conducted under nonsterile conditions) given in Table 21-3, the transmission of the algal suspension in each culture tube was read on a Coleman Junior II Spectrophotometer (Model 6/35; Coleman Instruments Division, Perkin Elmer Corp., 42 Madison St., Maywood, Illinois 60153; 8250 Mountain Sights Ave., Montreal 308, Quebec) at 678 nm by placing each tube directly into the spectrophotometer fitted with a 19 × 150 mm, Coleman no. 6-102 adapter. The tubes were then fitted again with the corresponding cotton plugs and returned into the bath. With some practice, it takes about 3–4 min for a group of 4 test tubes to be taken from the growth bath, read on the spectrophotometer, and returned to the bath.

The transmission, as it changed in the course of the growth experiment and read with an approximation to the nearest 0.25 %, is given for one of

TABLE 21-4. log_2 *O.D.* + 10 *against transmission* (T)

T	0	0.25	0.50	0.75	T	0	0.25	0.50	0.75
1	10.999	10.928	10.867	10.813	26	9.227	9.217	9.206	9.196
2	10.764	10.720	10.680	10.642	27	9.186	9.176	9.166	9.156
3	10.606	10.573	10.542	10.512	28	9.145	9.135	9.125	9.115
4	10.483	10.456	10.429	10.404	29	9.105	9.095	9.085	9.075
5	10.379	10.356	10.333	10.311	30	9.065	9.055	9.045	9.035
6	10.289	10.268	10.247	10.227	31	9.025	9.015	9.005	8.996
7	10.208	10.188	10.170	10.151	32	8.986	8.976	8.966	8.956
8	10.133	10.116	10.098	10.081	33	8.946	8.936	8.926	8.917
9	10.065	10.048	10.032	10.016	34	8.907	8.897	8.887	8.877
10	10.000	9.984	9.969	9.954	35	8.868	8.858	8.848	8.838
11	9.939	9.924	9.910	9.895	36	8.828	8.819	8.809	8.799
12	9.881	9.867	9.853	9.839	37	8.789	8.779	8.770	8.760
13	9.826	9.812	9.799	9.785	38	8.750	8.740	8.730	8.721
14	9.772	9.759	9.746	9.733	39	8.711	8.701	8.691	8.681
15	9.721	9.708	9.696	9.683	40	8.671	8.662	8.652	8.642
16	9.671	9.659	9.646	9.634	41	8.632	8.622	8.612	8.602
17	9.622	9.610	9.599	9.587	42	8.592	8.583	8.573	8.563
18	9.575	9.563	9.552	9.540	43	8.553	8.543	8.533	8.523
19	9.529	9.517	9.506	9.495	44	8.513	8.503	8.493	8.483
20	9.484	9.472	9.461	9.450	45	8.473	8.463	8.453	8.443
21	9.439	9.428	9.417	9.406	46	8.433	8.423	8.413	8.402
22	9.396	9.385	9.374	9.363	47	8.392	8.382	8.372	8.362
23	9.353	9.342	9.331	9.321	48	8.351	8.341	8.331	8.321
24	9.310	9.300	9.289	9.279	49	8.310	8.300	8.290	8.279
25	9.268	9.258	9.248	9.237	50	8.269	8.259	8.248	8.238

TABLE 21-4. $\log_2$ *O.D.* + 10 *against transmission* (T) (*cont.*)

T	0	0.25	0.50	0.75	T	0	0.25	0.50	0.75
51	8.227	8.217	8.206	8.196	76	6.933	6.916	6.898	6.881
52	8.185	8.174	8.164	8.153	77	6.863	6.845	6.827	6.808
53	8.142	8.132	8.121	8.110	78	6.790	6.771	6.752	6.733
54	8.099	8.088	8.078	8.067	79	6.714	6.694	6.675	6.655
55	8.056	8.045	8.034	8.023	80	6.635	6.614	6.594	6.573
56	8.012	8.000	7.989	7.978	81	6.552	6.531	6.509	6.488
57	7.967	7.956	7.944	7.933	82	6.466	6.443	6.421	6.398
58	7.922	7.910	7.899	7.887	83	6.375	6.351	6.328	6.303
59	7.876	7.864	7.852	7.841	84	6.279	6.254	6.229	6.204
60	7.829	7.817	7.805	7.793	85	6.178	6.151	6.125	6.098
61	7.781	7.770	7.757	7.745	86	6.070	6.042	6.013	5.984
62	7.733	7.721	7.709	7.697	87	5.955	5.925	5.894	5.863
63	7.684	7.672	7.659	7.647	88	5.832	5.799	5.766	5.732
64	7.634	7.622	7.609	7.596	89	5.698	5.663	5.627	5.590
65	7.583	7.570	7.557	7.544	90	5.553	5.514	5.475	5.434
66	7.531	7.518	7.505	7.491	91	5.393	5.350	5.307	5.262
67	7.478	7.465	7.451	7.437	92	5.215	5.168	5,118	5.068
68	7.424	7.410	7.396	7.382	93	5.015	4.961	4.904	4.846
69	7.368	7.354	7.340	7.325	94	4.785	4.722	4.656	4.587
70	7.311	7.297	7.282	7.267	95	4.515	4.439	4.359	4.275
71	7.252	7.238	7.223	7.208	96	4.186	4.091	3.989	3.881
72	7.192	7.177	7.162	7.146	97	3.763	3.636	3.497	3.343
73	7.131	7.115	7.099	7.083	98	3.171	2.977	2.753	2.488
74	7.067	7.051	7.034	7.018	99	2.165	1.748	1.162	0.160
75	7.001	6.984	6.967	6.950					

the test tubes in column 2 (Table 21-3). The O.D. (column 3) for the corresponding readings was obtained by using Eq. (4). The $\log_{10}$ O.D. (column 4) was obtained by using tables of common logarithms. To transform values of $\log_{10}$ O.D. into $\log_2$ (column 5), figures in column 4 were multiplied by the conversion factor, 3.32. To avoid possible errors connected with the usage of negative logarithms, an arbitrary figure 10 was added to all values of $\log_2$ O.D. This transformed all negative logarithms into positive ones (column 6). To facilitate plotting of the data, the initial optical density of the cell suspension and, therefore, the $\log_2$ O.D. + 10 at zero time was treated as equal to zero and the increases in $\log_2$ O.D. + 10 at the subsequent readings were obtained by subtracting from the values of $\log_2$ O.D. + 10 at each of the readings (column 6) the values of $\log_2$ O.D. + 10 at zero time. The Δ values (column 7) actually represent the number of doublings of O.D. at indicated time intervals from the beginning of observations (zero time).

To expedite calculations illustrated in great detail in Table 21-3, Table 21-4 was constructed in which the values of $\log_2$ O.D. + 10 (column 6 in Table 21-3) are directly related to transmission readings (column 2 in Table 21-3). By using Table 21-4, the intermediate steps (columns 3, 4, 5 in Table 21-3) are eliminated and the protocol of a growth experiment is shortened to contain the columns:

1. time, hours
2. transmission, *T* (read on colorimeter)
3. $\log_2$ O.D. + 10 (taken from Table 21-4)
4. accumulated increments of $\log_2$ O.D. + 10, Δ (obtained by subtracting the value of $\log_2$ O.D. + 10 at the initial zero time from each of the values of $\log_2$ O.D. + 10 obtained at subsequent readings).

In Table 21-4, values of $\log_2$ O.D. + 10 are given for transmission readings expressed as integrals and as fractions of the integrals. On most instruments, the transmission can be directly read to one unit (one percent transmission). Between the units of the transmission (integrals), transmission can be estimated with some approximation to one quarter (0.25%) of a unit. The transmission readings, as integrals, are given in Table 21-4 in the column '*T*' and the corresponding values of $\log_2$ O.D. + 10 for these integrals under the heading '*O*'. For the fractions between integrals, the $\log_2$ O.D. + 10 values are given in Table 21-4 in columns 0.25%, 0.50% and 0.75%. Thus, the example, for readings of the transmission:

51.0% the $\log_2$ O.D. + 10 is 8.227
51.25% the $\log_2$ O.D. + 10 is 8.217
51.50% the $\log_2$ O.D. + 10 is 8.206

51.75% the $\log_2$ O.D. + 10 is 8.196
52.0% the $\log_2$ O.D. + 10 is 8.185

To obtain the exponential growth rate, increments of $\log_2$ O.D. + 10 (column 7 in Table 21-3) are plotted against time as illustrated in Fig. 21-2 and a straight line is drawn to fit as many points as possible. Plotting of data and drawing of a straight line through a number of points, as illustrated in Fig. 21-2 serves several purposes.

First, it permits a judgment as to whether the exponential growth was observed during the growth experiment as well as to provide an indication of the duration of the exponential growth phase.

Second, it allows an improvement in the accuracy of measurements by disregarding small deviations from the expected strictly exponential increase. By drawing a straight line, a judgment can be made as to whether these deviations, caused by imperfections of the technique, are small enough to be disregarded in calculations of the exponential growth rate.

Third, larger one-directional deviations, which cause the curve at both ends to deflect from a straight line, indicate the lag phase (at the beginning of measurements) and the declining phase (at the end of the observations) of growth which must be excluded from the measurements of the exponential growth rate. By proper care of the culture used for inoculation and by using a heavier inoculum to start with a reasonably dense cell suspension, an attempt can be made to shorten or even to eliminate the lag phase of growth. Bending of the curve at the upper end, indicating a transition from the exponential phase to declining phase of growth, denote time (population density of the cell suspension) at which measurements must be discontinued.

The exponential growth rate is determined from the slope of the curve in Fig. 21-2. In this example, the exponential growth was observed from the first hour reading to the sixth hour reading. By using equation (2) and values as read in Fig. 21-2, the exponential rate is calculated as:

$$R_E = \frac{2.4-0.3}{\frac{6}{24}-\frac{1}{24}} = \frac{2.1}{\frac{5}{24}} = 10.1. \qquad (5)$$

A growth experiment usually consists of two or more replications with each represented by a cell suspension in a separate growth vessel. The change in transmission with time is recorded for all replications and the transmission readings for each of the replications are treated separately in the protocol of the growth experiment. It is recommended to plot data and to calculate values of R_E separately for each replication. Then, at a final stage, and only if the difference in values of R_E obtained for each

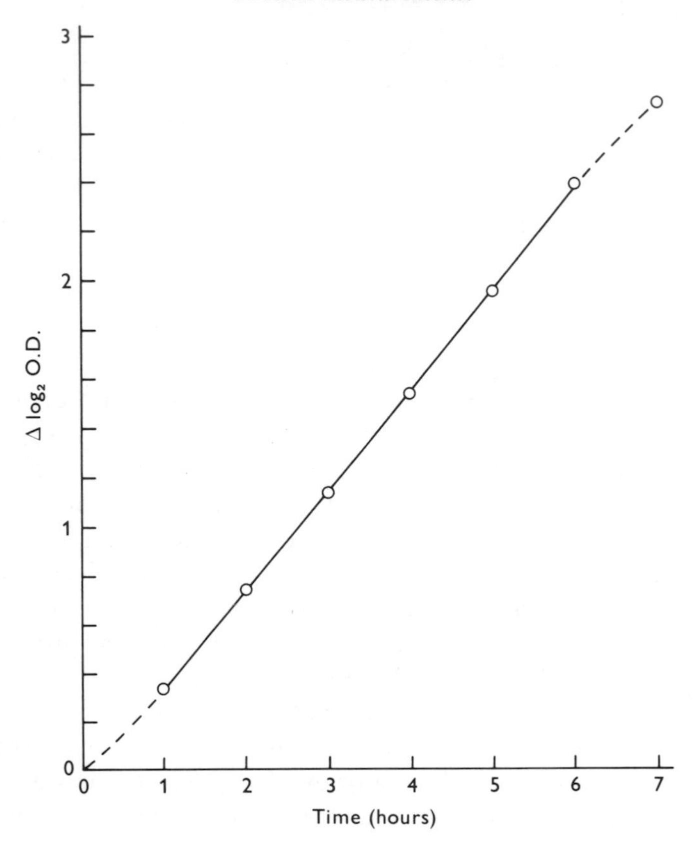

Fig. 21-2. Plotting of the growth experiment presented in Table 21-3.

replication is small, a value of R_E is calculated as an average of R_Es for individual replications.

The precision of measurements of optical density can be adversely affected by a number of factors listed below:

1. An inadequate mixing of the sample before it is used for O.D. measurements or sedimentation of cells in the process of measurements.

2. Poor optical properties of the vessels (test tubes) in which measurements are made. Optical unevenness of the glass can be checked by filling a number of vessels with water and determining variations in transmission.

3. Inadequate wavelength of light used for optical density measurements.

4. Poor adjustment of the instrument. The standardization must be done after a warming period and must be checked frequently during measurements.

The exclusion of all of these factors can be summarily ascertained by the technique described as a dilution curve.

VIII. DILUTION CURVE

In addition to the linearity of the exponential growth curve over an extended period of time, another major condition for using the increase in optical density in growth rate measurements is a proof that the increase in the O.D. (slope of the curve in Fig. 21-2) actually represents the increase in population density of a given cell suspension. Several factors affect the strict correspondence of growth measurements as increase in optical density to actual increase in cell mass. Of these factors the most important are: the make of the measuring device and its proper functioning (adjustment); the wavelength of the light source during measurements; properties of the organism subjected to measurements; and, the population density of the cell suspension used for measurements.

Each of these factors is complex. Without entering into a detailed discussion on the nature of these factors, a technique is offered to obtain a demonstration of the degree of the correspondence of the measurements of the exponential growth rate as increase in optical density to the actual (known in the given example in advance) increase in the population density of the suspension of the given organism.

A reasonably dense suspension of the organism is repeatedly diluted by half by adding at each dilution to the measured volume of the well-stirred cell suspension an equal volume of the medium. The dilutions obtained are: 1 (no dilution), $\frac{1}{2}$, $\frac{1}{4}$, $\frac{1}{8}$, $\frac{1}{16}$, $\frac{1}{32}$, $\frac{1}{64}$ and so on. With every dilution, the cell mass per volume unit of the suspension is reduced by half and, by moving in the opposite direction (from a more dilute to a denser suspension) the cell mass doubles. It remains to be proven that the optical density of the suspension increases correspondingly.

The transmission of each of the dilutions is determined with the available instrument. In the example given in Table 21-5 the transmissions of the suspensions of *Chlorella* were read on a Coleman Junior II Spectrophotometer at 438, 540, and 678 nm (columns 2, 5, and 8 in the Table 21-5). The $\log_2$ O.D. + 10 for each of the transmissions was obtained from Table 21-4 (columns 3, 6, and 9 in Table 21-5). The accumulated increments of the $\log_2$ O.D. + 10 (columns 4, 7, and 10 in Table 21-5 read in reverse beginning with the most diluted suspension) were then plotted against doublings of the cell mass (dilutions in reverse) in Fig. 21-3.

The slopes of the straight lines in Fig. 21-3 can be quantitatively evaluated by dividing the number of doublings in optical density by the corresponding number of dilutions in cell material. Thus, for example, the straight line drawn through points obtained by reading O.D. at 678 nm

TABLE 21-5. *Number of doublings in optical density for successive dilutions of the suspension of Chlorella*

	438 nm			540 nm			678 nm		
Dilutions	T	$\log_2$ O.D. + 10	Δ	T	$\log_2$ O.D. + 10	Δ	T	$\log_2$ O.D. + 10	Δ
(1)	(2)	(3)	(4)	(5)	(6)	(7)	(8)	(9)	(10)
1	—	—	—	8.75	10.081	6.738	1.50	10.867	7.696
$\frac{1}{2}$	2.50	10.680	6.405	27.25	9.176	5.833	10.00	10.000	6.829
$\frac{1}{4}$	14.75	9.733	5.458	52.00	8.185	4.842	30.00	9.065	5.894
$\frac{1}{8}$	37.75	8.760	4.485	72.25	7.177	3.834	54.75	8.067	4.896
$\frac{1}{16}$	60.75	7.793	3.518	85.00	6.178	2.835	73.75	7.083	3.912
$\frac{1}{32}$	77.00	6.863	2.588	91.50	5.307	1.964	85.50	6.125	2.954
$\frac{1}{64}$	87.00	5.955	1.680	95.25	4.439	1.096	92.50	5.118	1.947
$\frac{1}{128}$	92.50	5.118	0.843	97.00	3.763	0.420	96.00	4.186	1.015
$\frac{1}{256}$	95.75	4.275	0	97.75	3.343	0	98.00	3.171	0

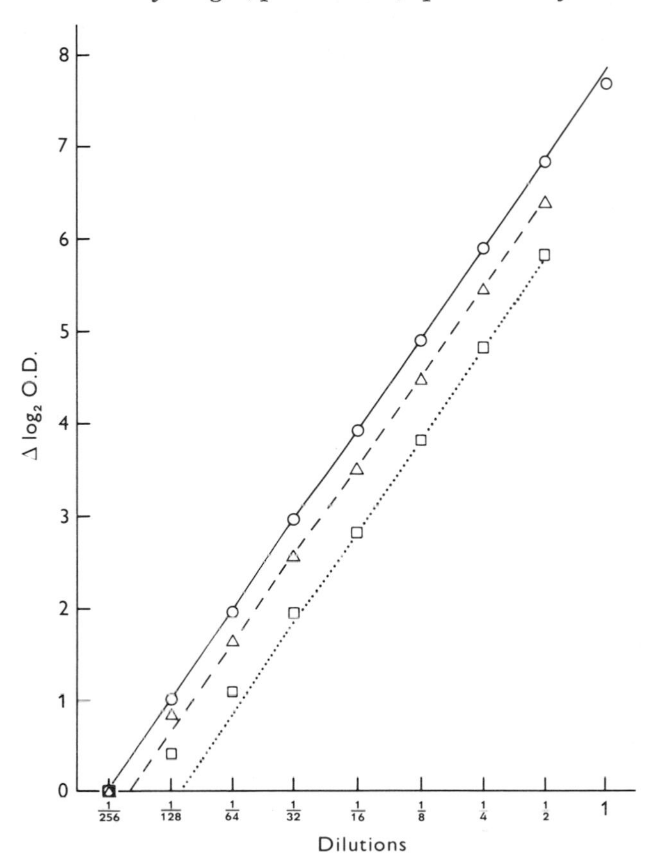

Fig. 21-3. Plotting of the experiment in which optical densities of successive dilutions of *Chlorella* were determined at different wavelengths (data of Table 21-5). Measurements: △ – at 438 nm; □ – at 540 nm; ○ – at 678 nm.

extends from 0 to 7.84. This corresponds to the actual number of doublings of cell material (number of dilutions) equal to 8. By dividing 7.84 by 8, a figure, 0.98, is obtained, indicating that for each actual doubling of cell material the number of doublings, as determined by the increase in O.D. read at 678 nm is 0.98.

Evaluation of the curves of Fig. 21-3 offered in Table 21-6 permits some comparisons. The slope of the line obtained for readings of O.D. at 438 nm referred to number of dilutions of cell material, that is the ratio Δ O.D./Δ P.D. (population density), equals 0.956. This means that the obtained value of the increase in O.D. is 4.4% lower than the actual increase in cell material. Compared to this, the ratios of Δ O.D. to Δ P.D. at 540 nm and 678 nm are correspondingly 1.0 and 0.98. Thus, for the

TABLE 21-6. *Ratio of the number of doublings in optical density (O.D.) to the known number of doublings in population density (P.D.) determined from Fig. 21-3*

Wavelength, nm	$\frac{\Delta\ \text{O.D.}}{\Delta\ \text{P.D.}}$	Deviation from theoretically expected
(1)	(2)	(3)
438	0.956	−4.4 %
540	1.000	0.0
678	0.982	−1.8 %

organism and conditions employed in the experiment illustrated in Fig. 21-3 and Table 21-6, measurements at 540 nm gave a perfect correspondence between the increase in O.D. and the known increase in population density of the cell suspension.

If 540 nm is chosen for measurements of O.D. as an indication of growth, attention must be drawn to the fact that the line drawn through the $\log_2$ O.D. obtained at 540 nm (Fig. 21-3) is straight and, therefore, the increase in optical density is strictly proportional to the increase in population density only at higher population densities. As seen in Fig. 21-3, the line connecting $\log_2$ O.D. at 540 nm is perfectly straight between readings obtained for dilutions $\frac{1}{16}$ to $\frac{1}{2}$. For population densities below that of the dilution $\frac{1}{16}$, the $\log_2$ O.D. curve begins to deviate from the straight line; the deviations are much more pronounced for dilutions $\frac{1}{64}$, $\frac{1}{128}$ and $\frac{1}{256}$. From Table 21-5 we find that the transmission for the dilution $\frac{1}{16}$ at 540 nm (column 5 in Table 21-5) is 85.0. Therefore, the initial population density during growth measurements, determined as increase in O.D. at 540 nm, must be at least equal or higher than that corresponding to the transmission 85.0. Otherwise, the bending of the $\log_2$ O.D. growth curve, often observable during the earlier portion of the growth experiments which we denote as a lag phase (Fig. 21-1), will actually be due not only to biological factors operating during and prior to observations, but also to the lack of proportionality of the O.D. measurements to the actual increase in population density.

Measurements of O.D. at 678 nm produce a $\log_2$ O.D. curve with a slope giving the ratio Δ O.D./Δ P.D. equal to 0.982. Thus, the exponential growth rate, as determined from O.D. measurements, is expected to be lower than the actual increase in population density by 1.8 %. This error for biological observations cannot be considered large. If desirable, the exponential growth rate, obtained as the increase in O.D. determined at

678 nm, can be corrected by adding 1.8% of its value. The advantage of determinations of exponential growth rate, as increase in O.D. measured at 678 nm, lies in the perfect linearity of the obtained $\log_2$ O.D. curve over a wide range of population densities, extending, in the example given in Fig. 21-3 and Table 21-5, to the extreme dilutions (1/256) characterized by such a high transmission (98.0) as indicated in column 8 of Table 21-5.

Factors which affect measurements of O.D. may change with a change in the organism under observation and in measuring device. Therefore, the dilution curve, as an evidence of applicability of optical density to measurement of population density, must be obtained every time when new factors affecting measurement of O.D. are introduced, and also from time to time even under seemingly unchanged conditions to ascertain that the strict proportionality of O.D. and P.D. continues to exist.

IX. REFERENCES

Buchanan, R. L. 1918. Life phases in a bacterial culture. *J. Infect. Dis.* **23**, 109–25.

Monod, J. 1942. *La Croissance des cultures bactériennes.* Hermann et Cie, Paris, France.

Monod, J. 1949. The growth of bacterial cultures. *Ann. Rev. Microbiol.* **3**, 371–94.

Sachs, J. V. 1873. *Lehrbuch der Botanik.* 3rd ed. Leipzig, Germany.

Shihira, I. and Krauss, R. W. 1965. *Chlorella. Physiology and Taxonomy of Forty-one Isolates.* Univ. Maryland, College Park. 96 pp.

Sorokin, C. and Krauss, R. W. 1958. The effect of light intensity on the growth rates of green algae. *Plant Physiol.* **33**, 109–13.

Winslow, C. E. A. and Walker, H. H. 1939. The earlier phases of the bacterial culture cycle. *Bacteriol. Rev.* **3**, 147–86.

22: Coulter Counter for phytoplankton

T. R. PARSONS*

Fisheries Research Board of Canada, Nanaimo, British Columbia, Canada

CONTENTS

* Present Address: Institute of Oceanography, University of British Columbia, Vancouver 8, British Columbia, Canada.

I. INTRODUCTION

In the following description it is assumed that the operator has at his disposal a Model B Coulter Counter (Coulter Electronics, Inc., Industrial Division, 590 W. 20th St., Hialeah, Florids 33010; 1814 Berkel Rd, Mississauga, Ontario) which has been set up by a company representative. It is also assumed that the operator has a basic knowledge of the effect of the various controls on the electronic cabinet and knows how to draw particles through the aperture, adjust the vacuum and clean the mercury column and electrodes when necessary.

The following instructions are abbreviated from an earlier publication (Sheldon and Parsons 1967*a*). For an account of other techniques, theory and accessory apparatus the reader is referred to this reference. In addition, certain sections have been modified from 'A Practical Handbook of Seawater Analysis' by Strickland and Parsons (1968).

II. EQUIPMENT

A. Coulter Counter Model B with a three position manometer switch permitting sample volumes of 0.05 ml, 0.5 ml or 2 ml. (Alternatively a line switch connected to a time clock can be installed by Coulter Electronics to permit greater variation in sample volumes.)

B. A selection of aperture tubes 30, 100, 200, and 560 μm diameter.

C. Two or three round bottomed flasks and a stirring motor.

D. Calibration material of various diameters

1. *Pollen* (Hollister-Stier Laboratories, P. O. Box 14197, Dallas, Texas):

a. paper mulberry (*Broussonetia papyrifera* (L.) Vent.) – 16 μm;
b. ragweed (*Ambrosia* sp.) – 19.5 μm;
c. pecan (*Carya illinoensis* (W. Wang) K. Koch) – 45 μm;
d. corn (*Zea mays* L.) – 90 μm.

2. *Plastic spheres* (Dow Chemical Co., 2030 Abbott Rd, Midland, Michigan 48640) 1 – 10 μm.

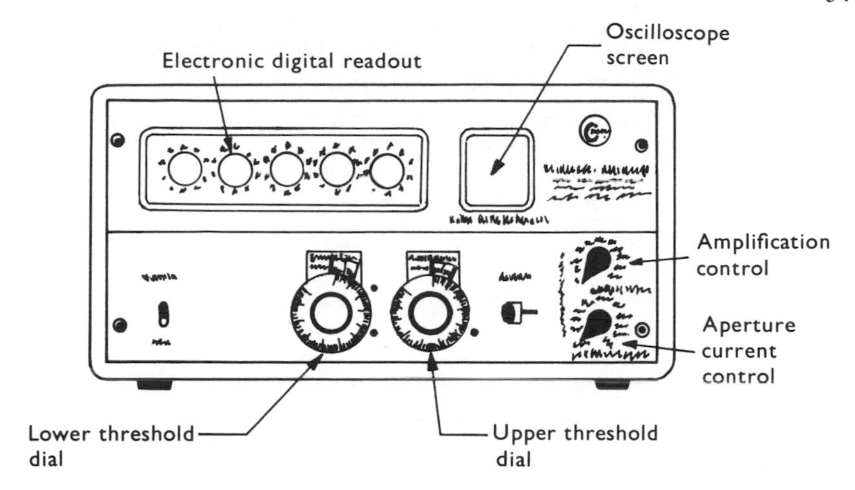

Fig. 22-1. Electronic cabinet of a Coulter Counter Model B showing controls referred to in the text.

III. CALIBRATION

Various controls on the electronic cabinet of the Coulter Counter Model B are shown in Fig. 22-1. In addition to these controls, a threshold mode switch will be found within the cabinet and a three position switch for different volumes of fluid will be found on the manometer. All other controls will be adjusted when the instrument is set up by the company representative and normally these do not have to be manipulated again during the experimental use of the instrument. Furthermore, the Model B Counter is stable over a temperature range of 5–25 °C and a salinity range of 8–40 ‰. For freshwater organisms, however, a sodium chloride concentration of 5 ‰ is suggested (Mulligan and Kingsbury 1968) and consequently the instrument calibration should be carried out in a solution with a NaCl concentration of 0.5 %.

A. Select particles having a diameter between 2 and 40 % of the aperture being calibrated. Suspend these in membrane filtered seawater (marine algae) or 0.5 % NaCl (freshwater algae). Particles should be suspended by shaking and left to separate and become fully hydrated overnight. The diameter of the material employed should then be checked with a microscope.

B. Set the threshold mode switch at *separate*; set the lower threshold at 5 and the upper threshold at 100. Adjust amplification and aperture current settings so that particles appear one-half to two-thirds of the way up the oscilloscope screen.

TABLE 22-1. *Particle concentrations (per ml) for calibration (A) and experiments (B), using different sized apertures*

Aperture size (μm)	Particles per ml	
	A[a]	B[a]
30	500 000	2 500 000
50	150 000	700 000
100	18 000	90 000
200	2 000	12 000
560	100	500

[a] A and B are at approximately the 2 and 10% coincidence counting levels.

C. Set the manometer switch to measure a 0.05 ml volume if a small aperture is being calibrated (i.e. 30 μm) or at 0.5 or 2 ml if a large aperture (i.e. 200 μm) is being calibrated. Count the number of particles in a known volume. Express this count as the number of particles/ml.

D. Adjust the number of particles in the saline sample so that there are less than the number of particles shown in column A of Table 22-1.

E. Set the threshold mode switch to *locked* and the upper threshold to 50. With the lower threshold set at 5, take two counts of the number of particles in a standard (0.05, 0.5 or 2 ml) volume of saline solution. Move the lower threshold to 10, keep the upper threshold at 50 and repeat the two counts. Continue to move the lower threshold up at intervals of 5 threshold settings, making two counts after each move and keeping the upper threshold set at 50.

F. Plot the results as shown in the example (Fig. 22-2) and determine the mode of the first peak. Subsequent peaks which may occur are usually due to two or more particles being stuck together.

G. Note amplification and aperture current settings. Divide the volume of the particles used for calibration by the threshold value for the mode.

The example shown in Fig. 22-2 is taken from Strickland and Parsons (1968) and shows the calibration of a 30 μm aperture using plastic spheres of 22.5 μm^3 volume. This calibration was obtained at an amplification of one-half and an aperture current of 1 (i.e., a sensitivity of 0.5). If these controls are changed so that their multiple (sensitivity) increases or decreases, each threshold setting will be changed proportionately, i.e., for a sensitivity of 4, each threshold setting is then equivalent to:

$$\frac{4}{0.5} \times 0.64 = 5.1 \ \mu m^3. \qquad (1)$$

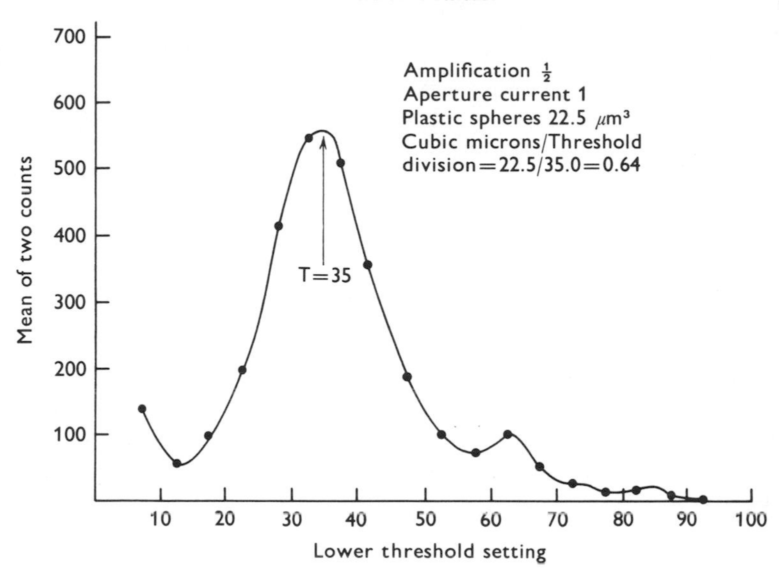

Fig. 22-2. Size distribution of plastic spheres used for the calibration of a 30 μm aperture (from Strickland and Parsons 1968).

IV. PHYTOPLANKTON GROWTH RATES

A. *Numerical increase in cells*

For single-celled, non-chain forming phytoplankton, the simplest measure of growth is to determine the increase in cell numbers over a specified time interval. This can be done by selecting an aperture tube such that the diameter of the phytoplankton is between 2 and 40% of the diameter of the aperture. The procedure followed is then the same as already described in III.B–D, preceding, except that in III.D, samples have to be diluted below the count level shown in column B of Table 22-1. If the phytoplankton sample is diluted the final concentration of cells, as measured with the Coulter Counter, must be corrected by the dilution factor. The growth rate of the organism is then calculated as shown in the following example.

A culture of *Monochrysis lutheri* (Chrysophyceae; Millport, strain 60, IUCC 1293) was counted at a sensitivity of 0.5 using a 50 μm tube; lower threshold at 5, upper threshold at 100. The population at these settings appeared approximately halfway up the oscilloscope screen and the mean count for 0.05 ml was 1.20×10^6 cells per ml. The sample was diluted by a factor of 10 using filtered culture medium, and recounted. The mean count on a 0.05 ml sample (corrected for dilution) was then 1.55×10^6 cells per ml (n_0). The count was repeated 10 h later (t) and the mean value was then 2.20×10^6 cells/ml (n_t). The growth constant (k) of the material was then calculated from the formula:

$$\frac{\log_{10} n_t - \log_{10} n_0}{t} = k\,(\mathrm{h})^{-1} \qquad (2)$$

and the generation time (T_g) in h:

$$T_g = \frac{0.3}{k}. \qquad (3)$$

In the preceding example, $k = 0.0125$ and $T_g = 24$ h.

B. *Increase in biomass*

1. Absence of detritus. For single-celled, and all chain-forming species (including field collections), it is recommended that increase in biomass be used as a measure of growth rather than increase in cell number. The following procedure is applicable to the measurement of growth rates of cultures, or of natural phytoplankton samples having a very low background of detrital material (i.e., biomass of phytoplankton to detritus greater than 5:1).

a. Select an aperture tube such that the particles being examined have a diameter between 2 and 40% of the aperture size. For mixed populations it may be necessary to use more than one aperture and sieve out larger plankters when using the smaller aperture tubes.

b. Set the sensitivity (aperture current × amplification settings) at the minimum setting for counting the smallest particles, and the lower threshold at the lowest level used for making counts (see V.A on determination of these settings). Set the upper threshold at 100.

c. Count the particles in a suitable volume of medium. Express the count as the number of particles/ml. Adjust the volume of the medium if the count exceeds the value shown in Table 22-1, column B.

d. Set the upper and lower thresholds at the required values for obtaining the number of particles in the size categories shown in Table 22-2 (see example in V.A) and make two counts at each sensitivity setting until the maximum sensitivity is reached.

e. Express the mean of two counts as the number of particles in a known volume of medium (e.g., 1 ml) and multiply each count by the geometric mean volume for each size interval. This figure is given in column 3 of Table 22-2. These results give amount as a logarithmic progression of particle diameter (see example in Fig. 22-3).

f. Remeasure the particle size spectrum after a set time interval and calculate the growth constant and generation time of the material, using

TABLE 22-2. *A grade scale based on 1 μm particle diameter in which particle volume varies by a factor of 2 in successive grades (from Sheldon and Parsons 1967a)*

Diameter (μm)	Volume (μm³)	Volume (μm³)	Diameter (μm)
1.00	0.52		
		0.73	1.12
1.26	1.04		
		1.47	1.41
1.58	2.08		
		2.94	1.78
2.00	4.19		
		5.92	2.24
2.52	8.38		
		11.8	2.82
3.18	16.8		
		23.8	3.57
4.00	33.5		
		47.4	4.49
5.04	67.0		
		94.7	5.66
6.34	131		
		189	7.12
8.00	268		
		379	8.98
10.1	536		
		758	11.3
12.7	1.07×10^3		
		1.52×10^3	14.3
16.0	2.15×10^3		
		3.03×10^3	18.0
20.2	4.29×10^3		
		6.07×10^3	22.6
25.4	8.58×10^3		
		12.1×10^3	28.5
32.0	17.2×10^3		
		24.3×10^3	35.9
40.3	34.3×10^3		
		48.6×10^3	45.3
50.8	68.7×10^3		
		97.1×10^3	57.0
64.0	137×10^3		
		194×10^3	71.9
80.6	275×10^3		
		388×10^3	90.5
102	549×10^3		

TABLE 22-2 (*cont.*)

Diameter (μm)	Volume (μm^3)	Volume (μm^3)	Diameter (μm)
		777×10^3	114
128	1.10×10^6		
		1.56×10^6	144
161	2.20×10^6		
		3.11×10^6	181
203	4.40×10^6		
		6.22×10^6	228
256	8.79×10^6		
		12.4×10^6	287
322	17.6×10^6		
		24.9×10^6	361
404	35.1×10^6		
		49.6×10^6	452
512	70.3×10^6		
		99.4×10^6	573
646	141×10^6		
		199×10^6	724
812	281×10^6		
		397×10^6	909
1020	562×10^6		

equations (2), (3), in A, preceding. Substitute the biomass of the material for the number of cells, n_t and n_o.

Fig. 22-3 and Table 22-3 have been taken from Sheldon and Parsons (1967*a*). The data represent an *in situ* light and dark bottle incubation during a heavy marine phytoplankton bloom. The data in Table 22-3 have been used to illustrate changes in the phytoplankton size spectrum shown in Fig. 22-3. In addition, relative growth constants have been calculated for various parts of the size spectrum (the size categories follow definitions given by Sheldon and Parsons 1967*b*) as well as for the total biomass of material at different depths. These latter growth constants are not absolute values for specific populations in the size spectrum since to some extent the populations overlap. However they represent a relative growth constant for a specific size group.

2. *Presence of detritus.* The procedure for measuring the growth of phytoplankton in the presence of appreciable detritus is the same as in the previous section (1, preceding), with the exception that several measurements of the particulate material must be made over a specified time interval. The treatment of the data is then the same as that described by

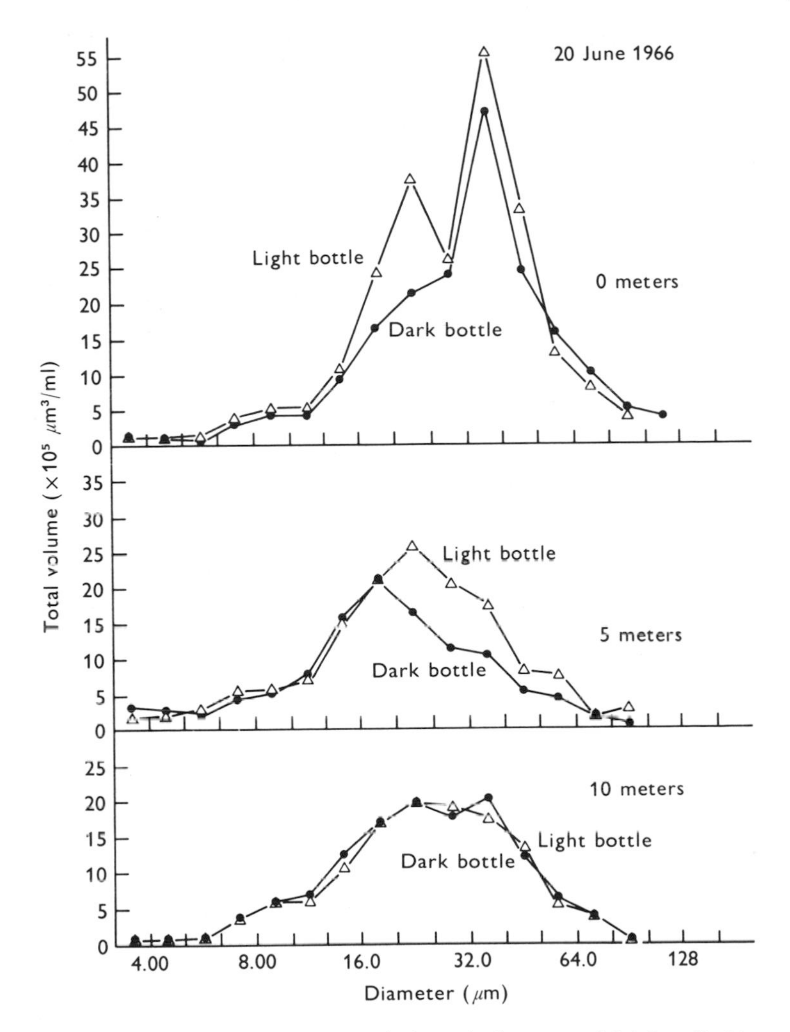

Fig. 22-3. Growth rates of suspended particulate material (after Sheldon and Parsons 1967*a*).

Cushing and Nicholson (1966). The theory behind this approach assumes that the phytoplankton growth rate remains constant over the period of study and that the level of detritus does not change. If this is true, then the expression:

$$\frac{V-\hat{D}}{V_0-\hat{D}} = \exp\left[\hat{k}(t-t_0)\right] \tag{4}$$

can be used to calculate the true growth constant (k), where V_0 is the total amount of particulate material at the beginning of the experiment (t_0), V is the amount of particulate material at time t and $\hat{D}$ is an estimate of the

TABLE 22-3. *Growth rates of suspended particulate material*

	Coulter Counter volumes ($\mu m^3 \times 10^6$/ml)					
Diameter (μm) ▶	0		5		10	
Depth (m)	Dark	Light	Dark	Light	Dark	Light
Nanoplankton region						
3.57	0.10	0.10	0.38	0.21	0.19	0.15
4.49	0.11	0.13	0.26	0.24	0.13	0.16
5.66	0.16	0.20	0.28	0.27	0.20	0.20
7.12	0.40	0.41	0.36	0.54	0.38	0.36
8.98	0.46	0.50	0.53	0.51	0.53	0.63
11.3	0.45	0.52	0.82	0.70	0.72	0.70
14.3	0.92	1.14	1.64	1.50	1.26	1.15
18.0	1.67	2.45	2.15	2.10	1.70	1.72
Microplankton region						
22.6	2.16	3.75	1.40	2.64	1.98	1.96
28.5	2.35	2.60	1.14	2.07	1.58	1.86
35.9	4.70	5.60	1.08	1.76	2.03	1.76
45.3	2.48	3.36	0.57	0.84	1.22	1.36
57.0	1.05	1.37	0.45	0.76	0.66	0.59
71.9	0.51	0.79	0.37	0.74	0.38	0.41
90.5	0.44	0.49	0.16	0.22	0.16	0.14
Total	17.96	23.41	11.59	15.10	13.12	13.15
Growth const. (days^{-1})	0.114		0.115		No growth	
Nanoplankton total	4.27	5.45	6.42	6.07	5.11	5.07
Growth const. (days^{-1})	0.106		No growth		No growth	
Microplankton total	13.69	17.96	5.17	9.03	8.01	8.08
Growth const. (days^{-1})	0.118		0.242		No growth	

Location: Saanich Inlet, B.C.
Date: 20 June 1966
Incubation: 24 h *in situ*.

true quantity of detritus (D). By substituting D back in the equation the true growth constant of the phytoplankton can be obtained.

The following example is taken from Sheldon and Parsons (1967*a*).

A sample of seawater collected in winter was incubated at 10 °C for several days. After an initial lag phase a small population of phytoplankton

TABLE 22-4. *Growth rate of a natural phytoplankton population*

	Coulter Counter volumes ($\mu m^3 \times 10^4$ ml)			
Time intervals (h) ▶	0	24	48	72
Diameter (μm) ▼				
7.12	3.9	6.3	14.0	34
8.98	5.0	8.5	32.0	91
11.3	4.0	7.5	32.5	106
14.3	3.8	6.1	17.2	52
18.0	3.3	4.4	5.8	15
22.6	3.1	5.0	3.5	6.2
28.5	2.5	5.0	5.8	9.8
35.9	4.4	5.1	11.0	26
45.3	0.7	4.2	10.0	35
57.0	—	—	1.8	20
71.9	—	—	—	18
90.5	—	—	—	5.4
Total	31	52	134	438

Location: Departure Bay
Date: 24 December 1966

Period of incubation: 72 h
Continuous illumination

developed and the growth of this population was followed for the next 72 h. The results are shown in Table 22-4. The calculation of the true growth constant of this population is also shown in Fig. 22-4 and as follows:

Calculation of $\hat{k}$

For $V_0 = 31 \times 10^4$ $V = 52 \times 10^4$ $t = 24$

$\hat{D}$	10	15	20	25	30	($\times 10^4$)
$V-\hat{D}$	42	37	32	27	22	($\times 10^4$)
$V_0-\hat{D}$	21	16	11	6	1	($\times 10^4$)
$\dfrac{V-\hat{D}}{V_0-\hat{D}}$	2	2.3	2.9	4.5	22	
$\hat{k} = \dfrac{1}{t}\log_{10}\left(\dfrac{V-\hat{D}}{V_0-\hat{D}}\right)$	0.0126	0.0148	0.0191	0.0274	0.0555	

Similarly for $V = 134 \times 10^4$

$\hat{D}$	10	15	20	25	30	($\times 10^4$)
$\hat{k}$	0.0160	0.0182	0.0213	0.026	0.042	

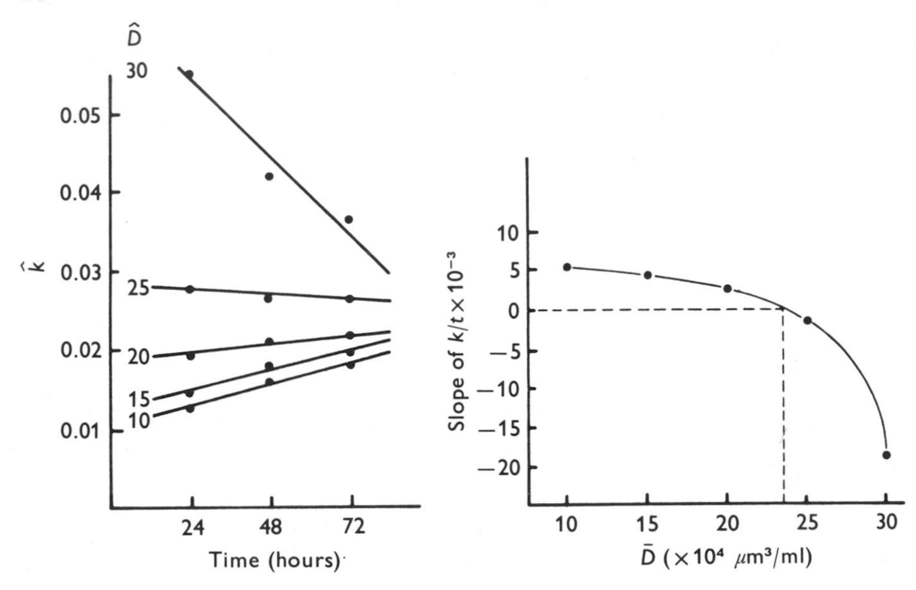

Fig. 22-4. Estimation of the quantity of detritus (after Sheldon and Parsons 1967*a*).

and for $V = 438 \times 10^4$

$\hat{D}$	10	15	20	25	30	$(\times 10^4)$
$\hat{k}$	0.0182	0.0195	0.0216	0.0256	0.0365	

The values of $\hat{k}$ at different times for different values of $\hat{D}$ are plotted in Fig. 22-4. The slopes of the lines are determined for:

$\hat{D}$	10	15	20	25	30	$(\times 10^4)$

and the slopes of $\hat{k}$ on t for 72 h unit time are:

0.0056 0.0047 0.0025 −0.0018 −0.019

The value of $\hat{D}$ at which there is no change in $\hat{k}$ with time is 23.5×10^4 μm^3/ml. The volume of detritus is, therefore, 23.5×10^4 μm^3/ml, and the original phytoplankton volume (V_0) is 7.5×10^4 μm^3/ml (i.e., 31.0–23.5 μm^3/ml).

The growth constant k_{10} $(h)^{-1}$ is $\dfrac{\log(134-23.5)-\log 7.5}{48} = 0.0245$

Therefore the generation time is $\dfrac{0.3}{0.0245} = 12.2$ h.

From Fig. 22-4 it may be seen that the population of phytoplankton developed two peaks, one at 11.3 μm diameter and the other at 45.3 μm diameter. The growth constant for the total population was estimated. As with the previous procedure (Table 22-3) it is possible to obtain separate growth constants for the two different phytoplankton populations shown in Table 22-4.

V. ADDITIONAL NOTES

A. The required sensitivity and threshold settings for determining a biomass spectrum of cellular material need only be determined once for each aperture, providing no subsequent changes are made to the electronics of the counter. From III, preceding, the particle volume corresponding to one threshold setting at a known sensitivity will have already been obtained. It is now necessary to determine the threshold settings (upper and lower) which will give two adjacent volumes in the left-hand column of Table 22-2. The following example serves to show how this is done:

With the 30 μm tube one threshold setting at a sensitivity of 4 was equal to 5.1 μm^3 (III.G, example). At this sensitivity, particles in the size range 134–268 μm^3 (Table 22-2) could be measured if the upper and lower thresholds were placed at:

$$T\,(\text{upper}) = \frac{268}{5.1} = 52.5, \tag{5}$$

$$T\,(\text{lower}) = \frac{134}{5.1} = 26.2. \tag{6}$$

Since the movement of amplification, or aperture current settings correspond to a 2× volume scale, the logarithmic scale of particle diameter given in Table 22-2 can now be followed by leaving the upper threshold at 52.5 and the lower threshold at 26.2 and changing the sensitivity by a factor of two each time a count is taken. In this instance the minimum sensitivity required in IV.B.1.a, preceding, will be found from the diameter in Table 22-2 corresponding to 2% of the aperture size (in this present example this would correspond to a sensitivity of approximately 0.125, i.e., aperture current setting of 1 and amplification of $\frac{1}{8}$). Conversely, the maximum sensitivity setting for any tube will be found by the same procedure from the diameter in Table 22-2 corresponding to 40% of the aperture size.

B. The precision of the Coulter Counter decreases with the number of particles counted, providing counts are being made below an acceptable level of coincidence (Table 22-1). Precision is easily checked by the operator but the following values serve as a guide to values which may be expected:

For a count of: 2000, range is ±65 (3.3%)
200, range is ±12 (6%)
70, range is ±11 (16%)
30, range is ±8 (27%)

VI. PROBLEMS

Various problems in the technical operation of a Coulter Counter are dealt with in an operation manual supplied with the instrument. Problems associated with the type of material being analyzed (especially natural samples) are often more difficult to diagnose. In general, however, some independent measure of biomass should be used in preliminary work in order to determine the relationship between volumes, as measured with the Coulter Counter, and absolute units of biomass, such as total organic carbon. A number of these relationships are given in Sheldon and Parsons (1967*a*, *b*).

Additional types of Particle Counters (micro) may be more suitable, such as the Π MC Particle Measurement Computer System (Millipore Corp., Ashby Rd, Bedford, Massachusetts 01730; 55 Montpellier Blvd, Montreal 379, Quebec); TGZ 3 Particle Size Analyzer (Carl Zeiss, 445 Fifty Ave., New York, New York 10018; 45 Valleybrook Dr., Don Mills 405, Ontario).

VII. REFERENCES

Cushing, D. H. and Nicholson, H. F. 1966. Method of estimating algal production rates at sea. *Nature* **212**, 310–11.

Mulligan, H. F. and Kingsbury, J. M. 1968. Application of an electronic particle counter in analyzing natural populations of phytoplankton. *Limnol. Oceanogr.* **13**, 499–506.

Sheldon, R. W. and Parsons, T. R. 1967*a*. *A Practical Manual on the Use of the Coulter Counter in Marine Science*. Coulter Electronics (Canada) Ltd. 66 pp.

Sheldon, R. W. and Parsons, T. R. 1967*b*. A continuous size spectrum for particulate matter in the sea. *J. Fish. Res. Bd Canada* **24**, 909–15.

Strickland, J. D. H. and Parsons, T. R. 1968. A practical handbook of seawater analysis. *Bull. Fish. Res. Bd Canada* 167. 311 pp.

23: Pigment analysis

EUGENE HANSMANN

Department of Biology, University of New Mexico, Albuquerque, New Mexico 87106

CONTENTS

I. OBJECTIVE

The objective of this chapter is to present a chemical method for estimating growth rates of algal cells in culture. Growth of an algal culture does not necessarily imply cell division, but cell division usually accompanies it. Growth, which is the addition of organic material to the cells can be measured directly by analyzing the standing stock or biomass of the culture. Several cellular components can be used as a measure of biomass, such as carbon, lipids, proteins, and plant pigments. Of these four cellular components, the pigments, primarily chlorophyll *a*, is most widely accepted.

II. RELATIONSHIP OF PIGMENTS AND GROWTH RATE

In using chlorophyll *a* as the biomass component, growth rates per unit biomass can be expressed by Eq. (1).

$$\left(\frac{dX}{dt}\right)\cdot\left(\frac{1}{x}\right) \tag{1}$$

where:

dX = change in chlorophyll *a* biomass in mg/m^3,

dt = change in time,

x = concentration (mg/m^3) of chlorophyll *a* at the beginning of the experiment.

It is very common in natural populations to express growth rates in terms of the rate of photosynthesis or carbon fixation per unit of chlorophyll *a* present. Equation (1) can be applied although the units are changed.

Besides chlorophyll *a* being used as an indicator of the growth rate of algal populations, there is evidence in the literature that chlorophyll *a* concentration correlates with cell numbers. Kobayasi (1961) in working with natural populations found a linear correlation between chlorophyll content and cell numbers. Eppley and Sloan (1966) indicated a strong correlation between growth rates (division/day) of *Dunaliella tertiolecta* (Chlorophyceae) and chlorophyll *a* per unit cell volume. A correlation coefficient of +0.903 was obtained. Unfortunately, R. W. Eppley (personal communication) indicated that this strong relationship does not hold for all populations. Axenic cultures of marine phytoplankton were grown at

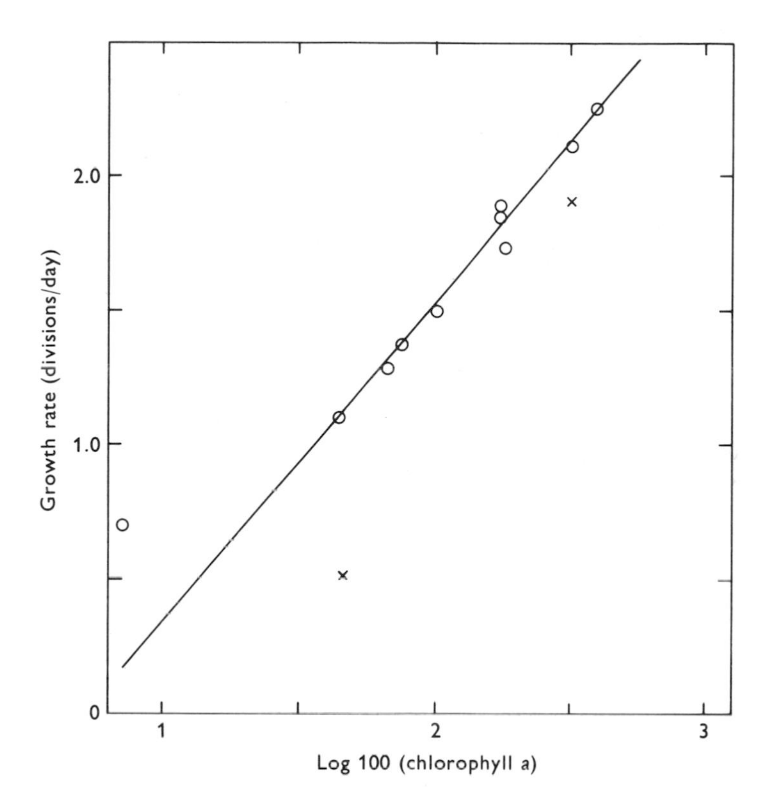

Fig. 23-1. Growth rate of *Dunaliella tertiolecta* at 20–21 °C, 0.07 cal cm^{-2} min^{-1} continuous light intensity. Reproduced with permission of *Physiologia Plantarum* (Eppley and Sloan 1966).

20–21 °C under constant light intensity of 0.07 cal/cm^2 min in aerated 5 l culture bottles. The culture medium was off-shore seawater enriched with nutrients and vitamins. Growth rates were expressed in the units of division/day obtained from serial cell counts carried out with a Coulter Particle Counter (see Chp. 22). Cell volume was also determined with this instrument. Photosynthetic pigment concentration was determined by extraction with 90% acetone and the use of the trichomatic equation (Parsons and Strickland 1963). Fig. 23-1 shows the relationship of chlorophyll *a* with growth rates of *D. tertiolecta*. Peters, Ball and Kevern (1968) showed a very good correlation between phytopigment units and the number of organisms found on artificial substrates in streams with correlation coefficients from 0.75–0.90 calculated for four stations.

III. EQUIPMENT

A. Membrane filtration equipment designed to hold 47 mm filter (Millipore Corp., Ashby Rd, Bedford, Massachusetts 01730; 55 Montpellier Bovd, Montreal 379, Quebec).

B. Membrane filters (HA, Millipore) pore size 0.45 μm, 47 mm diameter; or glass filter paper (CF/C Whatman). (Millipore filters have the advantage of completely dissolving in acetone, but they may leave a colloidal material even after centrifugation. Glass filter papers have the advantage of being much cheaper and they do not dissolve in acetone, although fibers may be introduced into the extract. Care should be taken to remove these fibers from solution during centrifugation. A more complete filtration may be possible with a Millipore filter.)

C. Graduated centrifuge tubes, glass-stoppered, capacity at least 10 ml.

D. Graduated cylinders, glass – various volumes.

E. Spectrophotometer (e.g., DU, DB, or DB-G, Beckman Instruments, Inc., 2500 Harbor Blvd, Fullerton, California 92634; 901 Oxford St., Toronto 18, Ontario).

F. Spectrophotometer cells, light path 10 mm, 100 mm.

G. Acetone, 90% reagent grade. Do not store, but make fresh for each run.

H. Optional – Magnesium carbonate (powdered), 0.1 g/l distilled water. Store in a stoppered Erlenmeyer flask. Mix thoroughly each time before use (see V.A, following).

IV. METHOD (CHLOROPHYLL *a*)

All of the chlorophyll and carotenoid pigments are extractable using 90% acetone. Although most pigments are usually disregarded except the ubiquitous chlorophyll *a*, other pigments may be useful in differentiating the various groups of algae in mixed cultures.

To a large extent the principles of extraction and analysis of pigment concentration are those developed by Richards and Thompson (1952). Since then modifications have been developed. Parsons and Strickland (1963, 1965) modified the technique and developed the equation which is commonly used today. The procedure used here is based on Parsons and Strickland (1965) with slight modifications (reproduced with permission of Information Canada).

This procedure is for laboratory cultures. Techniques used for the removal of zooplankton, which are included in field sampling, have been omitted. For analyses of natural collections consult Vollenweider (1969), Golterman (1969), and Strickland and Parsons (1968).

A. *Separation of cells*

1. Pour the sample into filter (preferably Millipore HA, pore size 0.45 μm). Depending on the volume or concentration of cells, more than one filter may be required to filter the sample as the filtering rate decreases as the filter becomes clogged.
2. Before removing the filter from the funnel, drain the filter completely dry by using a vacuum pump.

B. *Extraction*

Since pigments are very photosensitive, the extract should be exposed to the least amount of light possible during the preparation.
1. Place the filter with accumulated algae into centrifuge tube.
2. Add 5–10 ml 90% acetone. Note the exact amount of acetone (see 5, following).
3. Vigorously shake the centrifuge tube so the filter completely dissolves in the solvent.
4. Place the tubes in a dark refrigerator 20–24 h for complete extraction.
5. At the time the extraction is started, prepare an acetone blank with a clean Millipore filter. Use the same number of filters used for filtering the sample and the same quantity of acetone. This blank will be used as a standard.
6. After the extraction period, remove the samples from the refrigerator; allow them to warm to room temperature. If solvent has evaporated, build the volume to 10 ml by adding 90% acetone.
7. Centrifuge the sample and the blank for *ca.* 5–10 min at 3000–5000 rpm.

C. *Preparation of colorimeter*

The following technique is the procedure for the Beckman Spectrophotometer DB or DB-G (Beckman Manual 1967). Manuals included with other instruments give standardizing procedures.
1. Turn the instrument *idle* for several minutes. Satisfactory operation may be obtained in 2–3 min, but maximum stability is reached in *ca.* 1 h.
2. For operation in the visible range (320–760 nm) set the source selector in the visible position and turn on tungsten lamp. (Operation for setting the slit program is given in the operational manual.)
3. The analysis is based on comparing a sample of unknown transmission against a blank of 100% transmission (0.00 absorbance). Standardize the machine with the prepared blank as follows:

a. Place the opaque block (supplied with the instrument) in the sample compartment (*S*).

b. Turn the power switch to *On* position.

c. With the *Zero Adjust Control*, bring the meter needle to 0 scale line on the meter.

d. Return the switch to idle.

e. Rotate the wavelength dial to the desired wavelength.

f. Fill two cells with the prepared blank. (Do not touch the optical surfaces of the cells as smudges may absorb or reflect light. If spillage occurs while filling the cells, dry the outer surfaces with soft absorbent paper before inserting the cells into the sample compartment.)

g. Remove the opaque block.

h. Place the cells in the proper position (optical surfaces in line with the light beam) in the samples compartment; close the compartment cover.

i. Turn the power switch to *On* and with the *100% Adjust Control*, adjust the meter to read 100% transmission.

j. Discard the blank in one of the cells. This will be the cell used for each sample. This standardization against a blank should be performed at all wavelengths used in the analysis.

k. After discarding the blank, fill the cell with the extraction sample.

l. Place it in the sample compartment, seating the cell in the same orientation as in the previous step when 100% transmission was established; close the cover.

m. Turn the power switch to *On*; read the absorbance at the appropriate wavelengths for chlorophyll pigments (see D, following).

D. *Appropriate wavelengths*

The following equations (Table 23-1) have been proposed for the estimation of the concentration of chlorophyll *a* (C_a) in mg/l using measurements of absorbance (D) and a light path of 10 mm at the wavelength indicated in mu by the subscripts. Vernon (1960) indicates a probable error in the Richards and Thompson (1952) equation as great as 25% (Table 23-1; their equation 1). This may be attributed to incorrect specific absorption coefficient. Talling and Driver (1963) state that the specific absorption coefficients of the equation are based on those calculated prior to 1940. Unfortunately, considerable discrepancy exists for specific absorption coefficients of chlorophyll solutions made before 1940. Equation (2) by Parsons and Strickland (1965) is frequently used for chlorophyll *a* analysis with C in mg/l and measurements of absorbance (D):

TABLE 23-1. *Proposed equations for the estimation of chlorophyll a*

90% acetone	Richards and Thompson 1952 modifications:	$C_a = 15.6\,D_{665} - 2.0\,D_{645} - 0.8\,D_{630}$
	Odum *et al.* 1958	$C_a = 14.3\,D_{665}$
	Humphrey 1951	$C_a = 15.6\,D_{665} - 1.80\,D_{645} - 1.3\,D_{630}$
80% acetone	Vernon 1960	$C_a = 11.63\,D_{665} - 2.39\,D_{649}$
	MacKinney 1941	$C_a = 12.7\,D_{663} - 2.69\,D_{645}$
100%?	Holm 1954	$C_a = 9.78\,D_{662} - 0.99\,D_{644}$
acetone	Godnev and Sudnik 1958	$C_a = 10.0\,D_{662} - 0.8\,D_{643}$
Anhydrous	Comar and Zscheile 1942	$C_a = 9.93\,D_{660} - 0.777\,D_{642.5}$
ether	Koski 1959 as adapted by Norman 1957	$C_a = 10.68\,D_{663} - 0.9506\,D_{644}$
	Smith and Benitez 1955	$C_a = 10.1\,D_{662} - 1.01\,D_{644}$

Table reproduced from Talling and Driver (1963). See Talling and Driver for references.

$$\left.\begin{aligned} C_a &= 11.6\,D_{665} - 1.31\,D_{645} - 0.14\,D_{630}, \\ C_b &= 20.7\,D_{645} - 4.34\,D_{665} - 4.42\,D_{630}, \\ C_c &= 55.0\,D_{630} - 4.64\,D_{665} - 16.3\,D_{645}. \end{aligned}\right\} \tag{2}$$

The lower limits of detection for pigments by the method is approximately 0.02 mg/m³. The fluorometric approach to chlorophyll *a* analysis described by Yentsch and Menzel (1963) reduces the lower limits of detection to the extent that sub-sampling of algal cultures becomes practical.

Parsons and Strickland (1965) suggest that cells having a light path of 100 mm be used for the analysis. In the event of absorption values greater than 1.3, measurements should be made with 25 mm or 10 mm cells and the absorption values multiplied by 4 or 10 respectively to normalize the values to those expected with a 100 mm cell. One can also calculate pigment concentration by using only cells having a light path of 10 mm. More accuracy may be obtained with 100 mm cells especially if pigment concentrations are low. Similar equations to Eq. (2) are given for 100 mm cells by Strickland and Parsons (1968).

E. *Calculation of pigment*

Quantitative spectrophotometry, is based upon the fact that the absorbance of a substance is dependent on its concentration. If the absorbance is directly proportional to the concentration, the system is said to obey Beer's Law. The absorbance of a sample is also directly proportional to the length of the light path within the sample as stated in the Lambert Law, but because of the constant width of the sample cells, this factor cancels out in the calculation. Richards and Thompson (1952) found that at the wave-lengths used to calculate pigment concentration, that absorbancies below 0.8 were directly proportional to the concentration. Thus it becomes necessary to dilute samples having absorbancies greater than 0.8. The concentration of chlorophylls *a*, *b*, *c* in a known volume of liquid culture can be computed if the following data are known: (1) volume of acetone solvent in ml used for extraction; (2) volume of culture medium in liters. Knowing these values, the concentration of pigment in mg/m^3 can be computed from Eq. (3).

1. If cells of 100 mm light path are used or the factor is applied in using cells of a shorter light path (D, preceding), 10 ml of acetone solvent must also be used for extraction.

$$\text{Chlorophyll } a, b, c\ (\text{mg/m}^3) = \frac{(C)}{(V_c)} \tag{3}$$

where:

C = values obtained from Eq. (2),

V_c = original volume of culture in liters.

2. If cells of 10 mm light path are used exclusively, the pigment concentration (mg/m^3) can be computed from Eq. (4).

$$\text{Chlorophyll } a, b, c\ (\text{mg/m}^3) = \frac{(C)\,(V_a)}{(V_c)} \tag{4}$$

where:

V_a = volume of acetone solvent in 1 ml.

F. *Example*

This is a calculation for a chlorophyll *a* analysis using cells of 10 mm light path. The observed absorbancies were as follows:

$D_{665} = 0.057$, $D_{645} = 0.021$, $D_{630} = 0.010$;

chlorophyll $a = 11.6(0.057) - 1.31(0.021) - 0.14(0.010) = 0.632$ mg/l.

The concentration of chlorophyll *a* in a known volume of solvent can be converted to the concentration in a known volume of culture from Eq. (4). For example, if the solvent volume is 35.0 ml and the culture volume is 0.15 liters, the concentration of chlorophyll *a* (mg/m^3) would be as follows:

$$\text{Chlorophyll } a = (0.632)\,(35.0)/(0.15) = 147.5 \text{ mg/m}^3.$$

V. POSSIBLE PROBLEMS

A. *Degradation*

Moss (1967) indicates that the presence of a pigment degradation product having a visible absorption spectrum similar to that of chlorophyll is a possible source of error. The presence of this product (pheophytin *a*) is usually ignored in phytoplankton of open waters and may be ignored in laboratory cultures. The presence and absorption of this product in laboratory cultures can only be resolved by analyzing for its effect. The method of Moss (1967) can be used to analyze for this product. Richards and Thompson (1952) recommend adding magnesium carbonate to prevent acidity and the stimulation of pheophytin *a* in samples. Add *ca.* 1 ml of 0.01 % $MgCO_3$ solution to the sample before filtering (see IV.A, preceding).

B. *Sensitivity*

The lower limit of sensitivity of the method is approximately 0.2 mg/m^3 (Parsons and Strickland 1965). If pigment levels are in this range, the problem may be overcome by concentrating the extract. Some colloidal material may remain after the solution of the Millipore filter, even after centrifugation (Parsons and Strickland 1965). This is compensated for by the use of the acetone and clean Millipore blank prepared at the time of pigment extraction.

VI. OTHER USES

Chlorophyll *a* analysis is used extensively in both field and laboratory research. It has several applications in the field which are discussed and described by several authors including Golterman (1969), Strickland and Parsons (1968) and Vollenweider (1969).

VII. REFERENCES

Beckman Instrument, Inc. 1967. *Beckman Instrument Manual for Model DB Prism Spectrophotometer and Model DB-G Grating Spectrophotometer.* Fullerton, California. 54 pp.

Eppley, R. W. and Sloan, P. R. 1966. Growth rates of marine phytoplankton: Correlation with light absorption by cell chlorophyll *a. Physiol. Plant.* **19**, 47–59.

Golterman, H. L. 1969. *Methods for Chemical Analysis of Fresh Waters.* I.B.P. Handbook No. 8. Blackwell Scientific Publ., Oxford. 166 pp.

Kobayasi, H. 1961. Chlorophyll content in sessile algal community of Japanese mountain river. *Bot. Mag., Tokyo* **74**, 228–35.

Moss, B. 1967. A spectrophotometric method for the estimation of percentage degradation of chlorophylls to pheopigments in extracts of algae. *Limnol. Oceanogr.* **12**, 335–40.

Parsons, T. R. and Strickland, J. D. H. 1963. Discussion of spectrophotometric determination of marine plant pigments, with revised equation for ascertaining chlorophylls and carotenoids. *J. Mar. Res.* **21**, 155–63.

Parsons, T. R. and Strickland, J. D. H. 1965. Particulate organic matter. III.I Pigment analysis. III.I.I Determination of phytoplankton pigments. *J. Fish. Res. Bd Canada* **18**, 117–27.

Peters, J. C., Ball, R. C. and Kevern, N. R. 1968. An evaluation of artificial substrates for measuring periphyton production. *Michigan State Univ. Dept. Fish. Wildlife.* Tech. Rep. No. 1.

Richards, F. A. 1952. The estimation and characterization of plankton populations by pigment analysis. I. The absorprion spectra of some pigments occurring in diatoms, dinoflagellates and brown algae. *J. Mar. Res.* **11**, 147–55.

Richards, F. A. and Thompson, T. G. 1952. The estimation and characterization of plankton populations by pigment analysis. *J. Mar. Res.* **11**, 156–72.

Strickland, J. D. H. and Parsons, T. R. 1968. A Practical Handbook of Seawater Analysis. *Fish. Res. Bd Canada*, Ottawa. Bull. 167. 311 pp.

Talling, J. F. and Driver, D. 1963. Some problems in the estimation of chlorophyll *a* in phytoplankton. In Doty, M., ed., *Proc. Conference on Primary Productivity Measurements, Marine and Freshwater, 1961.* United States Atomic Energy Comm., Division of Technical Information. (TID-7633). 237 pp.

Vernon, L. P. 1960. Spectrophotometric determination of chlorophylls and pheophytins in plant extracts. *Anal. Chem.* **32**, 1144–50.

Vollenweider, R. A. 1969. *A Manual on Methods for Measuring Primary Production in Aquatic Environments.* I.B.P. Handbook No. 12. Blackwell Scientific Publ., Oxford. 213 pp.

Yentsch, C. S. and Menzel, D. W. 1963. A method for the determination of phytoplankton chlorophyll and phaeophytin by fluorescence. *Deep Sea Res.* **10**, 221–31.

NOTES ADDED IN PROOF

1. Barrett and Jeffrey (*Plant Physiol.* **39**, 44–7, 1964) recommend use of 100% acetone to prevent action by the degradation enzyme chlorophyllase.

2. The Parsons and Strickland equation (Eq. 2, p. 365) is based on earlier extinction coefficient of chlorophyll *c*. However, chlorophyll *c* is now known to have two components (Jeffrey, *Biochim. Biophys. Acta* **279**, 15–33, 1972). Purification of chlorophyll c_1 and c_2 indicates new extinction coefficients and new equations are being prepared (S. W. Jeffrey, personal communication). The chlorophyll *c* values are approximately half those presented in this chapter.

24: Intracellular growth rates

PAUL B. GREEN

Department of Biological Sciences, Stanford University, Stanford, California 94305

CONTENTS

I. PURPOSE

The immediate cause of the change in shape in any object is the pattern of activity in the various parts of the object. In plant cells this activity takes the form of expansion of surface. Knowledge of the rate and directionality of surface behavior in various parts of a single cell can reveal the major features of the mechanism of morphogenesis, be it an extension of a pre-existing form or a specific change in shape.

Once the detailed local behavior is known, its basis in local variations of physiology and structure can be meaningfully pursued. Unfortunately one cannot infer the pattern of local surface behavior from the changes in cell outline alone (see Fig. 24-1). A variety of patterns of local behavior can yield the same change in cell shape (Green 1965).

II. EQUIPMENT

Analysis requires subdivision of the growing zone. Ordinarily natural ornamentation is not present in growing regions within cells so artificial marks must be applied.

A. Traditional materials include carbon particles, grinding powders (Edmund Scientific Co., 555 Edscorp Bldg, Barrington, New Jersey 08007) and starch grains.

B. Spherical resin beads are often an improvement over the traditional materials because their position can be determined accurately even when somewhat out of focus (see Green, Erickson and Richmond 1970). Small beads 5–20 μm can be obtained when the size, less than 400 mesh, is specified (Bio-Rad Laboratories, 32nd and Griffin Ave., Richmond, California 94804). The beads obtained in the chloride form can be converted to the hydroxyl form as follows:

1. Suspend beads in 2 N KOH.
2. Plug the larger end of a Pasteur-type pipette with cleansing tissue.
3. Add some bead suspension to Pasteur-type pipette which filters the fluid through while the beads are retained on the cleansing tissue.
4. Treat the beads further with KOH.
5. Rinse the beads in distilled water.

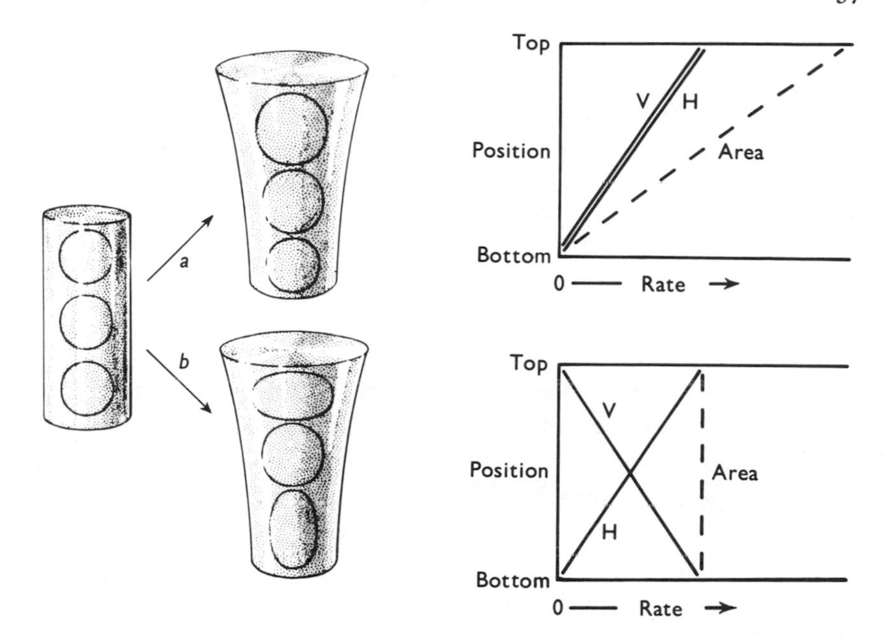

Fig. 24-1. Model morphogenetic system of a cylinder converted to two patterns (*a*, *b*) by surface deformation. In *a*, expansion is the same in all directions (isotropic) but gradient in rate of area expansion exists (see upper, right graph) with vertical (V) and horizontal (H) directions equal. In *b*, expansion in circumference predominates at top with expansion in length predominating at base, thus V and H are in antiparallel gradients (see lower, right graph).

C. Application can be made with a hair (glued to a convenient holder) either by hand for large cells, or by fine rod moved by a micromanipulator (de Fonbrune type, Aloe Scientific Co., St Louis, Missouri) on small cells.

D. In most instances it is convenient to follow displacements of marks by time-lapse photography. Data analysis can be carried out on prints from the film or from projected images. The Vanguard Motion Analyser (Vanguard Instrument Co., Melville, Long Island, New York) is suitable and can be connected to a key punch for computer manipulation of the data.

III. TEST ORGANISM

The Charophyceae, particularly *Nitella*, are convenient material. Their young internodes are naturally exposed and expand many fold. These plants have single apical cells which can be exposed for marking by dissection. For this a device to hold the plant, such as that described in Green (1968), is essential.

IV. METHOD

To account for shape change in radially symmetrical cells it generally suffices to know the growth behavior in various regions of the cell along two directions: the cell axis and circumference. When r is measured as the local relative rate of linear extension (relative elemental rate):

$$r = \frac{1}{x} \times \frac{dx}{dt} = \frac{d \ln x}{dt}, \tag{1}$$

where x is the distance between closely spaced marks, and t is time, the sum of the two perpendicular rates gives the relative rate of increase in surface area. The quotient of two perpendicular rates gives the directionality (anisotropy) if the quotient is other than one. If equal to one, growth is non-directional or isotropic. The formulae are rigorously correct only for marks 'infinitessimally' separated but are sufficient when used with closely spaced marks.

In practice r can be evaluated for the region between a pair of marks, over a period of time as:

$$r = \frac{\ln x_1 - \ln x_2}{t_2 - t_1}, \tag{2}$$

where x is the separation between two marks at times t_1 and t_2.

A. *Degree of surface subdivision*

The marks are closely enough spaced when the data show only a gradual difference in rate of growth between successive pairs of marks – or show no difference. When the behavior of pairs of marks shows no significant variation along the cell axis (or around the circumference) it is no longer necessary to sub-divide the growing district for measurement because fractional changes in the whole are the same as fractional changes in any sub-division.

B. *Frequency of measurement*

When extension rate is not the same down a growing axis of constant geometry, pairs of marks are displaced down the axis as they separate from each other. Thus their behavior reflects the activity of an ever enlarging district – which unfortunately gradually changes its position. Data must therefore be taken over short intervals. This places great demands on the precision of measurement – generally met only by time-lapse or other photography and subsequent dimensional analysis of the images.

C. *Relation to overall growth rate*

When the local relative rates, for successive regions down a given axis, are plotted as a function of distance down the axis, the profile shows the 'intensity' of the growth process down that axis. The area under such a curve is proportional to the rate of extension of the axis as a whole.

D. *Generality*

This mode of analysis applies equally to organs as well as cells; the validity is dependent primarily on the marks being closely spaced and frequently measured, so that any inhomogeneities of the growth axis do not greatly alter the data.

V. SAMPLE DATA, ANALYSES

Analyses of the type described have been made on both the internodes (Green 1954) and the apical cell (Green and King 1966) of *Nitella*. Castle (1958) has performed a similar intracellular analysis on the fungus *Phycomyces*. Much of the method derives from analyses done of the *Zea mays* root by Erickson and Sax (1956). An analysis of drug-induced changes in cell morphogenesis is found in Green, Erickson and Richmond (1970). Applications to tropisms in *Chara* (Charophyceae) are found in Sievers and Schröter (1971).

VI. PROBLEMS

The method will be difficult to apply to those algae which have slime sheaths because marks will be lost as slime is shed. The anion exchange resin beads will be less likely to stick to marine forms because the chloride in sea water will seriously compete with the cell wall anions (carboxyl groups).

VII. REFERENCES

Castle, E. S. 1958. The topography of tip growth in a plant cell. *J. Gen. Physiol.* **41**, 913–26.

Erickson, R. O. and Sax, K. 1956. Elemental growth rate of the primary root of *Zea mays*. *Proc. Amer. Phil. Soc.* **100**, 487–98.

Green, P. B. 1954. The spiral growth pattern of the cell wall in *Nitella axillaris*. *Amer. J. Bot.* **41**, 403–9.

Green, P. B. 1965. Pathways of cellular morphogenesis. *J. Cell Biol.* **27**, 343–63.

Green, P. B. 1968. Growth physics in *Nitella*: A method for continuous *in vivo* analysis of extensibility based on a micro-manometer technique for turgor pressure. *Plant Physiol.* **43**, 1169–84.

Green, P. B. and King, A. 1966. A mechanism for the origin of specifically oriented textures in development with special reference to *Nitella* wall texture. *Austral. J. Biol. Sci.* **19**, 421–37.

Green, P. B., Erickson, R. O. and Richmond, P. A. 1970. On the physical basis of morphogenesis. In Fredrick, J. F. and Klein, R. M., eds., Phylogenesis and Morphogenesis in the Algae. *Ann. New York Acad. Sci.* **175**, 712–31.

Sievers, A. and Schröter, K. 1971. Versuch einer Kausalanalyse der geotropischen Reaktionskette im *Chara*-Rhizoid. *Planta* (Berl.) **96**, 339–53.

Section v

Bioassay

25: Bioassay: biotin

A. F. CARLUCCI

Institute of Marine Resources, University of California, San Diego, La Jolla, California 92037

CONTENTS

I. INTRODUCTION

The most sensitive method yet available for the bioassay of biotin in seawater employs the Dinophyceae *Amphidinium carterae* (Carlucci and Silbernagel 1967). Radiocarbon uptake by starved cells for a specific exposure after a pre-incubation under controlled conditions is proportional to the dissolved biotin concentrations. If the incubation period is continued for several days cell numbers are proportional to biotin concentrations. The biotin concentrations of seawater are calculated using the same equation for either uptake or cell-counting data.

II. TEST ORGANISM

Amphidinium carterae (Haskins Laboratories, 165 Prospect St., New Haven, Connecticut 06520), marine dinoflagellate. Available from the culture collection of the Food Chain Research Group, Institute of Marine Resources, University of California, San Diego, La Jolla, California 92037.

III. EQUIPMENT

A. *Glassware*

All glassware should be cleaned with a detergent wash, treated with chromic–sulfuric acid for 12–24 h; rinsed thoroughly with deionized water; and baked for 12 h at *ca*. 260 °C in an oven.

1. 50 ml micro-Fernbach flasks with deLong necks (Bellco 607) with Morton stainless steel enclosures (Bellco B-25, Bellco Glass, Inc., 340 Edrudo Rd, Vineland, New Jersey 08360).
2. 125 ml Erlenmeyer flasks, sterile; plugged with cotton wool enclosed in cheesecloth.
3. 0.1, 1.0, 10, 25 ml graduated pipettes.
4. All-glass membrane filter unit, 1 l flask, for 47 mm diameter PH filter, 0.3 μm pore (Millipore Corp., Ashby Rd, Bedford, Massachusetts 01730; 55 Montpellier Blvd, Montreal 379, Quebec).
5. Filter funnel for 25 mm diameter HA filter, 0.45 μm pore (Millipore Corp.).

6. Fritted glass filter, bacteriological, UF grade.
7. Storage vessels for solutions and samples:
   a. glass bottle, 1 l,
   b. screw-cap test tubes, sterile,
   c. polypropylene bottles.

B. *Incubator*

The incubator should be uniformly illuminated with cool-white fluorescent bulbs preferably placed under the bioassay vessels.

Light should be of an intensity of 0.02 to 0.05 ly/min. Temperature should be 20 ± 1 °C.

C. *Radiocarbon counter*

Any suitable thin-end window, gas-flow Geiger counter may be used for determining radiocarbon incorporation by cells of the test alga (e.g., Nuclear-Chicago Corp, 2000 Nuclear Dr., Des Plaines, Illinois 60018; 260 Lesmill Rd, Don Mills, Ontario; Packard Instrument Co. Inc., 2200 Warrenville Rd, Downers Grove, Illinois 60515: Philips Electronic Instruments, 750 S. Fulton Ave., Mount Vernon, New York 10550; 116 Vanderhoof Ave., Toronto, Ontario).

D. *Cell counting*

Cell counts may be made with a hemacytometer (see Chp. 19, III.C.3, 4), or, if more convenient, with a counting instrument such as the Coulter Electronic Particle Counter (see Chp. 22). Cell counts are used to measure response to the vitamin in a sample if facilities for radiocarbon experiments are not available.

E. *Special reagents*

1. Vitamin-free seawater. Vitamins are removed from natural seawater by adsorption onto charcoal. Ten grams of Norit A (decolorizing carbon) are required for every liter of seawater.

a. Pretreat the charcoal by shaking 10 min in 500 ml 5% w/v NaCl solution prepared in high quality distilled water.

b. Filter the suspension through Whatman no. 1 filter.

c. Transfer the charcoal to a second 500 ml NaCl solution; repeat a and b twice.

d. After the third wash, add the charcoal to one liter seawater; shake for half hour.

e. Filter the seawater through Whatman no. 1 paper.
f. Filter seawater aseptically through a membrane filter (Millipore PH, 47 mm, 0.3 μm) into sterile container.
g. Store water aseptically.

2. *Nutrient solutions*

a. Chelated metals, nitrate, phosphate (N solution)

i. Dissolve in 100 ml distilled water: $CoCl_2.6H_2O$, 0.08 g; $CuSO_4.5H_2O$, 0.08.

ii. Dissolve in 800 ml distilled water: $FeCl_3.6H_2O$, 0.2 g; $ZnSO_4.7H_2O$, 0.06; $MnSO_4.H_2O$, 0.12; $Na_2MoO_4.H_2O$, 0.03; Na_2EDTA (disodium ethylenediaminetetraacetate), 1.2.

iii. Add 1.0 ml of solution i (preceding) to solution ii (preceding). Dilute to 900 ml with distilled water.

iv. Adjust pH of solution iii to *ca.* 7.5 with dilute NaOH.

v. Add to solution iv: KNO_3, 10 g; KH_2PO_4, 1.4, distilled water to 1 l.

vi. Autoclave solution for 15 min at 15 lb/in^2 (see Chp. 12, II.C.2.c); store in glass bottle in dark. This is the nutrient solution N.

b. Vitamin solution (V solution)

i. Dissolve in 100 ml distilled water cyanocobalamin (B_{12}), 10 mg; thiamine.HCl, 10.

ii. Dilute 10 ml of preceding vitamin solution to 100 ml with distilled water.

iii. Sterilize vitamin solution (ii, preceding) through fritted glass filter (UF grade); store in sterile screw-cap test tubes at −20 °C. This is the vitamin solution V.

3. *Radioactive carbonate.* Prepare enough volume of a radioactive carbonate ($Na_2{}^{14}CO_3$) solution so that each vessel receives 1 ml. Add the $Na_2{}^{14}CO_3$ to 5 % w/v NaCl in distilled water. The activity of each milliliter of solution should be about 1 μCi. Sterilize by autoclaving at 15 lb/in^2 for 15 min (see Chp. 12, II.C.2.c).

F. *Algal inoculum*

1. To each of 3 sterile 125 ml Erlenmeyer flasks, add 50 ml of vitamin-free seawater. Each flask is plugged with cotton enclosed in cheesecloth.
2. Add to each flask 0.25 ml of nutrient solution N (E.2.a, preceding) and 0.05 ml of vitamin solution V (E.2.b, preceding).

3. Add to one flask 0.05 ml of the undiluted standard biotin solution (see IV.C.1.f, following). This flask contains the complete medium.
4. Inoculate the complete medium (3, preceding) with 1 ml of actively growing dense culture of *Amphidinium carterae*.
5. Incubate 7 days.
6. Transfer 0.5 ml to one flask containing biotin-free medium (2, preceding).
7. Incubate second flask 7 days.
8. Transfer 5.5 ml from second culture (7, preceding) to other biotin-free medium flask.
9. Incubate this last transfer 6 days. At the end of this time the cells will be sufficiently starved of biotin and physiologically active. The cell concentration in this flask should be about 5×10^5/ml and the medium free of excess biotin.

IV. METHOD

A. *Sample collection, storage*

Any suitable clean sampler may be used. Samples should be filtered immediately after collection, if possible, through a HA membrane filter (Millipore HA, 25 mm, 0.45 μm). Store the filtrate in a sterile polypropylene bottle at $-20\,^\circ$C. A sample of 50 ml is sufficient.

B. *Procedure*

1. Sample preparation. Thaw the samples. If any question arises as to sterility, filter through membrane filter (Millipore HA, 25 mm, 0.45 μm) just prior to assay. Add duplicate aliquots of between 5 and 20 ml for each sample, depending on biotin content, to 50 ml micro-Fernbach flasks. It may be necessary to dilute the sample if its biotin content exceeds 5 ng (10^{-9} g)/l. Bring the volume of each assay flask to 20 ml with vitamin-free seawater.

2. *Nutrient addition*

a. Add to each flask 0.1 ml of nutrient solution N (III.E.2.a, preceding) and 0.02 ml of vitamin solution V (III.E.2.b, preceding). To avoid adding these solutions by separate aliquots, suitable amounts may be mixed together just prior to use and addition made with one aliquot.

b. To one of each duplicate flask of sample, add a known amount of biotin to serve as an internal standard (see Calibration, C.2, following). Internal standardization is necessary to account for the inhibitory properties of seawater.

3. *Inoculation.* To each assay flask add *ca.* 0.5 ml inoculum of *A. carterae* (III.F.9, preceding). The initial concentration in each flask should be *ca.* 10^4 cells/ml. If the inoculum cell concentration is higher or lower than 5×10^5 cells/ml, add to each flask proportionally less or greater than 0.5 ml. The inoculum aliquot should be in the range of 0.3–0.7 ml/flask.

4. *Incubation, analyses.* Incubate all flasks for 94 h in the incubator (see III.B, preceding). Proceed as follows for either cell counts (a) or radiocarbon uptake (b).

a. Cell counts – determine the number of cells/ml in each flask after a further incubation of 3 days. Use a hemacytometer (Chp. 19, III.C.3, 4) or Coulter Counter (Chp. 22).

b. Radiocarbon uptake

i. Add 1 ml of ^{14}C-labeled carbonate solution (containing 1 μCi of activity) to each flask at 2 min intervals; mix contents well; replace the flask in the incubator for 2 h.

ii. Collect the cells in each flask on a membrane filter (HA Millipore, 25 mm, 0.45 μm). Wash the sides of the flask with filtered seawater and scrub with a policeman. Collect the cells in the wash on the filter also. (Arrange times of addition of radiocarbonate and filtration so that each sample receives 2 h of incubation with the isotope.)

iii. Count the activity of the cells on the filter with the gas-flow Geiger counter.

5. *Calculation of vitamin content.* Read the apparent biotin concentration in each sample from a calibration curve prepared by plotting radiocarbon uptakes or cell numbers in the external standards (see C, Calibration, following) against biotin concentrations. Let this apparent biotin concentration be A ng/l. Read the concentration of biotin of the sample plus internal standard from the same calibration curve. Let this concentration of biotin be B ng/l. Calculate the concentration of biotin in the sample from the expression:

$$\text{ng biotin/l} = \left(\frac{A}{B-A}\right) \times \left(\frac{20}{v}\right), \qquad (1)$$

where v is the number of milliliters of sample originally taken for the analysis.

C. *Calibration*

1. *Standard biotin solution*

a. Dissolve 10 mg pure crystalline biotin in 100 ml high quality distilled water.

b. Sterilize the solution by passing it through a fritted glass filter (UF grade).

c. Store filtrate in 10 ml portions at $-20\,^{\circ}$C.

d. Dilute 1 ml of this solution to 100 ml with distilled water in volumetric flask.

e. Take 1 ml of dilute solution and again dilute 1 to 100 ml.

f. Dilute 5 ml of this last dilution to 25 ml. This final solution has a biotin concentration of 2000 pg (10^{-12} g)/ml. Refer to this solution as A and make further dilutions in vitamin-free seawater as follows:

5 ml of solution A to 10 ml (B)
4 ml of solution A to 10 ml (C)
3 ml of solution A to 10 ml (D)
2 ml of solution A to 10 ml (E)
1 ml of solution A to 10 ml (F)
0.5 ml of solution A to 10 ml (G)

2. *Internal standards.* Add to each of the duplicate samples, 0.1 ml of solution F (1.f, preceding). The concentration of biotin in the flask will be 1 ng/l. The internal standard is analyzed with the sample containing no added biotin and is used for the calculations described earlier (B.5).

3. *External standards.* With each series of samples being assayed prepare 7 micro-Fernbach flasks containing 20 ml of vitamin-free seawater supplemented with nutrients and vitamins (same as samples, B.2, preceding). To one flask make no further addition and to each of the other flasks add 0.1 ml of solutions G–B, respectively. The concentrations of added biotin will be 0, 0.5, 1, 2, 3, 4, and 5 ng/l, respectively. Inoculate and incubate these external standards along with the samples being assayed.

V. SAMPLE DATA

From either the cell counts of radiocarbon uptake data the concentrations of biotin in a seawater sample is determined as follows.

Assume the apparent concentration of biotin in a sample is 0.65 ng/l and its internal standard is 1.50 ng/l. In Eq. (1); $A = 0.65$ ng/l and $B = 1.50$ ng/l. The amount of biotin in the sample if 5 ml were used in the assay is:

$$\text{ng biotin/l} = \left(\frac{0.65}{1.50-0.65}\right) \times \left(\frac{20}{5}\right) \tag{2}$$

$$= 3.0 \text{ ng biotin/l.}$$

VI. CAPABILITIES (MARINE)

A. *Range*

0.2 to 5 ng biotin/l.

B. *Precision*

1. At the 3 ng biotin/l level – the correct value lies in the range:

   Mean of n determinations $\pm 0.28/n^{\frac{1}{2}}$ ng biotin/l.

2. At 1 ng biotin/l level – the correct value lies in the range:

   Mean of n determinations $\pm 0.09/n^{\frac{1}{2}}$ ng biotin/l.

VII. LIMITATIONS

A. *Inhibition of seawater to test alga*

If the inhibition to *A. carterae* by the sample is greater than 30–35% (determined by the amount of the added 1 ng/l biotin recovered in the internal standard; i.e., $1-(B-A)\times 100\%$), the results of the assay become unreliable. The assay should be repeated using a greater sample dilution.

B. *Adaptation for freshwater systems*

The *A. carterae* assay for biotin in seawater may be applied to freshwater systems if enough seawater salts are added to the sample so that its salinity will be in the range of 27–35‰. Generally it is sufficient to dilute one part freshwater with 3 parts of vitamin-free seawater. The sensitivity of the freshwater assay, therefore, will be about one-fourth that of the seawater assay. The less sensitivity for freshwater systems still render it an adequate assay for most waters. For greater sensitivity the assay could employ as a test organism a biotin-requiring freshwater alga and the above procedure modified accordingly, i.e., alga-starvation times, pre-incubation durations, radiocarbon exposure times, etc.

VIII. ACKNOWLEDGMENTS

I am grateful to Susan Silbernagel for excellent technical assistance in the development of the assay.

This work was supported, in part, by the Marine Life Research Program, Scripps Institution of Oceanography's component of the California

Cooperative Oceanic Fisheries Investigation, a project sponsored by the State of California, and, in part, by the United States Atomic Energy Commission, Contract No. AT(11-1) GEN 10, P.A. 20.

IX. REFERENCE

Carlucci, A. F. and Silbernagel, S. B. 1967. Bioassay of seawater. IV. The determination of dissolved biotin in seawater using ^{14}C uptake by cells of *Amphidinium carterae*. *Can. J. Microbiol.* **13**, 979–86.

26: Bioassay: cyanocobalamin

A. F. CARLUCCI

Institute of Marine Resources, University of California, San Diego, La Jolla, California 92037

CONTENTS

I. INTRODUCTION

Although there are several methods available for the bioassay of dissolved cyanocobalamin (vitamin B_{12}) in seawater, the most sensitive is the one described by Carlucci and Silbernagel (1966). The marine Bacillariophyceae, *Cyclotella nana*, is the assay organism. Radiocarbon uptake by starved cells of the test alga with controlled conditions are proportional to the concentrations of the dissolved vitamin in the samples. If tracer techniques are not convenient, the samples are incubated for a total of 5 days after which time cell numbers are proportional to dissolved vitamin concentrations. The amount of cyanocobalamin in each sample is calculated using the same equation for either the uptake or cell-counting data.

II. TEST ORGANISM

Cyclotella nana (13-1), a marine centric diatom. Available from the culture collection of the Food Chain Research Group, Institute of Marine Resources, University of California, San Diego, La Jolla, California 92037.

III. EQUIPMENT

A. *Glassware*

See Chp. 25, III.A.

B. *Incubator*

See Chp. 25, III.B.

C. *Radiocarbon counter*

See Chp. 25, III.C.

D. *Cell counting*

See Chp. 25, III.D.

E. *Special reagents*

1. Vitamin-free seawater. See Chp. 25, III.E.1.

2. Nutrient solutions

a. Chelated metals, nitrate, and phosphate (see Chp. 25, III.E.2.a).
b. Sodium silicate (S solution)
i. Dissolve in 1 l distilled water either

$Na_2SiO_3.5H_2O$ 10 g

or

$Na_2SiO_3.9H_2O$ 14

ii. Sterilize silicate solution S through fritted glass filter (UF grade); store in sterile polypropylene container.
c. Hydrochloric acid
i. Make a 0.1 N solution
ii. Determine by titration the amount necessary to neutralize 10 ml of silicate solution S (preceding b.i). If X ml is the volume of HCl, dilute $100X$ ml to 1 l.
iii. Sterilize HCl solution through fritted glass filter (UF grade); store in sterile polypropylene bottle.
d. Vitamin solution (V solution)
i. Dissolve in 100 ml distilled water

thamine.HCl 10 mg
biotin 10

ii. Dilute 10 ml of preceding solution to 100 ml with distilled water.
iii. Sterilize vitamin solution (ii, preceding) through fritted glass filter (UF grade); store in screw-cap test tubes at −20°C. This is vitamin solution V.

3. Radioactive carbonate. See Chp. 25, III.E.3.

F. *Algal inoculum*

1. Add 50 ml of vitamin-free seawater to each of 3 sterile 125 ml Erlenmeyer flasks. Plug the flasks with cotton enclosed in cheesecloth.
2. Add to each flask 0.25 ml of nutrient solution N (E.2.a, preceding); 0.3 ml of silicate solution S (E.2.b, preceding); 0.3 ml of HCl (E.2.c, preceding); and 0.05 ml of vitamin solution V (E.2.d, preceding).
3. Add to one flask 0.05 ml of the undiluted standard cyanocobalamin solution (see IV.C.1, following). This flask contains the complete medium.
4. Inoculate the complete medium (3, preceding) with 1 ml of actively growing culture of *Cyclotella nana*.

5. Incubate for 3 days.
6. Transfer 0.5 ml to one of the flasks containing the cyanocobalamin-free medium (2, preceding).
7. Incubate second flask 4 days.
8. Transfer 0.5 ml to the other flask containing cyanocobalamin-free medium.
9. Incubate 3 days. The culture is in log phase of growth, almost entirely stripped of vitamin B_{12}, and physiologically active. The cell concentration in this flask should be between 0.5 and 2×10^6 cells/ml and the medium free of cyanocobalamin.

IV. METHOD

A. *Sample collection, storage*

See Chp. 25, IV.A.

B. *Procedure*

1. Sample preparation. Thaw the samples. If any question arises concerning their sterility, filter through membrane filter (Millipore HA, 25 mm, 0.45 μm) just prior to the assay. Add duplicate aliquots, between 5 and 20 ml for each sample, to the 50 ml micro-Fernbach flasks. It is necessary to dilute the sample if its cyanocobalamin content exceeds 3 ng (10^{-9} g)/l. Bring the volume of each flask to 20 ml with vitamin-free seawater.

2. Nutrient addition. See III.E.2, preceding, for solutions.

a. Add to each flask 0.1 ml of nutrient solution N; 0.12 ml of silicate solution S; 0.12 ml of HCl; and, 0.02 ml of vitamin solution V. To avoid adding these solutions as many separate aliquots, suitable amounts may be mixed together just prior to use and addition be made with one aliquot.

b. To one of each duplicate flask of sample, add a known amount of cyanocobalamin to serve as an internal standard (see C.2, following). Internal standardization is necessary to account for the inhibitory properties of seawater.

3. Inoculation. To each assay add *ca.* 0.25 ml inoculum of *C. nana* (III.F.9, preceding). The initial concentration in each flask should be *ca.* 10^4 cells/ml. If the inoculum cell concentration is higher or lower than 1×10^6 cells/ml, add to each flask proportionally less or greater than 0.25 ml. The inoculum aliquot should be in the range of 0.2–0.5 ml/flask.

4. Incubation, analyses. Incubate all flasks for 46 h in the light incubator (see III.B, preceding). Proceed as follows for either cell counts (a), or radiocarbon uptake (b).

a. Cell counts – determine the number of cells/ml in each flask after a further incubation of 3 days. Use a hemacytometer (Chp. 19, III.C.3, 4) or, an electronic particle counter such as the Coulter Counter (Chp. 22).

b. Radiocarbon uptake

i. Add 1 ml of ^{14}C-labeled carbonate solution, containing 1 μCi of activity, to each flask at 2 min intervals; mix contents well, replace the flask in the incubator for 2 h.

ii. Collect the cells in each flask on a membrane filter (Millipore HA, 25 mm, 0.45 μm). Wash the sides of the flask with filtered seawater, scrubbing with a rubber policeman. Collect the cells in the wash on the filter also. (Arrange the times of addition of radiocarbonate and filtration so that each sample receives 2 h incubation with the isotope.)

iii. Count the activity in the cells with the gas-flow Geiger counter.

5. Calculation of vitamin content. Read the apparent cyanocobalamin concentration in each sample from a calibration curve prepared by plotting radiocarbon uptakes or cell numbers in the external standards (see C, Calibration, following) against vitamin concentrations. Let this apparent cyanocobalamin concentration be A ng/l. Read the concentration of vitamin in the sample plus internal standard from the same calibration curve. Let this concentration of cyanocobalamin be B ng/l. Calculate the concentration of cyanocobalamin in the sample from the expression:

$$\text{ng cyanocobalamin/l} = \left(\frac{A}{B-A}\right) \times \left(\frac{20}{v}\right), \qquad (1)$$

where v is the number of milliliters of sample originally taken for the analysis.

C. *Calibration*

1. Standard cyanocobalamin solution

a. Dissolve 11 mg pure crystalline cyanocobalamin in 100 ml high quality distilled water. (Use 11 mg of vitamin instead of 10 mg to allow for the presence of *ca.* 10% of water of crystallization.)

b. Sterilize the solution by passage through a fritted glass filter (UF grade).

c. Store in 10 ml portions in screw-cap test tubes at −20 °C.

d. Dilute 1 ml of this solution to 100 ml with distilled water.

e. Dilute 1 ml of this dilute vitamin solution to 100 ml with distilled water.

f. Make a final dilution of 1 ml to 10 ml with distilled water. In this final solution, 1 ml = 1 ng cyanocobalamin. Let this solution be A. Make further dilutions in vitamin-free seawater as follows:

6 ml of A to 10 ml (solution B)
4 ml of A to 10 ml (solution C)
2 ml of A to 10 ml (solution D)
8 ml of D to 10 ml (solution E)
4 ml of D to 10 ml (solution F)
2 ml of D to 10 ml (solution G)
1 ml of D to 10 ml (solution H)

2. *Internal standards.* Add to each duplicate sample, 0.1 ml of solution D (1.f, preceding). The concentration of cyanocobalamin in this flask will be 1 ng/l. The internal standard is analyzed along with the sample which contains no added vitamin and is used for the calculations described earlier (B.5).

3. *External standards.* With each series of samples being assayed prepare 8 micro-Fernbach flasks containing 20 ml of vitamin-free seawater supplemented with all the solutions (same as samples, B.2, preceding). Make no further addition to one flask. To each of the seven remaining flasks add 0.1 ml of solutions H–B above, respectively. The concentrations of added vitamin B_{12} will be 0.1, 0.2, 0.4, 0.8, 1.0, 2.0, and 3.0 ng/l, respectively. The external standards are then treated in the same manner as the samples.

V. SAMPLE DATA

The amount of cyanocobalamin in a sample is calculated from either the cell-count or radiocarbon uptake data as follows.

As an example, assume that the apparent concentration in a sample is 0.75 ng cyanocobalamin/liter and that its internal standard is 1.50 ng cyanocobalamin/liter. In Eq. (1) A = 0.75 ng/l and B = 1.50 ng/l. If a 5 ml aliquot of sample was used then the concentration of cyanocobalamin is:

$$\text{ng cyanocobalamin/liter} = \left(\frac{0.75}{1.50-0.75}\right) \times \left(\frac{20}{5}\right),$$
$$= 4 \text{ ng/l}.$$

VI. CAPABILITIES (MARINE)

A. *Range*

0.05 to 3 ng cyanocobalamin/liter.

B. *Precision*

1. At the 1.5 ng cyanocobalamin/liter level – the correct value lies in the range:

$$\text{Mean of } n \text{ determinations } \pm 0.5/n^{\frac{1}{2}} \text{ ng cyanocobalamin/liter.}$$

2. At the 0.2 ng cyanocobalamin/liter level – the correct value lies in the range:

$$\text{Mean of } n \text{ determinations } \pm 0.02/n^{\frac{1}{2}} \text{ ng cyanocobalamin/liter.}$$

VII. LIMITATIONS

A. *Inhibition of seawater to test alga*

If the inhibition of the seawater sample to *C. nana* is greater than 30–35% (as determined from the recovery of the added cyanocobalamin to the internal standard) then the results of the assay become unreliable. The assay should be repeated using a greater sample dilution.

B. *Adaptation for freshwater systems*

The *C. nana* assay for cyanocobalamin in seawater may be used to determine the concentration of this vitamin in freshwater systems if enough seawater salts are added to the sample so that its salinity will be in the range of 27–35‰. Generally it is sufficient to dilute one part of the sample with 3 parts of vitamin-free seawater. The sensitivity of the assay will be, therefore, about one-fourth that of the seawater assay. The less sensitivity for freshwater systems still render the *C. nana* assay adequate for vitamin B_{12} determinations in these waters.

C. *Response to cyanocobalamin analogues*

C. nana, clone 13–1, responds to a number of cyanocobalamin analogues as well as cyanocobalamin (Carlucci and Silbernagel 1966). For an assay which gives the major (over 90%) response to cyanocobalamin *C. nana*, clone 3H (available from Woods Hole Oceanographic Institution, Woods Hole, Massachusetts 02543) should be used (Guillard 1968).

VIII. ACKNOWLEDGMENTS

I am grateful to Susan Silbernagel for excellent technical assistance in the development of the assay.

This work was supported, in part, by the Marine Life Research Program, Scripps Institution of Oceanography's component of the California Cooperative Oceanic Fisheries Investigation, a project sponsored by the State of California, and, in part, by the United States Atomic Energy Commission, Contract No. AT(11-1)GEN 10, P.A. 20.

IX. REFERENCES

Carlucci, A. F. and Silbernagel, S. B. 1966. Bioassay of seawater. 1. A ^{14}C-uptake method for the determination of concentrations of vitamin B_{12} in seawater. *Can. J. Microbiol.* **12**, 175–83.

Guillard, R. R. L. 1968. B_{12} specificity of marine centric diatoms. *J. Phycol.* **4**, 59–64.

NOTE ADDED IN PROOF

Cyclotella nana has been referred to *Thalassiosira pseudonana*.

27 Bioassay: thiamine*

KENNETH GOLD

Osborn Laboratories of Marine Sciences,
Seaside Park, Coney Island, Brooklyn, New York 11224

CONTENTS

* Supported by a contract with the U.S. Atomic Energy Commission, reference number NYO-3658-23.

I. OBJECTIVE

Bioassays for thiamine are recommended where the concentration of thiamine is too low to be measured by the standard chemical method (Horwitz 1965). Where an investigator elects to use a bioassay technique, it must be remembered that an organism's growth response may be to: (a) thiamine analogues; (b) thiamine breakdown products; or (c) the intact thiamine molecule. The investigator should also bear in mind that thiamine is rapidly decomposed or inactivated, at high alkalinity (Wagner and Folkers 1964); by elevated temperatures in seawater (Gold, Roels and Bank 1966); and by its being complexed with certain heavy metals (Gold 1968).

In spite of the acknowledged short-comings inherent in the bioassay method, it is often the only way to determine the concentration of the vitamin in a natural product. Thus, three species of algae are available, which, with very little difficulty or previous investigator experience, can be used to measure the thiamine content of fluids, tissues, and sediments (after appropriate extraction of the vitamin where required). These three species are: the Dinophyceae *Crypthecodinium cohnii* (marine); and Chrysophyceae *Monochrysis lutheri* (marine), and *Ochromonas danica* (freshwater).

The assays employing *M. lutheri* and *O. danica* have been described by Carlucci and Silbernagel (1966) for *M. lutheri* and by Baker *et al.* (1964) for *O. danica*.

Details of the assay for thiamine using the colorless dinoflagellate *Crypthecodinium cohnii* are described here. Attributes of the organism include a heterotrophic mode of nutrition eliminating the need for special lighting and allowing rapid and abundant growth (Provasoli and Gold 1962), and an ability to survive long maintenance periods on agar slants (Keller, Hutner and Keller 1968). This assay lacks the sensitivity of the *M. lutheri* method; thus, it is recommended: (1) for detection of high concentrations of thiamine such as found in sediments and biological tissues; (2) in studies of thiamine chemistry in salt solutions; and (3) in comparative nutritional studies along the lines proposed by Droop (1958) with respect to algal utilization of thiamine precursors. (*C. cohnii* responds to the thiazole moiety, while *M. lutheri* responds to the pyrimidine fraction of the thiamine molecule.)

II. EQUIPMENT

A. *Materials*

Few specialized instruments are needed for thiamine assays using *C. cohnii*. All glassware is routinely baked at 200 °C for several hours before use.

1. Water bath (preferred), or constant temperature incubator, maintaining the optimum growth temperature of 30 °C.
2. Spectrophotometer (Spectronic 20 spectrophotometer, Bausch and Lomb Inc., Analytical Systems Division, 820 Linden Ave., Rochester, New York 14625; 1790 Birchmount Rd, Scarborough, Ontario) with the wavelength set at 660 nm.
3. Borosilicate, screw-cap test tubes, 125 × 20 mm (for new caps, autoclave several times for 15 min; or boil 3 h in distilled water to remove toxic compounds).
4. Borosilicate flasks with screw-caps for cultures and solutions, 1 l.
5. Borosilicate pipettes, sterile, 5 ml sereological recommended and assorted sizes.
6. Borosilicate, volumetric flask, 100 ml.
7. Borosilicate bottles, for stock biotin.
8. Glass filter and funnel (Reeve Angel, 934AH).

B. *Special reagents*

1. Metals mix. See Table 27-1, footnote 3.

2. Hydrochloric acid (0.01 N). Dilute 0.85 ml of concentrated HCl to 1 l with distilled water.

3. Stock thiamine solution (1 mg/ml). Dissolve 100 mg of thiamine hydrochloride in 100 ml of 0.01 N HCl. This solution is stable for extended periods when stored in the refrigerator. For the highest degree of accuracy use volumetric glassware and a thiamine standard, available from the US Pharmacopeia (USP Reference Standards, 46 Park Row, New York, New York 10016).

4. Working thiamine solution (1 μg/ml). Dilute stock thiamine solution 1:1000 with 0.01 N HCl. This is thiamine solution T.

5. Biotin solution (1 μg/ml). An ampoule containing 25 μg biotin/ml water (ICN Nutritional Biochemicals Division, 26201 Miles Rd, Cleveland, Ohio 44128) is smashed in a bottle; dilute the contents with 24 ml distilled water.

TABLE 27-1. *Complete basal medium (E6) for thiamine bioassays using Crypthecodinium cohnii (heterotrophic Dinophyceae)*

Salt solution[1]	1 l	$(NH_4)_2SO_4$	50 mg
Glutamic acid[2]	1.5 g	Biotin[4]	1 μg
Metals mix[3]	3 ml	Thiamine.HCl[5]	1 mg
Na_2 glycerophosphate.$5\frac{1}{2}H_2O$	0.15 g	Tris[6]	3 g
K_2HPO_4	10 mg	pH before and after	
Dextrose	3 g	autoclaving 6.4–6.6[7]	

N.B.: Chemical preservatives should not be added to the medium or to any of the solutions used to prepare it. Solutions should be kept frozen, refrigerated or prepared fresh before use.

[1] Salt solutions should be filtered (glass fiber filters, Reeve Angel, 934 AH). DO NOT FILTER THE MEDIUM after it is prepared. A slight precipitate may be visible, but it will go into solution on autoclaving.

a. Filtered seawater can be used as the salt solution for a routine maintenance of *C. cohnii.*

b. Thiamine-free seawater can be prepared in two ways. The fastest and simplest way is to autoclave the seawater for 15 min (Gold 1968). Thiamine is not destroyed on autoclaving in the complete basal medium as a result of low pH and glutamic acid protection. An alternative way to remove thiamine is to use activated charcoal. In this laboratory, powdered charcoal (Merck, NF) was found to be a non-toxic source of charcoal. It was added directly to the seawater, shaken or stirred for 30 min; removed by filtration. A useful filter because of its high flow rate is the Reeve Angel glass fiber filter 934 AH.

[2] Glutamic acid is dissolved in 1 l of salt solution at the start of the preparation of the basal medium.

[3] Metals mix. A solution is prepared which can be stored for extended periods frozen, or kept refrigerated for several months without adding preservatives. Preparation is as follows:

Dissolve 1 g Na_2EDTA (ethylenediaminetetraacetic acid) in 80 ml distilled water by adding just enough NaOH to completely dissolve the Na_2EDTA. When the Na_2EDTA is in solution, adjust volume to 100 ml with distilled water. (Failure to dissolve the Na_2EDTA completely, results in a flocculent brown precipitate on addition of the metals.)

Add the following as solids (dissolving each in turn before adding the next):

$FeCl_3.6H_2O$	50 mg	$ZnCl_2$	10 mg
H_3BO_3	1 g	$CoCl_2.6H_2O$	5 mg
$MnCl_2.4H_2O$	150 mg		

Adjust the pH to 6.5 with concentrated HCl. The final solution is clear and straw-colored. A color change to pink occurs on freezing, but the potency appears to remain unchanged.

[4] Biotin. Add 1 ml biotin solution (see text, II.B.5) to basal medium.

[5] Thiamine hydrochloride. Omit for thiamine-free basal medium. Additions of thiamine are made only to prepare the external and internal standard curves. (See text, IV, for details.)

[6] Tris is tris(hydroxymethyl)aminomethane.

[7] The pH before autoclaving is adjusted with HCl or NaOH as required. The pH after autoclaving should be 6.6. Higher values than pH 6.8 should be avoided.

III. TEST ORGANISM

Crypthecodinium cohnii is available from me, or the American Type Culture Collection (ATCC 30021).

IV. METHOD

The method is based on the growth responses of the thiamine-depleted alga to graded doses of the vitamin. It is first grown in thiamine-depleted basal medium, then transferred to medium containing the vitamin or suspected to contain it. After a suitable period ranging from 48–96 h, growth in the presence of an unknown amount of vitamin is compared with growth on the standard curve prepared from known amounts of thiamine.

Extraction of thiamine is required where this method is used to assay for vitamin activity in sediments or biological materials. Though I have no personal experience with extraction of thiamine from natural products, the methods of extraction suggested by Horwitz (1965) or Baker *et al.* (1964) should suffice.

A. *Basal medium*

The basal medium used in these tests is shown in Table 27-1.

B. *Inoculum*

Crypthecodinium cohnii is routinely maintained at room temperature in the medium in Table 27-1. When prepared with filtered seawater as the salt solution it is designated Medium E6. Subcultures are prepared 3 times weekly by transferring 1 ml of a dense, actively-growing, motile, stock culture into 10 ml fresh medium. In this way, cultures are kept in excellent physiological condition for the assay.

Inoculum for the assay is prepared by transferring 0.1 ml of the stock culture into 200 ml of thiamine-free medium (Table 27-1, omitting the thiamine.HCl solution) contained in a 1 l screw-cap flask. Growth progresses in the flask for 3 days at which time the cells should be thiamine-depleted. The resultant optimum inoculum usually contains an estimated 80–90 % motile cells.

Keller, Hutner and Keller (1968) prepared agar-solidified stock medium for *C. cohnii*. There are no experimental data on slant cultures for use in thiamine depletion; cells that were resuspended in complete liquid medium resumed motility within a few days after inoculation. Experimentalists are encouraged to test the agar slant method as a sensible alternative to thiamine depletion in liquid medium.

C. *Procedure*

1. External standard (dose-response curve)

a. Prepare 1 l thiamine-free basal medium using seawater that has been treated to remove the vitamin (see Table 27-1, footnote 1b).

b. Set aside 100 ml of thiamine-free basal medium. Add to it 1 ml of thiamine hydrochloride solution T containing 1 μg/ml of the vitamin. This solution T should be prepared immediately before use by diluting the stock solution as described earlier (II.B.4).

c. Dispense the thiamine-enriched basal medium (b, preceding) into the test tubes as follows: 9, 8, 6, 4, 2, 1 ml. The minimum number of replicate tubes in each set should be 3. Dilute each thiamine-enriched portion with thiamine-free medium (a, preceding) so that the final volume in each test tube is 9 ml. Dispense 9 ml of thiamine-free medium into each of 3 additional test tubes. This is the vitamin-free control. After inoculation, the final concentration of thiamine.HCl in each tube will be (μg/l): 9, 8, 6, 4, 2, 1, 0.

2. Internal standard. An internal standard may be desired as a way to detect the presence of factors that interfere with the bioassay (i.e., inhibition or enhancement). To prepare this solution, proceed as in 1.b, preceding, using seawater medium prepared without charcoal treatment (Table 27-1, footnote 1b). Dispense this thiamine-enriched seawater, and dilute it with thiamine-free medium as in 1.c, preceding. For a more extensive discussion of the use of an internal standard, see Gold (1964) and Carlucci and Silbernagel (1966).

3. Unknown. Prepare two sets of basal medium as follows with the seawater to be tested.

a. Batch *a* – basal medium with test seawater treated with charcoal to remove thiamine (Table 27-1). This will be used to dilute the test seawater (b.iii, following).

b. Batch *b*

i. Basal medium with test seawater not treated to remove thiamine (Table 27-1, omitting footnote 1b).

ii. Dispense into test tubes as follows: 9, 8, 6, 4, 2, 1 ml.

iii. Dilute test samples (ii, preceding) with batch *a* (a, preceding) so that final volume in each test tube is 9 ml.

4. *Sterilization.* Autoclave external standard (1.a, preceding); internal standard (2, preceding); unknown (3, preceding) for 15 min (see Chp. 12, II.C.2.c). (The low pH and presence of high concentration of glutamic acid prevents thiamine destruction.) Cool the medium to room temperature or 30 °C.

5. *Inoculation.* Aseptically inoculate all tubes with 1 ml of thiamine-depleted *Crypthecodinium cohnii* using a 5 ml sterile serological pipette as follows. Draw 5 ml of culture medium into the pipette and rapidly dispense 1 ml portions into each of 4 test tubes. Return the final 1 ml portion to the inoculum flask, mix well and repeat the procedure until all of the test tubes have been inoculated. Rapid delivery of culture medium results in accumulation of cells in the final 1 ml in the pipette. This procedure helps reduce the uneven inoculum that otherwise results. An alternative is to inoculate each tube separately with the contents of a 1 ml pipette, but the risks of contamination are far greater.

6. *Incubation.* Incubate the cultures at 30 °C. Cells tend to accumulate at the miniscus, so it is desirable (but not required) to mix the cultures gently once or twice each day. Qualitative growth responses to thiamine are observable within 24 h; a 72–96 h incubation period is recommended where quantitative results are sought.

D. *Harvesting, analysis*

1. At the end of the incubation period, cells are killed by adding a drop of 37% formaldehyde solution to each test tube. Live cells tend to form 'streams' in the medium, thereby obscuring spectrophotometer results. Killed cells stay in suspension long enough to make an accurate measurement.

2. Calibrate the instrument to zero absorbance at 660 nm, using an uninoculated portion of basal medium. Absorbance should be checked frequently, and the instrument readjusted if there is detectable drift in the meter.

3. Mix the cultures well before dispensing portions into the spectrophotometer cells. Avoid bubbles. Read absorbance as soon as the instrument meter comes to rest.

4. Calculate the mean for each concentration of vitamin; plot on linear graph paper, where: concentration of thiamine, is the abscissa; absorbance, the ordinate.

5. Comparison of the absorbance of the unknown with the external standard curve is used to determine concentration of thiamine.

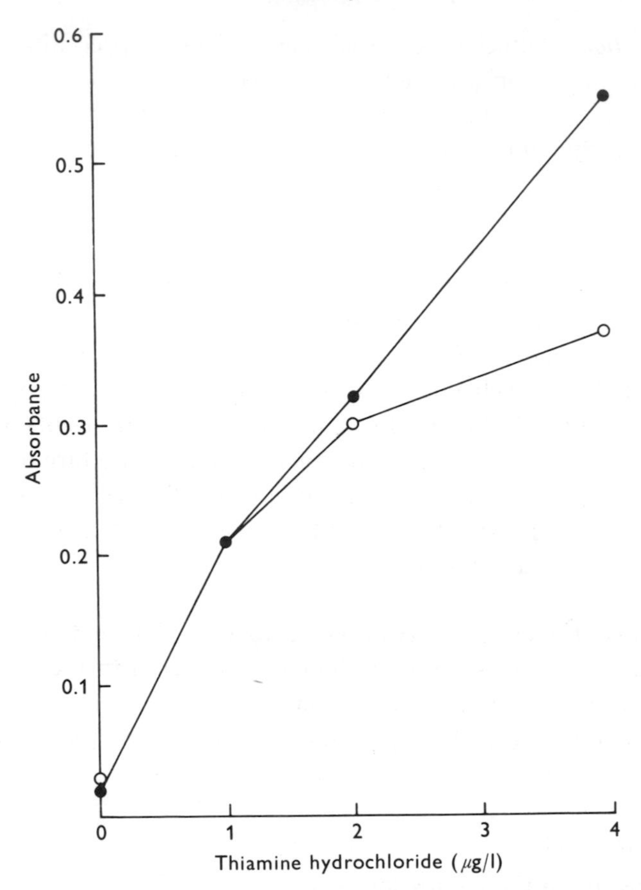

Fig. 27-1. An external standard curve plotted over range 0–4 μg thiamine/liter. ○ 48 h incubation; ● 96 h incubation.

V. SAMPLE DATA

An example of a standard curve is shown in Fig. 27-1. The longer incubation time (96 h) is recommended for quantitative bioassay. Shorter times may suffice for qualitative purposes.

VI. ALTERNATIVE BIOASSAY

A ^{14}C assimilation method is described by Carlucci and Silbernagel (1966). Briefly, the method consists of inoculating the thiamine-starved pigmented *Monochrysis lutheri* into an external standard, internal standard, and unknown. After 48–72 h incubation in the light, a portion of $NaH^{14}CO_3$ is added to each culture, and the cells exposed to the radioisotope for 2 h.

Cells are collected on filters, dried, and the incorporated radioactivity measured. ^{14}C assimilation is proportional to the amount of thiamine present in the culture. The useful range of concentrations detected by this method was 2–100 ng/l. Concentrations of thiamine could be read directly from the response curve plotted for the external standard data, or could be calculated mathematically using data derived from the internal standard.

Cell counts, instead of incorporated radioactivity, could be used to bioassay seawater with *M. lutheri*. An incubation period of 120 h was recommended by the authors.

This technique is similar to that described for biotin and cyanocobalamin in the preceding chapters (Chps. 25, 26).

VII. REFERENCES

Baker, H., Frank, O., Fennelly, J. J. and Leevy, C. M. 1964. A method for assaying thiamine status in man and animals. *Amer. J. Clin. Nutr.* **14**, 197–201.

Carlucci, A. F. and Silbernagel, S. B. 1966. Bioassay of seawater II. Methods for the determination of concentrations of dissolved vitamin B_1 in seawater. *Can. J. Microbiol.* **12**, 1079–89.

Droop, M. R. 1958. Requirement for thiamine among some marine and supra-littoral protista. *J. Mar. Biol. Ass. U.K.* **37**, 323–9.

Gold, K. 1964. A microbiological assay for vitamin B_{12} in seawater using radiocarbon. *Limnol. Oceanogr.* **9**, 343–7.

Gold, K. 1968. Some factors affecting the stability of thiamine. *Limnol. Oceanogr.* **13**, 185–8.

Gold, K., Roels, O. A. and Bank, H. 1966. Temperature dependent destruction of thiamine in sea water. *Limnol. Oceanogr.* **11**, 410–13.

Horwitz, W., ed. 1965. *Official Methods of Analysis of the Association of Official Agricultural Chemists*, 10th ed. pp. 758–60. Assn. Official Agric. Chem., Washington, D.C.

Keller, S. E., Hutner, S. H. and Keller, D. 1968. Rearing the colorless marine dinoflagellate *Cryptothecodinium cohnii* for use as a biochemical tool. *J. Protozool.* **15**, 792–5.

Provasoli, L. and Gold, K. 1962. Nutrition of the American strain of *Gyrodinium cohnii*. *Arch. Mikrobiol.* **42**, 196–203.

Wagner, A. F. and Folkers, K. 1964. *Vitamins and Enzymes*. Interscience, New York. 532 pp.

28: Bioassay: vitamin content of algae

K. V. NATARAJAN

Greater Hartford Community College,
34 Sequassen Street, Hartford, Connecticut 06106

CONTENTS

I. OBJECTIVES

Considerable attention is being focused on the distribution and ecological importance of vitamins in aquatic systems. We have considerable knowledge of the significance of cyanocobalamin, thiamine and biotin in the marine environment and scanty information in freshwater systems. Major investigations have been on the assay of vitamins in the dissolved state. The presence of vitamins in algae and the possible liberation into the surrounding medium during the metabolic processes, or death and decay indicate that algae also may be playing a major part in the vitamin cycles. The interest in utilization of algae as human and animal food has induced investigations on the vitamin content of algae. The scope of the problem of vitamin content of algae is enormous considering the broad range of algal groups which are involved in the aquatic food chain. It is hoped this chapter will stimulate more work on the significance of vitamins in algae in terms of their nutritive value as well as their contribution in the general productivity of the aquatic environment.

Most of the known vitamins occur in algae although more work has been done of the B-complex vitamins. Vitamins A and C are reported present in algal materials with chemical assays but reliable microbiological assays have yet to be developed. Thus microbiological assay methods will be discussed only for the B-complex vitamins.

Reliable algal bioassays have not been developed for many vitamins except in the case of cyanocobalamin, thiamine and biotin. The bacterial and fungal assays available are more advantageous in terms of specificity, sensitivity, rapidity, and require less expensive laboratory equipment. It is possible that better algal bioassays will be developed for many vitamins; until then it is advisable and imperative to use the microorganisms recommended here which have been proved and tested to give dependable results. These include several bacteria and two fungi.

Lactobacillus spp. have been used for the bioassay of thiamine, riboflavin, niacin, pantothenic acid, biotin and cyanocobalamin because of fast growth, non-pathogenic nature, easy maintenance and culture, and requirement for the B-complex vitamins. The other microorganisms (*Streptococcus faecalis* for folic acid, *Saccharomyces carlsbergensis* for pyridoxine and inositol,

Neurospora crassa for choline) are selected for their wide use and reliability.

II. EQUIPMENT

A. *General equipment, glassware*

The equipment needed for bioassays is standard in most laboratories; such as an autoclave, refrigerator and freezer for storage of cultures and stock solutions, centrifuge, pH meter, test tube shakers, a large supply of 15 × 150 mm or 25 × 200 mm culture tubes with stainless steel or aluminum closures, racks, inoculating needles, pipettes, etc.

Required specialized equipment includes:

1. Spectrophotometer or colorimeter plus cuvettes, matched tubes, filters, etc., for measuring turbidity (see Chp. 21).
2. Water bath or incubator, at 30–37 °C (± 0.5 °C).
3. Rotary evaporator.
4. Vacuum desiccator.
5. Amber glass containers for storing light-sensitive compounds.

All glassware used in vitamin assays should be specially cleaned to eliminate contamination problems. The sensitivity of the organisms are such that organic impurities and left over impurities of acids and detergents in the glassware will enhance or inhibit the organisms. The procedure here is recommended. Soak in hot detergent, then wash in hot water followed by cold water. Rinse many times in distilled water. Dry in oven. Sometimes it is advisable to fill the washed glassware with distilled water and autoclave for 15 min before drying in a hot air oven; other times, soaking in 75% sulfuric or chromicsulfuric acid for a few hours after detergent cleaning. In this instance care should be taken to remove all acid from the glassware by repeated rinsing in tap water followed by distilled water (but see Chp. 2, II.C.2). In general select a suitable method of cleaning that results in consistent assays and that the nature of the glassware does not become a variable in the interpretation of the results.

B. *Reagents*

The use of dehydrated media is recommended to avoid the cumbersome and time consuming preparation of media, and possible introduction of undesirable variables and discrepancies between assays. Most of the dehydrated culture media are economical and can be obtained from Difco Laboratories (P.O. Box 1058A, Detroit, Michigan 48232), BBL (BioQuest Division, Becton, Dickinson and Co., P.O. Box 243, Cockeysville, Maryland 21030; 2464 South Sheridan Way, Clarkson, Ontario), Fisher Scientific Company (711 Forbes Ave., Pittsburgh, Pennsylvania 15219; 8555 Devonshire Rd, Montreal 307, Quebec). The media employed are listed in methods section (IV.A, B, following) when each assay organism is discussed. Synthetic media recommended by Freed (1966) and György and Pearson (1967*b*) can also be used. Besides the dehydrated culture media, the following reagents are required which should be made from reagent grade chemicals.

1. Sodium chloride solution (saline)

a. Dissolve 9.0 g NaCl in distilled water; dilute to 1 l.
b. Separate in 10 ml quantities in screw-cap test tubes.
c. Autoclave for 15 min (see Chp. 12, II.C.2.c).
d. Store in the refrigerator.

2. Sodium acetate buffer (1 %)

a. Dissolve 1 g anhydrous CH_3COONa in 80 ml distilled water.
b. Adjust pH to 4.5 with acetic acid.
c. Make to 100 ml.

3. Sodium acetate (2.5 M). Dissolve 205 g anhydrous CH_3COONa in distilled water; dilute to 1 l.

4. Sodium acetate. Prepare 2 N as needed for extraction (IV.C.1, following).

5. Sulfuric acid. Prepare 0.1 N, 0.44 N, 1 N, 2 N H_2SO_4 in suitable quantities as needed for extraction (IV.C.1, 3, 4, 6, following).

6. Hydrochloric acid. Prepare 0.1 N, 1 N, 2 N, 3 N, 6 N HCl in suitable quantities as needed for extraction (IV.C.1, 2, 5, 7, 8, following).

7. Sodium hydroxide. Prepare 0.01 N, 0.1 N, 1 N, 2 N, 3 N, 6 N, 10 N NaOH in suitable quantities as needed for extraction (IV.C.3, 4, 6, 7, 8, 9, following).

8. Sodium cyanide (0.01 %). Dissolve 10 mg of NaCN in distilled water; dilute to 100 ml (IV.C.10, following).

9. Takadiastase. Prepare fresh at the time of extraction (IV.C.1, 9, following). This is the registered trademark of an *Aspergillus oryzae* diastase, obtainable from Parke-Davis and Co., Detroit, Michigan. (Also available as α-Amylase, A6630, Sigma Chemical Co., P.O. Box 14508, St Louis, Missouri 63178.)

10. Mylase P. Prepare fresh at the time of extraction (IV.C.5, following), obtainable from ICN Nutritional Biochemicals, 26201 Miles Rd, Cleveland, Ohio 44128; Wallerstein Laboratories, New York, New York,

11. Phosphate-citrate buffer. A 0.02 M phosphate–0.01 M citrate buffer, at pH 4.6, may be prepared (see IV.C.10, following) or preferably make the buffer as follows:

a. 0.5 M citric acid – dissolve 10.5 g citric acid monohydride in 100 ml distilled water.

b. 0.5 M sodium phosphate – dissolve 7.1 g Na_2HPO_4 in 100 ml water.

c. Mix 21.3 ml 0.5 M citric acid; 37.4 ml 0.5 M Na_2HPO_4; 200 ml water.

III. TEST ORGANISMS

Table 28-1 lists the microorganisms and information concerning them. Most are recommended by the Association of Vitamin Chemists, and the Association of Official Agricultural Chemists. Freeze-dried cultures can be obtained from the American Type Culture Collection (Order Department, 12301 Parklawn Dr., Rockville, Maryland 20852). The freeze-dried cultures are revived following the procedure given with the cultures and maintained in the prescribed media. Cultures are also available from the National Collection of Industrial Bacteria (Chemical Research Laboratory, Teddington, Middlesex, England) or Institute for Fermentation (Takeda Pharmaceutical Industries, Jusonischinocho Higashinyodogawa-ku, Osaka, Japan).

IV. METHODS

A. *Stock cultures*

1. Lactobacillus fermenti, L. casei, L. arabinosus, L. leichmannii. Stock cultures of lactobacilli are maintained in Bacto-Lactobacilli Agar AOAC (Difco) made according to instructions accompanying the medium. (In summary dissolve 48 g of medium in 1 liter distilled water, using heat. Transfer 10 ml medium to culture tubes; autoclave for 15 min (see Chp. 12, II.C.2.c). Allow the medium to solidify in an upright position; store in refrigerator.)

Incubate stab cultures in triplicate at 35–37 °C for 18–48 h until growth is seen; store at 2–8 °C in refrigerator. Incubation time and transfer intervals vary slightly with respect to each species and should be followed, as in Table 28-1. One of the triplicate tubes is used for transfer purposes and the other two for preparation of inocula.

TABLE 28-1. *List of microorganisms and culture conditions*

Vitamin	Organism	Sensitivity	Medium for stock culture (Difco)	Incubation period h, °C	Transfer intervals
Thiamine	*Lactobacillus fermenti* 36 (ATCC 9338)	0.5–3 ng/ml	Lactobacilli Agar AOAC	24–48 35–37	Monthly
Riboflavin	*Lactobacillus casei* (ATCC 7469)	2.5–15 ng/ml	Lactobacilli Agar AOAC	24–48 35–37	Monthly
Niacin	*Lactobacillus arabinosus* 17–5 (ATCC 8014)	5–30 ng/ml	Lactobacilli Agar AOAC	24–48 35–37	Monthly
Pyridoxine	*Saccharomyces carlsbergensis* 4228 (ATCC 9080)	0.5–5 ng/ml	Lactobacilli Agar AOAC	18–24 25–30	Weekly
Pantothenic acid	*Lactobacillus arabinosus* 17–5 (ATCC 8014)	2.5–20 ng/ml	Lactobacilli Agar AOAC	18–24 35–37	Weekly
Biotin	*Lactobacillus arabinosus* 17–5 (ATCC 8014)	2.5–50 pg/ml	Lactobacilli Agar AOAC	24–48 35–37	Monthly
Inositol	*Saccharomyces carlsbergensis* 4228 (ATCC 9080)	0.1–1 μg/ml	Lactobacilli Agar AOAC	18–24 25–30	Weekly
Choline	*Neurospora crassa* 34486 (ATCC 9277)	0.125–1.25 μg/ml	Neurospora Culture Agar	72 25–28	Monthly
Folic acid	*Streptococcus faecalis* (ATCC 8043)	0.2–1 ng/ml	Lactobacilli Agar AOAC	18–24 35–37	Weekly
Cyanocobalamin	*Lactobacillus leichmannii* (ATCC 7830)	2.5–25 pg/ml	Lactobacilli Agar AOAC	24–28 37	Weekly

The stock cultures may be carried in Bacto-Micro Assay Culture Agar (Difco), also obtained in a dehyrated form, and made using similar procedures described earlier.

2. *Saccharomyces carlsbergensis.* Stock cultures are maintained in slants of Bacto-Lactobacilli Agar AOAC (Difco) or Bacto-Micro Assay Culture Agar (Difco) (see 1, preceding).

Incubate the slants in triplicate at 25–30 °C for 18–24 h; store in the refrigerator.

3. *Neurospora crassa.* Stock cultures are maintained in slants of Bacto-Neurospora Culture Agar (Difco). Dissolve 65 g of medium in 1 l distilled water using heat; distribute 10 ml quantities in culture tubes and autoclave 15 min. Make slants of the tubes and store in the refrigerator until use.

Incubate agar slant cultures in triplicate at 25–28 °C for 2–4 days until sporulation occurs; store in the refrigerator.

4. *Streptococcus faecalis.* Stock cultures are maintained in stabs of Bacto-Lactobacilli Agar AOAC (Difco) or Bacto-Micro Assay Culture Agar (Difco) (see 1, preceding).

Incubate the stabs at 35–37 °C for 18–24 h; store in the refrigerator.

B. *Inoculum*

1. *Lactobacillus fermenti, L. casei, L. arabinosus, L. leichmannii.* For bioassay of thiamine, riboflavin, niacin, pantothenic acid, biotin and cyanocobalamin. Prepare inoculum for assay by transferring cells from a stock culture into 10 ml of Bacto-Lactobacilli Broth AOAC (Difco) or Bacto-Micro Inoculum Broth (Difco); incubate 16–24 h at 35–37 °C. Aseptically centrifuge; decant supernatant liquid; resuspend cells in 10 ml of sterile saline solution (II.B.1, preceding). Use one drop of this suspension to inoculate each assay tube.

2. *Saccaromyces carlsbergensis*

a. Pyridoxine assay – prepare inoculum by transferring cells from stock culture to 10 ml of single strength Bacto-Pyridoxine Y Medium (Difco) containing 1 ng/ml each of pyridoxal hydrochloride, pyridoxamine hydrochloride, and pyridoxine hydrochloride. Incubate 18–24 h at 25–30 °C; centrifuge aseptically; decant supernatant liquid; resuspend cells in 10 ml sterile single strength Bacto-Pyridoxine Y Medium. Use one drop of this suspension to inoculate each assay tube.

b. Inositol assay – inoculate stock culture into 10 ml of Bacto-Inositol Assay Medium (Difco) containing 10 μg inositol. Incubate 18–24 h at 30 °C; centrifuge; decant supernatant; resuspend cells in 10 ml sterile saline (II.B.1, preceding). Use one drop suspension to inoculate each assay tube.

3. Neurospora crassa. For bioassay of choline. Suspend a loopful of spores from a 48 h slant culture in 10 ml sterile saline. Use one drop of spore suspension for each assay flask.

4. Streptococcus faecalis. For bioassay of folic acid. Transfer cells from stock culture to 10 ml Bacto-Lactobacilli Broth AOAC (Difco) or Bacto-Micro Inoculum Broth (Difco). Incubate 18-24 h at 35–37 °C; centrifuge; resuspend cells in 10 ml sterile saline. Use one drop of suspension to inoculate assay tube.

C. *Extraction procedures*

In the extraction of algal samples for the analyses of vitamins different procedures have been used by different investigators. The procedures selected here are convenient and generally acceptable. Consult the references at the end of the chapter if difficulties arise. The following guidelines should be valid to all extraction procedures.

Wash the large algae to remove all adhering growth, then dry in hot air at less than 100 °C; grind in a mill; store in the dark. Microalgae can be concentrated by centrifugation and dried as before or freeze-dried.

When using reagents for extraction it is advisable to add volumes in milliliters of not less than 10 times the dry weight of the sample in grams.

At the end of extraction procedures the make-up of the final volume will depend on the quantity of the vitamin present in the original sample. A rough estimate of the vitamin in sample is usually necessary and extraction procedures have to be slightly modified depending on the amount of vitamin. All volumes are made with distilled water in volumetric flasks.

Before use of an extract for assay it may have to be filtered through Whatman no. 40 paper and a subsample diluted further so that the final concentration of vitamins is in the assay range.

1. Thiamine

a. Weigh 1 g of algal sample; suspend in 15 ml 0.1 N H_2SO_4 or 0.1 N HCl.
b. Heat in a boiling water bath 30 min, with frequent mixing (or auto-

clave 15 min, see Chp. 12, II.C.2.c). The pH should be *ca.* 1.5 during digestion; if it rises, add dilute acid.

c. Cool; adjust pH to 4.0–4.5 with 2 N CH_3COONa.

d. Add 100 mg takadiastase; incubate at 37 °C for 24 h.

e. Adjust pH to 6.5–6.6; dilute to suitable volume (100 ml); filter, if necessary.

2. *Riboflavin*

a. Weigh 1 g of algal sample; add 50 ml of 0.1 N HCl.

b. Autoclave 15 min (see Chp. 12, II.C.2.c).

c. Cool; adjust pH to 4.5.

d. Make to convenient volume (100 ml); filter. Adjust pH to 6.8.

3. *Niacin*

a. Weigh 1 g of algal sample; add 10 ml of 1 N H_2SO_4.

b. Autoclave 30 min (see Chp. 12, II.C.2.c).

c. Adjust pH to 6.8 with 1 N NaOH.

d. Make to 25–50 ml; filter.

4. *Pyridoxine*

a. Hydrolyse 1 g of algal sample with 50 ml of 0.44 N H_2SO_4 by autoclaving at 20 lb/in^2 for 1 h.

b. Adjust pH to 5.0–5.2 with 3 N NaOH.

c. Make to 200 ml; filter.

d. Store in an amber glass bottle.

5. *Pantothenic acid*

a. Weigh 1 g of algal sample; suspend in 50 ml of distilled water.

b. Adjust pH to 6.8–7.0.

c. Autoclave 15 min. (see Chp. 12, II.C.2.c).

d. Cool; add 0.1 g Mylase P; buffer with 2 ml 2.5 M CH_3COONa.

e. Adjust pH with HCl to 4.8.

f. Incubate at 50 °C for 12–24 h.

g. Dilute sample to 50 ml.

h. If interfering fatty materials are present, adjust pH to 4.5–4.8 with 0.1 N HCl; filter.

6. *Biotin*

a. Weigh 1 g of algal sample into 50 ml flask; add 25 ml 2 N H_2SO_4.

b. Autoclave for 2 h. (see Chp. 12, II.C.2.c).
c. Cool; adjust pH to 6.8–7.0 with 2N NaOH.
d. Dilute to proper volume; filter, if necessary.

7. Inositol

a. Hydrolyse 0.5 g of algal sample by refluxing with 50 ml of 6N HCl for 48 h.
b. Concentrate hydrolysate to dryness in vacuum on rotary evaporator.
c. Add 10 ml water to residue; reevaporate mixture.
d. Repeat preceding step (c).
e. Remove the remaining amounts of HCl by placing flask with residue in vacuum desiccator containing a vessel of NaOH pellets.
f. Dissolve the residue in *ca.* 40 ml water; adjust pH to 7.0 with 6N NaOH.
g. Make to 50 ml; filter.

8. Choline

a. Hydrolyse 0.5 g of algal sample by refluxing with 50 ml 3N HCl for 18–24 h.
b. Concentrate hydrolysate as in inositol extract (7, preceding). (Direct neutralization of HCl cannot be used because of the toxicity of large amounts of salts for the test organisms.)

9. Folic acid

a. Weigh 1 g of algal sample into 50 ml flask; add 5 ml of 1 % CH_3COONa buffer, and 100 mg of takadiastase.
b. Incubate under toluene, at 37 °C for 24 h.
c. Heat in boiling water bath for 5 min; cool; adjust to pH 6.6–6.8 with 0.1 N NaOH.
d. Make to 100 ml; filter.

10. Cyanocobalamin

a. Weigh 1 g of sample; add 20 ml 0.02M phosphate–0.01M citrate buffer (pH 4.6). (Can use 0.1M acetate buffer at pH 5.2, or 0.1M phosphate buffer at pH 6.0).
b. Add 0.1 ml of 0.01 % NaCN.
c. Autoclave for 15 min.
d. Filter; make to 50 ml.

TABLE 28-2. *List of assay media, incubation and readings*

Assay media (Difco)	Vitamins	Organisms (ATCC no.)	Assay incubation period (h)	Assay incubation temp (°C)	Assay reading (μm)[a]
Thiamine assay	Thiamine	*L. fermenti* 9338	16–18	35–37	640
Riboflavin assay	Riboflavin	*L. casei* 7469	18–24	35–37	640
Niacin assay	Niacin	*L. arabinosus* 8014	16–18	35–37	640
Pyridoxine Y	Pyridoxine	*S. carlsbergensis* 9080	22	25–30	640
Pantothenate AOAC	Pantothenic acid	*L. arabinosus* 8014	16–18	35–37	640
Biotin assay	Biotin	*L. arabinosus* 8014	16–20	35–37	640
Inositol assay	Inositol	*S. carlsbergensis* 9080	18–24	30	640
Choline assay	Choline	*N. crassa* 9277	72	25	weigh
Folic TE	Folic acid	*S. faecalis* 8043	16–18	35–37	640
B_{12} assay	Cyanocobalamin	*L. leichmannii* 7830	16–24	37	640

[a] Can be read for 540–660 μm.

D. *Vitamin stocks, standards, samples*

It is recommended that all vitamin salts be dried to constant weight and stored in the dark in a desiccator over P_2O_5 or concentrated H_2SO_4 for at least 24 h. The last stock solution is the working standard which is usually made immediately before the assay. All solutions are made with glass-distilled water, unless otherwise noted, in a volumetric flask, and stored in the refrigerator.

If possible U.S.P. reference standard vitamins should be used which can be obtained from the United States Pharmocopeia, 4630 Montgomery Ave., Bethesda, Maryland 20014. U.S.P. reference standards of thiamine hydrochloride (200 mg), riboflavin (500 mg), niacinamide (500 mg), pyridoxine hydrochloride (200 mg), calcium pantothenate (200 mg), folic acid (250 mg), cyanocobalamin (1.5 g) are available at a uniform service charge of *ca.* $10.00 (U.S.) each.

All the assay media required for the different vitamin assays are given in Table 28-2. They are obtainable from Difco in a dehydrated form and are rehydrated following the instructions given in the respective bottles.

All the standards and samples are made in duplicate.

1. Thiamine

a. Stock solutions

i. (100 μg/ml). Dissolve 100 mg thiamine.HCl in distilled water; make volume to 1 l.

ii. (1 μg/ml). Dilute 1 ml of i to 100 ml with distilled water.

iii. (10 ng/ml – working standard). Dilute 1 ml of ii to 100 ml with distilled water.

The thiamine stock solution can also be made in 25% ethyl alcohol or in 0.01 N HCl if desired.

b. Standard preparation (total volume = 10 ml)

standard curve	(ng)	0	5	10	20	30	40	50
use:								
working standard	(ml)	0.0	0.5	1.0	2.0	3.0	4.0	5.0
distilled water	(ml)	5.0	4.5	4.0	3.0	2.0	1.0	0.0
thiamine assay medium	(ml)	5.0	5.0	5.0	5.0	5.0	5.0	5.0

c. Sample preparation (total volume = 10 ml)

use:					
extract	(ml)	1	2	3	4
distilled water	(ml)	4	3	2	1

thiamine assay medium	(ml)	5	5	5	5

Sterilize by steaming at 100 °C for 15 min (see Chp. 12, II.C.2.b).

2. *Riboflavin*

a. Stock solutions

i. (25 μg/ml). Dissolve 25 mg riboflavin in *ca.* 750 ml distilled water and 1.2 ml glacial acetic acid; warm to hasten solution. Cool; make to 1 l.

ii. (10 μg/ml). Dilute 40 ml of i to 100 ml with distilled water.

iii. (100 ng/ml – working standard). Dilute 1 ml of ii to 100 ml with distilled water. Store all stock solutions in dark in a refrigerator.

b. Standard preparation (total volume = 10 ml)

standard curve	(ng)	0	25	50	75	100	150	200
use:								
working standard	(ml)	0.0	0.25	0.5	0.75	1.0	1.5	2.0
distilled water	(ml)	5.0	4.75	4.5	4.25	4.0	3.5	3.0
riboflavin assay medium	(ml)	5.0	5.0	5.0	5.0	5.0	5.0	5.0

c. Sample preparation (total volume = 10 ml)

use:					
extract	(ml)	0.5	1.0	1.5	2.0
distilled water	(ml)	4.5	4.0	3.5	3.0
riboflavin assay medium	(ml)	5.0	5.0	5.0	5.0

Sterilize by autoclaving for 10 min.

3. *Niacin*

a. Stock solutions

i. (100 μg/ml). Dissolve 50 mg niacin in distilled water; make volume to 500 ml.

ii. (100 ng/ml working standard). Dilute 1 ml of i to 1 l with distilled water.

b. Standard preparation (total volume = 10 ml)

standard curve	(ng)	0	50	100	200	300	400	500
use:								
working standard	(ml)	0.0	0.5	1.0	2.0	3.0	4.0	5.0
distilled water	(ml)	5.0	4.5	4.0	3.0	2.0	1.0	0.0
niacin assay medium	(ml)	5.0	5.0	5.0	5.0	5.0	5.0	5.0

c. Sample preparation (total volume = 10 ml)

use:					
extract	(ml)	1	2	3	4
distilled water	(ml)	4	3	2	1
niacin assay medium	(ml)	5	5	5	5

Sterilize by autoclaving for 10 min.

4. Pyridoxine

a. Stock solutions

i. (100 μg/ml). Dissolve 50 mg pyridoxine.HCl in distilled water; make to 500 ml.

ii. (1 μg/ml). Dilute 1 ml of i to 100 ml with distilled water.

iii. (10 ng/ml – working standard). Dilute 1 ml of ii to 100 ml with distilled water.

b. Standard preparation (total volume = 10 ml)

standard curve	(ng)	0	5	10	15	20	30	40
use:								
working standard	(ml)	0.0	0.5	1.0	1.5	2.0	3.0	4.0
distilled water	(ml)	5.0	4.5	4.0	3.5	3.0	2.0	1.0
pyridoxine assay medium	(ml)	5.0	5.0	5.0	5.0	5.0	5.0	5.0

c. Sample preparation (total volume = 10 ml)

use:					
extract	(ml)	1	2	3	4
distilled water	(ml)	4	3	2	1
pyridoxine assay medium	(ml)	5	5	5	5

Sterilize by steaming at 100 °C for 10 min.

5. Pantothenic acid

a. Stock solutions

i. (50 μg pantothenic acid/ml). Dissolve 54.4 mg Ca.pantothenate in 500 ml distilled water. Add 10 ml of 0.2 N acetic acid and 100 ml of 0.2 N sodium acetate; make to 1 l with distilled water.

ii. (5 μg/ml). Dilute 10 ml of i to 100 ml with distilled water.

iii. (50 ng/ml – working standard). Dilute 1 ml of ii to 100 ml with distilled water.

b. Standard preparation (total volume = 10 ml)

standard curve	(ng)	0	25	50	75	100	150	200
use:								
working standard	(ml)	0.0	0.5	1.0	1.5	2.0	3.0	4.0
distilled water	(ml)	5.0	4.5	4.0	3.5	3.0	2.0	1.0
pantothenic acid assay medium	(ml)	5.0	5.0	5.0	5.0	5.0	5.0	5.0

c. Sample preparation (total volume = 10 ml)

use:					
extract	(ml)	1	2	3	4
distilled water	(ml)	4	3	2	1
pantothenic acid assay medium	(ml)	5	5	5	5

Sterilize by autoclaving 10 min.

6. *Biotin*

a. Stock solutions

i. (50 μg/ml). Dissolve 50 mg crystalline d-biotin in 50% ethyl alcohol; make to 1 l.

ii. (1 μg/ml). Dilute 5 ml of i to 250 ml with 50% ethyl alcohol.

iii. (10 ng/ml). Dilute 5 ml of ii to 500 ml with 50% ethyl alcohol.

iv. (200 pg/ml – working standard). Dilute 2 ml of iii to 100 ml with distilled water.

b. Standard preparation (total volume = 10 ml)

standard curve	(pg)	0	50	100	200	300	400	500
use:								
working standard	(ml)	0.0	0.25	0.5	1.0	1.5	2.0	2.5
distilled water	(ml)	5.0	4.75	4.5	4.0	3.5	3.0	2.5
biotin assay medium	(ml)	5.0	5.0	5.0	5.0	5.0	5.0	5.0

c. Sample preparation (total volume = 10 ml)

use:					
extract	(ml)	1	2	3	4
distilled water	(ml)	4	3	2	1
biotin assay medium	(ml)	5	5	5	5

Sterilize by autoclaving for 10 min.

7. *Inositol*

a. Stock solutions

i. (2 mg/ml). Dissolve 200 mg inositol in distilled water; make to 100 ml.

ii. (2 μg/ml – working standard). Dilute 1 ml of i to 1 l with distilled water.

b. Standard preparation (total volume = 10 ml)

standard curve	(μg)	0	1	2	4	6	8	10
use:								
working standard	(ml)	0.0	0.5	1.0	2.0	3.0	4.0	5.0
distilled water	(ml)	5.0	4.5	4.0	3.0	2.0	1.0	0.0
inositol assay medium	(ml)	5.0	5.0	5.0	5.0	5.0	5.0	5.0

c. Sample preparation (total volume = 10 ml)

use:					
extract	(ml)	1	2	3	4
distilled water	(ml)	4	3	2	1
inositol assay medium	(ml)	5	5	5	5

Sterilize by autoclaving for 5 min.

8. *Choline*

a. Stock solutions

i. (500 μg/ml). Dissolve 500 mg anhydrous choline.Cl in distilled water; make to 1 l.

ii. (5 μg/ml – working standard). Dilute 1 ml of i to 100 ml with distilled water.

b. Standard preparation (total volume = 20 ml)

standard curve	(μg)	0	2.5	5	10	15	20	25
use:								
working standard	(ml)	0.0	0.5	1.0	2.0	3.0	4.0	5.0
distilled water	(ml)	10.0	9.5	9.0	8.0	7.0	6.0	5.0
choline assay medium	(ml)	10.0	10.0	10.0	10.0	10.0	10.0	10.0

c. Sample preparation (total volume = 20 ml)

use:					
extract	(ml)	1	2	3	4
distilled water	(ml)	9	8	7	6

choline assay medium	(ml)	10	10	10	10

Sterilize by autoclaving 10 min.

9. *Folic acid*

a. Stock solutions

i. (100 μg/ml). Dissolve 50 mg folic acid in 30 ml of 0.01 N NaOH; add 300 ml distilled water. Adjust pH to 7–8 with dilute HCl; make to 500 ml with distilled water.

ii. (1 μg/ml). Dilute 10 ml of i with 500 ml distilled water; adjust pH to 7–8 with dilute HCl; make to 1 l with distilled water.

iii. (200 ng/ml). Dilute 200 ml of ii with 500 ml of distilled water; adjust pH to 7–8 with dilute HCl; make to 1 l with distilled water.

iv. (2 ng/ml – working standard). Dilute 1 ml of iii to 100 ml with distilled water.

b. Standard preparation (total volume = 10 ml)

standard curve	(ng)	0	2	4	6	8	10
use:							
working standard	(ml)	0.0	1.0	2.0	3.0	4.0	5.0
distilled water	(ml)	5.0	4.0	3.0	2.0	1.0	0.0
folic TE medium	(ml)	5.0	5.0	5.0	5.0	5.0	5.0

c. Sample preparation (total volume = 10 ml)

use:					
extract	(ml)	0.5	1.0	1.5	2.0
distilled water	(ml)	4.5	4.0	3.5	3.0
folic TE medium	(ml)	5.0	5.0	5.0	5.0

Sterilize by autoclaving for 5 min.

10. *Cyanocobalamin*

a. Stock solutions

i. (100 ng/ml). Dissolve 50 μg cyanocobalamin in 25 % ethyl alcohol; make to 500 ml with 25 % ethyl alcohol.

ii. (100 pg/ml). Dilute 1 ml of i to 1 l with 25 % ethyl alcohol.

iii. (10 pg/ml – working standard). Dilute 10 ml of ii to 100 ml with distilled water.

All solutions are stored in dark in the refrigerator.

b. Standard preparation (total volume = 10 ml)

standard curve	(pg)	0	10	20	30	50	70	100	150
use:									
working standard	(ml)	0	1*	2*	3*	5*	0.7†	1.0†	1.5†
distilled water	(ml)	5	4	3	2	0	4.3	4.0	3.5
cyanocobalamin assay medium	(ml)	5	5	5	5	5	5.0	5.0	5.0

* Solution iii (working standard). † Solution ii.

c. Sample preparation (total volume = 10 ml)

use:					
extract	(ml)	1	2	3	4
distilled water	(ml)	4	3	2	1
cyanocobalamin assay medium	(ml)	5	5	5	5

Sterilize by autoclaving for 5 min.

E. *Incubation, reading, calculations*

Incubation periods and temperature of incubation vary slightly with individual assays which are summarized in Table 28-2.

To measure the final growth in the case of *Lactobacillus* spp. assays, two methods are used – titration of acid produced, and turbidimetry. Turbidimetry is recommended here because of the short incubation period, convenience and accuracy. A few precautions are necessary in turbidimetric assays. Because of the short incubation period, all the tubes in one assay should be incubated for the same length of time, though the exact time of incubation is varied slightly. Uniform temperature of the tubes must also be maintained. After the incubation all tubes are either steamed for 5 min to stop growth or refrigerated until the assay is read. Before reading shake the tubes to suspend the bacteria uniformly, then let settle *ca.* 20–30 s to remove air bubbles. The culture is transferred to cuvettes and read. The reading of each set of tubes should be done as quickly as possible with the sterile medium used for blank.

Assays using *Saccharomyces carlsbergensis* and *Streptococcus faecalis* are also read turbidimetrically using the fore-mentioned procedures.

Neurospora crassa assays are carried out with weighing. Remove the mycelial growth from the liquid medium as well as from the sides of the flask with the use of a wire loop or a rubber policeman. Place on a filter paper and remove all excess water. Transfer the mycelium to a beaker and dry it in an oven at 90 °C for 2 or more h; weigh until constant weight is obtained.

Plot the values of the optical density readings (or with *Neurospora crassa*, the weight of the mycelium) against the concentration of the vitamins in the standard tubes to make a standard curve. Read the vitamin concentration of the unknown from this standard curve. With the unknown sample, more than three different concentrations should be used such that at least three are in the region of the standard curve. Discard values which are ± 10 % from the mean of the duplicate tubes. From the three sets of values obtained for the extracted sample calculate the average concentration/ml of the extract. Then calculate the vitamin content in the sample using the following:

$$\text{vitamin/g sample} = \frac{(Y) \times (\text{volume of extract (ml)})}{\text{wt of sample (g)}} \times \text{dilution factor} \quad (1)$$

where:

Y is the average concentration of vitamin/ml of sample.

Depending on the amount of sample used the quantities can be expressed in grams or milligrams of sample.

V. SAMPLE DATA

A wide range of concentration of vitamins in algae have been reported in the literature (in μg/g dry weight): thiamine, 0.2–24.5; riboflavin, 0.2–51.1; niacin, 1–240; pyridoxine, 0.14–10.4; pantothenic acid, 0.25–150; biotin, 0.018–2.5; inositol, 55–2370; choline, 24–4885; folic acid, 0.05–47; cyanocobalamin 0.024–2.8. The vitamin content of algal materials so far analyzed compares favorably with the conventional vegetable and animal food resources. Table 28-3 gives the range of vitamin content of a few representative species belonging to the major algal phyla, assayed with the organisms recommended.

VI. POSSIBLE PROBLEMS

A. *Specificity*

The microorganisms used give the best reproducible results under the conditions employed here. The major disadvantage in any microbiological assay is the tendency for the organisms to respond to substances other than the particular vitamins, especially breakdown products. In the extraction procedures employed the conjugated or bound vitamins are liberated into the free form which is then assayed. The amount and type of material liberated depend on the extraction procedures. Regarding the organisms used which are selected for their maximum response the following points

TABLE 28-3. *Vitamin content of representative algal species (μg/g dry weight)*

Alga . . . Vitamins	*Chlorella pyrenoidosa* (Chlorophyceae)	*Chaetoceros simplex* (Bacillariophyceae)	*Ulva pertusa* (Chlorophyceae)	*Fucus spiralis* (Phaeophyceae)	*Rhodymenia palmata* (Rhodophyceae)
Thiamine	10.6–24.5	3.15	0.90	0.40	6.32
Riboflavin	27.5–51.1	5.33	2.83	10.00	5.00
Niacin	—	62.30	7.50	22.80	16.91
Pyridoxine	—	1.84	—	—	0.14
Pant. acid	3.2–16.1	29.50	2.35	23.00	4.30
Biotin	1.8–2.5	1.75	0.22	0.06	0.18
Inositol	1950–2370	—	330	—	—
Choline	—	—	61	—	—
Folic acid	6.4–28.3	2.10	0.12	1.91	—
Cyanocobalamin	—	0.047	0.063	0.08	0.028
References	Pratt and Johnson 1965, 1966	Kanazawa 1969	Kanazawa 1963	Teeri and Bieber 1958	Kanazawa, Saito and Idler 1966

should be noted. *Lactobacillus arabinosus* in the niacin assay measures niacin, niacinamide, and nicotinuric acid; and in the pantothenic assay measures the free pantothenic acid. *L. fermenti* responds to intact thiamine but not to thiazole or pyrimidine moieties. *Saccharomyces carlsbergensis* in the pyridoxine assay measures the total activity namely pyridoxine, pyridoxal, and pyridoxamine. *Streptococcus faecalis* will not respond to low potency materials, although it gives reproducible results. With regard to cyanocobalamin, *L. leichmannii* responds to total cobalamins and deoxyribosides.

B. *Sensitivity*

The sensitivity of the microorganisms used in the bioassays may vary slightly depending on the conditions in the laboratory as well as the inoculum, medium, temperature, pH, etc. To maintain the sensitivity of the organisms during indefinite subculturing it is advisable to grow them in minimal medium containing all the substances necessary for growth, including the needed vitamin. Sometimes when the organisms lose their sensitivity with prolonged use, it is better to start with a new culture. Algal bioassays are more sensitive but have not been developed for many vitamins (see Chps. 25–7).

C. *Standard curve*

The response of the organisms to the standards may vary from assay to assay and as such it is advisable to run a standard curve with each set of samples. The working standards should also be prepared fresh at the start of each assay. Variations in the linearity of the standard curve, generally can be traced to human errors. When there is too much variability in the dose-response curve, satisfactory statistical methods can be used, if applicable. Standard curves will vary from time to time which is inherent in microbiological assays.

D. *Incubation*

Incubation periods are not too critical as long as the whole set of samples and standards are treated uniformly. Temperature in the incubators, whether hot air incubators or water baths, should be maintained constant (± 0.5 °C). Care should be taken that all the assay tubes are maintained in a uniform environment. Shaking and aeration are not necessary but sometimes, especially in assays using *Saccharomyces*, it may be useful. If the inoculum is too thick and gives erratic results, it may have to be diluted.

E. *Interfering substances*

In microbiological assays many variables are involved. The concept of comparing the extracted samples with reference to purified standards has its own drawbacks. Chemicals, glassware, cotton which contains thiamine, biotin, folic acid and para-aminobenzoic acid, etc., may be sources of trouble. A number of compounds chemically related to the individual vitamins may also elicit confusing responses. Various precursors (e.g., desthiobiotin with biotin; pyridoxine with pyridoxamine and pyridoxal); moieties (e.g., pyrimidine and thiazole in thiamine; pantoic acid and β-alanine in pantothenic acid); biologically active forms as in biotin (α- and β-biotin); substitution products; derivatives; etc., may also influence bioassays. Discrepancies in assays due to antagonisms and inhibitions have also to be considered. These properties are to be taken into account in the analyses of the results. As better extraction procedures and more versatile microorganisms are discovered the assay procedures have to be changed. The present status of our knowledge of vitamins is improving every day and to keep up with this knowledge better techniques have to be selected and applied.

VII. ALTERNATIVE TECHNIQUES

A. *Algal*

Reliable algal bioassay methods are available for thiamine, biotin, and cyanocobalamin (Chps. 25–7). These can be used in the bioassay of extracted samples. The major problem in the use of algal bioassays will be the high concentration of vitamins in the extracted samples and the high sensitivity of the algae. For the samples to fall in the assay range of algal bioassays the samples may have to be diluted 10–100 fold.

For cyanocobalamin, four organisms are available and have been used extensively. The bacteria *Lactobacillus leichmannii* and *Escherichia coli* and the Euglenophyceae *Euglena gracilis* Z strain (IUCC 753), and *Ochromonas malhamensis* (Chrysophyceae). Sometimes it is advisable to use one or more organisms and compare the results in extensive investigations because of the many analogues of cyanocobalamin present in algal materials. The *L. leichmannii* assay measures the total cobalamins and deoxyribosides; *E. coli* responds to total cobalamins but is not suitable unless the extracted samples are further purified before assay due to its response to methionine and factor B; *O. malhamensis* responds to true cyanocobalamin as does *Euglena gracilis* which is similar in response to *L. leichmannii*.

B. *Acidimetric*

Acidimetric methods (titration) are used in bioassays involving *Lactobacillus*. *Lactobacillus* spp. produce lactic acid as a metabolic product, in proportion to the concentration of the vitamin in the medium, which can be measured by titration with alkali. If titration is to be used the assay tubes should be incubated for 60–72 h to allow for the maximum production of acid. After incubation the contents of the tubes are quantitatively transferred to a flask, the tubes rinsed with 10 ml of distilled water, mixed, and contents titrated with 0.1 N NaOH using a few drops of 0.1 % bromothymol blue as an indicator. The appearance of a green color at *ca*. pH 6.8 is taken as the end point. A pH meter can also be used and titration carried out to a pH of 7.0 Standard curves are plotted using milliliters of 0.1 N NaOH against concentrations of vitamin.

C. *Chemical*

For many vitamins (vitamins A, C, D, E, K) only chemical and animal assays are available. With reference to B-complex vitamins besides microbiological assays, chemical methods are available for thiamine, riboflavin, niacin, pyridoxine, pantothenic acid and folic acid. Chemical assays involve spectrofluorometry, spectrophotometry, ion-exchange, gas–liquid chromatography, thin-layer chromatography, paper chromatography, column chromatography, etc. The advantages of chemical assays are simplicity, and are less time consuming and require lesser number of reagents although in some instances they are complicated. The disadvantages are they are less specific and less sensitive than microbiological assays and may give erroneous results due to interference by biologically inactive materials present in the sample or introduced during the various extraction procedures. Chemical assays are used especially where the vitamins are expected to be present in large quantities in the extracted materials.

D. *Agar plate*

Agar plate methods are simple and quick but less sensitive. They are tremendously advantageous in cases of analyses of a large number of samples. The basic principle involves diffusion of the vitamins through the agar medium and can be adapted from crude to more sophisticated and quantitative assays.

In the cup plate method a number of cups of definite diameter are bored (with the use of glass cylinders or specially designed apparatus) in an agar plate made with the proper assay medium and seeded with the vitamin requiring organisms. Aliquots of standards and extracted samples in different dilutions are pipetted into the cups as quickly as possible and the plates are incubated overnight at 37 °C. The vitamins in the cups diffuse out into the surrounding agar producing a zone of growth proportional to the concentration. At the end of the incubation period the growth zones are measured either with a ruler or needle-pointed vernier calipers.

In other variations of the plate method, paper discs or pads soaked with the vitamin standards and samples are placed on the seeded agar plates and growth zones measured after incubation.

E. *Bioautography*

The principles involved in bioautography are simple and involve a combination of agar plate method and chromatography. The technique consists of preparing agar plates with the proper assay medium and seeded with the vitamin requiring bacterium. Chromatography of the extracts and standards are carried out using the proper solvent systems, and the dried chromatograms cut into pieces and placed on the seeded agar plates, incubated and growth zone measured. Depending on the individual needs this method is applicable with proper modifications and made semi-quantitative.

F. *Animals*

Animal assays which are less quantitative, costly and slow were the only assays available during the initial stages of vitamin research and before the discovery of more sophisticated chemical and microbiological assays. They can be used in laboratories where animal facilities and skilled technical help are available. Animal assays are measured in terms of: (1) growth; (2) reaction time based upon vitamin deficiency and recovery; (3) other graded responses which vary with vitamin doses; or (4) all-or-none vitamin assays. The animals commonly used are rats, chicks, guinea pigs, birds, etc. Crude algal vitamin assays can easily be carried out, where facilities and technical know-how are available, by grinding dried algal samples and using them for feeding experiments.

VIII. REFERENCES

Association of Official Agricultural Chemists. 1965. In Horwitz, W., ed., *Official Methods of Analysis*, 10th ed., chapter 39, Vitamins and other nutrients, pp. 752–86. Association of Official Agricultural Chemists, Washington, D.C.

Bolinder, A. E. and Larsen, B. 1961. Studies on the microbiological determination of niacin in some marine algae. *Acta Chem. Scand.* **15**, 823–38.

Daisley, K. W. 1969. Monthly survey of vitamin B_{12} concentrations in some waters of the English Lake District. *Limnol. Oceanogr.* **14**, 224–8.

Difco Technical Information. 1969. *Media for the Microbiological Assay of Vitamins and Amino Acids.* Difco Laboratories, Detroit, Michigan. 72 pp.

Ericson, L. E. and Carlson, B. 1953. Studies on the occurrence of amino acids, niacin, and pantothenic acid in marine algae. *Ark. Kemi* **6**, 511–22.

Freed, M. 1966. *Methods of Vitamin Assay*, 3rd ed. Interscience Publ., New York. 424 pp.

György, P. and Pearson, W. N. 1967*a*. *The Vitamins – Chemistry, Physiology, Pathology, Methods*, Vol. 6, 2nd ed. Academic Press, New York. 338 pp.

György, P. and Pearson, W. N. 1967*b*. *The Vitamins – Chemistry, Physiology, Pathology, Methods*, Vol. 7, 2nd ed. Academic Press, New York. 354 pp.

Hutner, S. H., Cury, A. and Baker, H. 1958. Microbiological assays. *Anal. Chem.* **30**, 849–67.

Ikawa, M., Borowski, P. T. and Chakravarti, A. 1968. Choline and inositol distribution in algae and fungi. *Appl. Microbiol.* **16**, 620–3.

Kanazawa, A. 1962. Studies on the vitamin B-complex in marine algae I. On vitamin contents. *Mem. Fac. Fish. Kagoshima Univ.* **10**, 38–69.

Kanazawa, A. 1963. Vitamins in algae. *Bull. Jap. Soc. Sci. Fish.* **29**, 713–31.

Kanazawa, A. 1969. On the vitamin D of a diatom, *Chaetoceros simplex*, as the diet for the larvae of marine animals. *Mem. Fac. Fish. Kagoshima Univ.* **18**, 93–7.

Kanazawa, A., Saito, A. and Idler, D. R. 1966. Vitamin B in dulse (*Rhodymenia palmata*). *J. Fish. Res. Bd Canada* **23**, 915–16.

Kavanagh, F. 1963. *Analytical Microbiology.* Academic Press, New York. 707 pp.

Natarajan, K. V. 1968. Distribution of thiamine, biotin, and niacin in the sea. *Appl. Microbiol.* **16**, 366–9.

Natarajan, K. V. and Dugdale, R. C. 1966. Bioassay and distribution of thiamine in the sea. *Limnol. Oceanogr.* **11**, 621–9.

Pratt, R. and Johnson, E. 1965. Production of thiamine, riboflavin, folic acid, and biotin by *Chlorella vulgaris* and *Chlorella pyrenoidosa*. *J. Pharm. Sci.* **54**, 871–4.

Pratt, R. and Johnson, E. 1966. Production of pantothenic acid and inositol by *Chlorella vulgaris* and *C. pyrenoidosa*. *J. Pharm. Sci.* **55**, 799–802.

Snell, E. E. 1950. Microbiological methods in vitamin research. In György, P., ed., *Vitamin Methods.* **1**, 327–505. Academic Press, New York.

Strohecker, R. and Henning, H. M. 1965. *Vitamin Assay – Tested Methods.* Verlag Chemie, GmbH, Weinheim/Bergstr. 360 pp.

Teeri, A. E. and Bieber, R. E. 1958. B-complex vitamins in certain brown and red algae. *Science* **127**, 1500.

Subject Index

(*Italic numbers indicate main reference to subject.*)

Author Index

(*Italics numbers indicate author citation in full; * indicates chapter author.*)

Taxonomic Index